# The Environmental Microbiome

The environment includes an almost infinite variety of microorganisms and their products. The human microbiome is influenced by this environmental microbiome and its interconnectedness, which recognizes these environmental elements as part of a complex ecosystem impacting human health. The recognition that humans co-evolved with microbes has contributed to the acknowledgment that different ecological communities of commensal, symbiotic, and pathogenic microorganisms share and contribute to our body. Microbiomes include species across all major kingdoms, including viruses as well as archaea, bacteria, and microbial eukaryotes. Recent research is rapidly strengthening the appreciation that associated microorganisms make essential contributions to human health and environmental well-being. This volume is an exploration of the environmental microbiome.

**Sunil Thomas** is a Research Professor at the Lankenau Institute for Medical Research in Wynnewood, Pennsylvania, where he leads interdisciplinary work spanning cell biology, microbiology, and translational medicine. His research focuses on next-generation immunotherapies for chronic conditions such as inflammatory bowel disease and Alzheimer's disease, with particular emphasis on how the human microbiome influences disease progression and treatment response.

Over his career, he has contributed to infectious disease research, vaccine development, and viral structural biology, helping advance diagnostics, therapeutics, and preventive strategies. He earned his Ph.D. in Environmental Biotechnology from Cochin University of Science and Technology in India and completed postdoctoral training at the Mount Sinai School of Medicine. He has authored over 70 peer-reviewed publications, contributed to books on vaccines and infectious diseases, and holds multiple patents.

# The Environmental Microbiome

Edited by
Sunil Thomas

CRC Press
Taylor & Francis Group
Boca Raton London New York

CRC Press is an imprint of the
Taylor & Francis Group, an **informa** business

Designed cover image: Sunil Thomas

First edition published 2027
by CRC Press
2385 NW Executive Center Drive, Suite 320, Boca Raton FL 33431

and by CRC Press
4 Park Square, Milton Park, Abingdon, Oxon, OX14 4RN

*CRC Press is an imprint of Taylor & Francis Group, LLC*



ISBN: 978-1-041-06936-2 (hbk)
ISBN: 978-1-041-06937-9 (pbk)
ISBN: 978-1-003-63795-0 (ebk)

DOI: 10.1201/9781003637950

Typeset in Joanna
by Newgen Publishing UK

# PREFACE

Microorganisms represent the earliest and most enduring forms of life on Earth. Long before the emergence of complex multicellular organisms, microbial life shaped the planet's atmosphere, soils, waters, and biogeochemical cycles. Even today, all eukaryotic life – including plants, animals, and humans – remains fundamentally rooted in microbial ancestry. Modern eukaryotic cells themselves are conserved amalgamations of ancient microbial partnerships, a testament to the central role microorganisms have played in the evolution of life.

Higher organisms cannot exist in isolation from microorganisms. From plants drawing nutrients through symbiotic soil microbes to animals relying on gut bacteria for digestion, microorganisms are indispensable partners in survival. The collective community of microorganisms that inhabit a host or environment is known as the microbiome. These microbial communities are not passive inhabitants; rather, they form dynamic, interactive systems that profoundly influence the health, development, and resilience of their hosts.

The microbiome – comprising trillions of bacteria, archaea, fungi, viruses, and protozoa – acts much like a vital organ. In humans and animals, it supports digestion, synthesizes essential vitamins, regulates metabolism, trains the immune system to distinguish friend from foe, and influences inflammation, behavior, mood, and disease susceptibility. In plants, the microbiome enhances nutrient uptake, improves stress tolerance, and protects against pathogens. Microbial diversity is a defining feature of a healthy microbiome, conferring stability and adaptability to changing conditions.

Beneficial microorganisms produce a wide array of secondary metabolites that are critical for maintaining physiological balance and overall well-being. When this balance is disrupted – a condition known as dysbiosis – the consequences can be severe. Dysbiosis has been linked to infections, allergies, autoimmune diseases, metabolic disorders, reduced crop productivity, and even neurological and behavioral conditions. Understanding the mechanisms that govern microbial balance is therefore essential for advancing health, agriculture, and environmental sustainability.

Recent breakthroughs in high-throughput sequencing, metagenomics, and bioinformatics have revolutionized microbiome research. These technologies allow rapid identification and functional analysis of microbial communities associated with different hosts, diets, environments, and lifestyles. What once took decades to uncover can now be achieved in a fraction of the time, opening new frontiers in medicine, nutrition, agriculture, ecology, and biotechnology.

The six interconnected volumes in the following list present a comprehensive exploration of microbial life across biological systems and ecosystems:

Volume 1: The Human Microbiome

Volume 2: Diet and the Microbiome

Volume 3: Animal Microbiomes

Volume 4: Plant Microbiomes

Volume 5: Insect Microbiomes

Volume 6: The Environmental Microbiome

Together, these volumes provide an integrated perspective on how microbial communities shape life on Earth.

I extend my sincere and heartfelt gratitude to all the authors who contributed chapters to this book series. Their expertise, dedication, and scholarly insight form the foundation of these volumes. Each contribution reflects a commitment to scientific rigor and a shared passion for advancing knowledge in the rapidly evolving field of microbiome research.

The collective efforts of these authors have enriched the depth, breadth, and quality of this series, transforming it into a comprehensive and authoritative resource. Their willingness to share ideas, collaborate across disciplines, and invest their time and intellect has been invaluable. This work stands as a testament to their contributions, and I am deeply thankful for their role in bringing this series to fruition.

This year marks the 100th anniversary of the Lankenau Institute for Medical Research. The anniversary represents decades of unwavering commitment to scientific inquiry, education, and the advancement of medical knowledge. The Lankenau Institute for Medical Research has played a pivotal role in shaping and supporting my scientific endeavors. Its culture of intellectual rigor, encouragement, and dedication to research excellence has provided an environment in which curiosity could thrive, and ideas could evolve into meaningful scientific contributions.

This six-volume Microbiome book series is presented as a tribute to the Lankenau Institute for Medical Research. It stands as a testament to the Institute's enduring legacy, its profound impact on generations of researchers, and its mission to advance science for the betterment of human health and society. Through this work, I honor its past achievements, acknowledge its present influence, and celebrate its continued inspiration for future scientific discovery.

# CONTRIBUTORS

CHETANA AKHAND
Environmental Bioprocesses, CSIR-National Environmental Engineering Research Institute (NEERI)
Nagpur, Maharashtra, India

RASHI BAMROTWAR
Environmental Bioprocesses, CSIR-National Environmental Engineering Research Institute (NEERI)
Nagpur, Maharashtra, India

Academy of Scientific & Innovative Research (AcSIR)
Ghaziabad, Uttar Pradesh, India

SEJAL BHAIRAM
Environmental Bioprocesses, CSIR-National Environmental Engineering Research Institute (NEERI)
Nagpur, Maharashtra, India

SANDEEP CHOVATIYA
Department of Biosciences
Saurashtra University
Gujarat, India

NISHANT A. DAFALE
Environmental Bioprocesses, CSIR-NationalEnvironmental Engineering Research Institute (NEERI)
Nagpur, Maharashtra, India

Academy of Scientific & Innovative Research (AcSIR)
Ghaziabad, Uttar Pradesh, India

VALÉRIA MAIA DE OLIVEIRA
Microbial Resources Division, Pluridisciplinary Research Center for Chemistry, Biology and Agriculture – CPQBA
University of Campinas – UNICAMP
Paulínia, São Paulo, Brazil

PRERNA DIWAN
Department of Microbiology, Ram Lal Anand College
University of Delhi
New Delhi, India

MANAR EL SAMAK
Department of Microbiology and Immunology
Faculty of Pharmacy
Suez Canal University
Ismailia, Egypt

PAULO ADEMAR AVELAR FERREIRA
Federal University of Santa Maria
Santa Maria, Rio Grande do Sul, Brazil

GISELLE GOMES MONTEIRO FRACETTO
Federal Rural University of Pernambuco
Recife, Pernambuco, Brazil

RAKESH KUMAR GUPTA
Department of Microbiology, Ram Lal Anand College
University of Delhi
New Delhi, India

SUNILA HOODA
Department of Microbiology, Ram Lal Anand College
University of Delhi
New Delhi, India

GABRIELLE MARQUES INÁCIO
Microbial Resources Division, Pluridisciplinary Research Center for Chemistry, Biology and Agriculture – CPQBA
University of Campinas – UNICAMP
Paulínia, São Paulo, Brazil

HASNAA L. KAMEL
Department of Microbiology and Immunology
Faculty of Pharmacy
Sinai University – Kantara
Ismailia, Egypt

CHARMY R. KOTHARI
Department of Biotechnology
Christ College
Rajkot, Gujarat, India

RAMESH K. KOTHARI
Department of Biosciences
Saurashtra University
Rajkot, Gujarat, India

SHASHI PRABHA KUMARI
Department of Microbiology, Ram Lal Anand College
University of Delhi
New Delhi, India

LAWRENCE L. LIVORNESE JR.
Division of Infectious Disease
Lankenau Medical Center
Wynnewood, Pennsylvania, USA

BHUMIKA MAHARANA
Environmental Bioprocesses, CSIR-National Environmental Engineering Research Institute (NEERI)
Nagpur, Maharashtra, India

Academy of Scientific & Innovative Research (AcSIR)
Ghaziabad, Uttar Pradesh, India

KELLY HIDALGO MARTINEZ
Microbial Resources Division, Pluridisciplinary Research Center for Chemistry, Biology and Agriculture – CPQBA
University of Campinas – UNICAMP
Paulínia, São Paulo, Brazil

ACACIO APARECIDO NAVARRETE
Graduate Program in Environmental Sciences
University of Brazil, Campus Fernandópolis
Fernandópolis, São Paulo, Brazil

TALLES RENAN SOUZA NICOLETTI
Graduate Program in Environmental Sciences
University of Brazil, Campus Fernandópolis
Fernandópolis, São Paulo, Brazil

RAJESH K. PATEL
Department of Biosciences
Veer Narmad South Gujarat University
Surat, Gujarat, India

LENORE PIPES
University of Hawaii at Manoa
Honolulu, Hawaii, USA

HINA H. RADADIYA
Department of Biosciences
Saurashtra University
Rajkot, Gujarat, India

ABDELRAHMAN M. SEDEEK
Department of Microbiology and Immunology
Faculty of Pharmacy
Galala University
Suez, Egypt

RIDDHI SINGH
Environmental Bioprocesses, CSIR-National Environmental Engineering Research Institute (NEERI)
Nagpur, Maharashtra, India

Academy of Scientific & Innovative Research (AcSIR)
Ghaziabad, Uttar Pradesh, India

SAMAR M. SOLYMAN
Department of Microbiology and Immunology
Faculty of Pharmacy
Suez Canal University
Ismailia, Egypt

DHRUVI SUVAGIYA
Department of Biosciences
Saurashtra University
Gujarat, India

SUNIL THOMAS
Lankenau Institute for Medical Research
Wynnewood, Pennsylvania, USA

Department of Gastroenterology
Lankenau Medical Center
Wynnewood, Pennsylvania, USA

DINH MINH TRAN
Institute of Biotechnology and Environment
Tay Nguyen University
Dak Lak, Vietnam

CHAPTER 1

# Environmental Microbiome

*Manar El Samak, Abdelrahman M. Sedeek, Hasnaa L. Kamel, and Samar M. Solyman*

## 1.1 INTRODUCTION

Environmental microbiomes represent the diverse and complex communities of microorganisms, including bacteria, archaea, fungi, viruses, and microeukaryotes, that inhabit diverse ecosystems across our planet (Nathani et al., 2020). These microbial communities constitute one of the most abundant and diverse forms of life on Earth, with an estimated $10^{12}$ distinct microbial species existing in environments ranging from terrestrial soils and aquatic systems to extreme habitats such as deep-sea hydrothermal vents and hypersaline lakes (Locey & Lennon, 2016; Thompson et al., 2017). The environmental microbiome encompasses not only the microorganisms themselves but also their collective genetic material, metabolic activities, and the complex network of interactions that govern ecosystem functioning (Fierer, 2017).

The significance of environmental microbiomes extends far beyond their remarkable abundance and diversity. Microorganisms serve as the primary drivers of global biogeochemical cycles, mediating the transformation of essential elements including carbon, nitrogen, phosphorus, and sulfur (Falkowski et al., 2008; Huang et al., 2024). Through their metabolic activities, environmental microbes regulate atmospheric composition, soil fertility, water quality, and climate patterns, making them indispensable for maintaining planetary ecosystem stability and resilience (Zhao et al., 2023). Furthermore, these communities possess extraordinary metabolic versatility, enabling them to perform unique biochemical transformations under conditions that would be lethal to most multicellular organisms (Nagar et al., 2023).

At the same time, the environmental microbiome serves as a vast reservoir of genetic and functional diversity, harboring an estimated 99% of all microbial genes on Earth (Schloss & Handelsman, 2003). This genetic wealth represents an untapped resource for biotechnological applications, including the discovery of novel enzymes, bioactive compounds, and metabolic pathways with potential applications in medicine, agriculture, and environmental remediation (Ferrer et al., 2016; Handelsman, 2004). Recent estimates suggest that environmental microbiomes contain millions of previously unknown genes, many of which encode proteins with unique catalytic properties and biotechnological potential (Sunagawa et al., 2015). The exploration of this diversity has been shaped by more than a century of scientific investigation, during which methodological advances have profoundly expanded our understanding of microbial life.

The study of microbial communities has progressed through several paradigm shifts. In the late 19th century, Sergei Winogradsky and Martinus Beijerinck established the first principles of microbial ecology. Winogradsky's demonstration of chemolithotrophy revealed that microbes could harness inorganic compounds for energy, while Beijerinck articulated the principle of microbial ubiquity, "everything is everywhere, but the environment selects", and developed enrichment culture methods that uncovered metabolic

DOI: 10.1201/9781003637950-1

specialists (Madsen, 2005; O'Malley, 2008). These pioneering efforts laid the conceptual foundation for understanding microbial contributions to ecosystem functioning, even though culture-based techniques of the time captured less than 1% of microbial taxa. Despite this limitation, culture-dependent studies yielded fundamental discoveries, including microbial nitrogen fixation, sulfur oxidation, and methane metabolism.

The molecular revolution of the late 20th century transformed this field by opening a window into uncultured diversity. Carl Woese's development of ribosomal 16S ribosomal ribonucleic acid (rRNA) phylogenetics provided the basis for reconstructing microbial evolutionary relationships and revealed the existence of the domain archaea, reshaping the tree of life (Woese & Fox, 1977). Polymerase chain reaction (PCR) and DNA sequencing expanded access to environmental DNA (eDNA), allowing researchers to explore "microbial dark matter" that could not be cultivated (Fang et al., 2023). By the early 2000s, metagenomic approaches enabled the recovery of complete genomes directly from environmental samples. Landmark studies such as Tyson and colleagues' genome reconstructions from acid mine drainage and the Global Ocean Sampling Expedition demonstrated the vast scale of microbial diversity and extensive horizontal gene transfer within natural populations (Rusch et al., 2007; Tyson et al., 2004; Venter et al., 2004). Global initiatives such as the Earth Microbiome Project and Tara Oceans have since cataloged microbial communities across thousands of ecosystems, linking taxonomic composition to environmental gradients and establishing global patterns of microbial biogeography (Gilbert et al., 2014).

Advances in high-throughput sequencing have given rise to the omics era, in which microbiomes can be studied as integrated systems (Knight et al., 2018). Marker-gene metabarcoding provides broad taxonomic resolution across large datasets, while shotgun metagenomics reveals the genomic potential of entire communities (Quince et al., 2017). Assembly of metagenome-assembled genomes and the application of long-read sequencing technologies, including Oxford Nanopore and PacBio, have expanded access to uncultured organisms, many of which represent novel lineages with unique metabolic pathways (Nurk et al., 2017). Between 2020 and 2025, the field of environmental microbiology has experienced rapid advances. Multi-omics integration now allows comprehensive analysis of genetic potential, gene expression, protein production, and metabolite profiles, providing unparalleled resolution of microbial function (Hasin et al., 2017; Mallick et al., 2021). Metatranscriptomics and metaproteomics extend analysis to functional activity, revealing how communities respond dynamically to environmental fluctuations (Franzosa et al., 2015). Metabolomics offers direct insight into the chemical repertoire produced by microbial metabolism, bridging the gap between genetic potential and ecological phenotype (Caesar et al., 2021). Stable isotope probing has further enabled the tracing of nutrient fluxes within communities, while network inference and machine learning approaches are increasingly employed to uncover hidden ecological patterns and predict functional interactions (Neufeld et al., 2007) in complex datasets, allowing predictive modeling of microbiome responses to climate change, pollution, and disturbance (Gloor et al., 2017; Topçuoğlu et al., 2020). Single-cell genomics and microfluidic cultivation methods have enabled access to previously uncultured microorganisms, uncovering novel lineages with distinctive metabolic properties, as shown in Table 1.1 (Blainey, 2013). eDNA has emerged as a powerful tool for biodiversity monitoring and ecosystem surveillance, offering a non-invasive means of detecting microbial shifts and tracking antimicrobial resistance (Thomsen & Willerslev, 2015).Together, these methods provide a multi-layered systems biology framework that links microbial diversity to function and positions microbiome research at the intersection of ecology and biotechnology.

Building on these advances, researchers have increasingly turned to the applied potential of microbiomes as sources of new bioactive chemistry and as targets for rational engineering. Contemporary bioprospecting integrates cultivation innovations, sequencing, and analytical chemistry to unlock novel natural products from environmental microbiomes. Techniques such as diffusion chambers and the iChip enable recovery of previously uncultivable taxa by mimicking native environments (Sedeek et al., 2022). Shotgun metagenomics and genome mining tools like antiSMASH allow systematic detection of biosynthetic gene clusters (BGCs), guiding discovery toward high-value candidates (Blin et al., 2025; Kenshole et al., 2021; J. Zheng et al., 2025). Metabologenomics, which couples genomic

TABLE 1.1

*Advanced techniques for environmental microbiome analysis: Capabilities and applications*

| Technique | Primary capability | Key applications in environmental microbiology | Advantages | Challenges |
|---|---|---|---|---|
| **16S rRNA amplicon sequencing** | Taxonomic profiling (bacteria, archaea, fungi) | Assessing pollution, contamination, and diversity in soil, water, and air samples | Cost-effective; targets highly conserved genes | Only targets one gene; cannot identify all functional capabilities |
| **Shotgun metagenomics** | Comprehensive taxonomic profiling (all microbes); functional potential (all genes present) | Discovering new organisms, exploring dynamic populations, and agricultural and ecological remediation studies | Captures all genomes; provides functional insights | Computationally intensive; complex data structure |
| **Metatranscriptomics** | Active gene expression; functional profile (what genes are active) | Studying the ecology of microorganism-mediated processes; analyzing gene expression under varying conditions (e.g., soil, sludge, rhizospheres) | Provides real-time active functions; offers insights into unculturable microbes | High ribosomal RNA abundance; mRNA instability; difficulty differentiating host from microbial RNA; limited reference databases |
| **Metaproteomics** | Protein synthesis; direct functional information (what proteins are produced) | Applications in bioremediation, carbon cycling, bioenergy, and understanding host-microbe crosstalk | Provides direct functional information; proteins are signature molecules of cellular processes | Challenges in protein extraction from complex samples; lack of genomic sequences for many microbes |
| **Microfluidics** | High-throughput single-cell isolation; controlled cultivation; precise manipulation of microenvironments; study of cell–cell and host-microbe interactions | New taxa discovery; screening for antibiotics and industrial enzymes; studying antibiotic resistance; biofilm studies; organ-on-a-chip models | High-throughput; lower cost; versatile; precise control; single-cell resolution; enables *in situ* cultivation | High cost and time for mass production of specific chips; accessibility issues; technology still in infancy |
| **Stable isotope probing** | Powerful technique in microbial ecology that links the identity of microorganisms to their function in an ecosystem without the need for cultivation | Useful to identify active members in microbial communities in bioremediation and biogeochemical cycles | Culture independent, in situ technique | Cost and availability |

predictions with high-resolution metabolomics, enables dereplication and structural prediction, while synthetic biology through clustered regularly interspaced short palindromic repeats (CRISPR)-based cluster activation, promoter refactoring, and heterologous expression has unlocked cryptic pathways and expanded structural diversity (Ameruoso et al., 2022; Z. Liu et al., 2021; Xu & Wright, 2019). Synthetic biology approaches are also applied to microbiome engineering, enabling targeted interventions such as pollutant degradation, soil fertility enhancement, and biosensor development (McCarty & Ledesma-Amaro, 2019). At the same time, bioprospecting continues to reveal new enzymes, antibiotics, and metabolites, while ecosystem-based biotechnologies increasingly harness entire microbial consortia to perform complex transformations for bioremediation, wastewater treatment, and bioenergy production (Pieper & Reineke, 2000; Ragozzino et al., 2025; Singh et al., 2016). In parallel, interest has expanded from the discovery of individual products toward the design of microbial consortia with predictable and useful properties.

The rational engineering of microbial communities has emerged as a frontier in applied microbiome research. Omics technologies provide detailed knowledge of community structure, while computational models simulate interactions and predict emergent behaviors, essential for the design of synthetic microbial communities (SynComs) (Karkaria et al., 2021; Y.-Q. Li et al., 2023). Integration of metabolomics with metagenomics enables the identification of ecological niches and nutrient limitations, guiding SynCom optimization for applications in agriculture, bioremediation, and biotechnology (Raajaraam & Raman, 2024; Shayanthan et al., 2022). These strategies extend the impact of environmental microbiome science from exploration of natural diversity toward predictive engineering and translational applications.

These discoveries have also highlighted the relevance of environmental microbiomes to the One Health framework, which emphasizes the interdependence of environmental, animal, and human health (Li & Zhang, 2023). Microbial communities safeguard ecosystem services critical for human well-being, such as nutrient cycling, pollutant degradation, and water purification, and they form the backbone of engineered processes such as wastewater treatment (Graves, 2022). Conversely, they can also serve as reservoirs for pathogens and antimicrobial resistance genes (Bengtsson-Palme et al., 2018). The environmental resistome, representing the natural diversity of resistance determinants, illustrates how ecosystems provide both evolutionary origins and contemporary reservoirs for antimicrobial resistance (D'Costa et al., 2011; Maestre-Carballa et al., 2022). Anthropogenic pressures, including widespread antibiotic use in medicine, agriculture, and aquaculture, have accelerated the spread of resistance through horizontal gene transfer, raising urgent concerns about environmental reservoirs as pathways for AMR dissemination (Larsson & Flach, 2022; Timms et al., 2023). The COVID-19 pandemic further underscored the importance of understanding microbial ecology in pathogen emergence, persistence, and spillover, reinforcing the need to integrate environmental microbiome research into health surveillance systems (P. Zhou et al., 2020).

Ecological and evolutionary principles provide a conceptual framework for understanding how microbiomes are structured and how they function (Nemergut et al., 2013). Community assembly is driven by both deterministic processes, such as environmental filtering and species interactions, and stochastic processes, including dispersal and ecological drift (Kayiranga et al., 2024). Classic theories of niche differentiation and neutral dynamics both contribute to explaining microbial community patterns, and contemporary approaches increasingly integrate these perspectives (Stegen et al., 2012). Successional dynamics play a central role in microbial ecology: primary succession shapes communities in newly formed habitats such as volcanic substrates and retreating glaciers, while secondary succession occurs in disturbed but previously colonized environments (Walker & Del Moral, 2003). Microbial communities are also structured by metabolic interdependencies. Cross-feeding interactions, or syntrophy, allow metabolic byproducts of one organism to fuel the growth of another, creating networks of interdependence that stabilize communities under resource-limited conditions (Morris et al., 2013). The spatial organization of microbial populations further influences ecosystem processes. Many microbes form structured biofilms, which confer protection, promote nutrient retention, and facilitate intercellular signaling (Flemming & Wuertz,

2019). Advanced imaging techniques have revealed microscale heterogeneity within these structures, highlighting the importance of spatial gradients in shaping ecological dynamics (Darch et al., 2012).

The growing recognition of environmental microbiomes as key components of conservation and restoration has expanded their relevance to applied ecology (Young et al., 2005). Microbial communities accelerate soil formation, enhance vegetation establishment, and buffer ecosystems against environmental stress (Bardgett & Putten, 2014). Microbial indicators are emerging as sensitive measures of ecosystem health, capable of detecting degradation earlier than conventional methods and guiding proactive interventions (de Vries et al. 2018). At the same time, microbiome research raises ethical and policy challenges. Questions of bioprospecting, benefit-sharing, and intellectual property are central to debates over biodiversity governance, particularly under international frameworks such as the Nagoya Protocol (Unit, 2025). The rapid rise of microbiome engineering and synthetic biology introduces issues of biosafety, ecological risk, and societal oversight (Torres et al., 2016). Equitable access to environmental genomic resources, data sovereignty, and the integration of microbiome science into

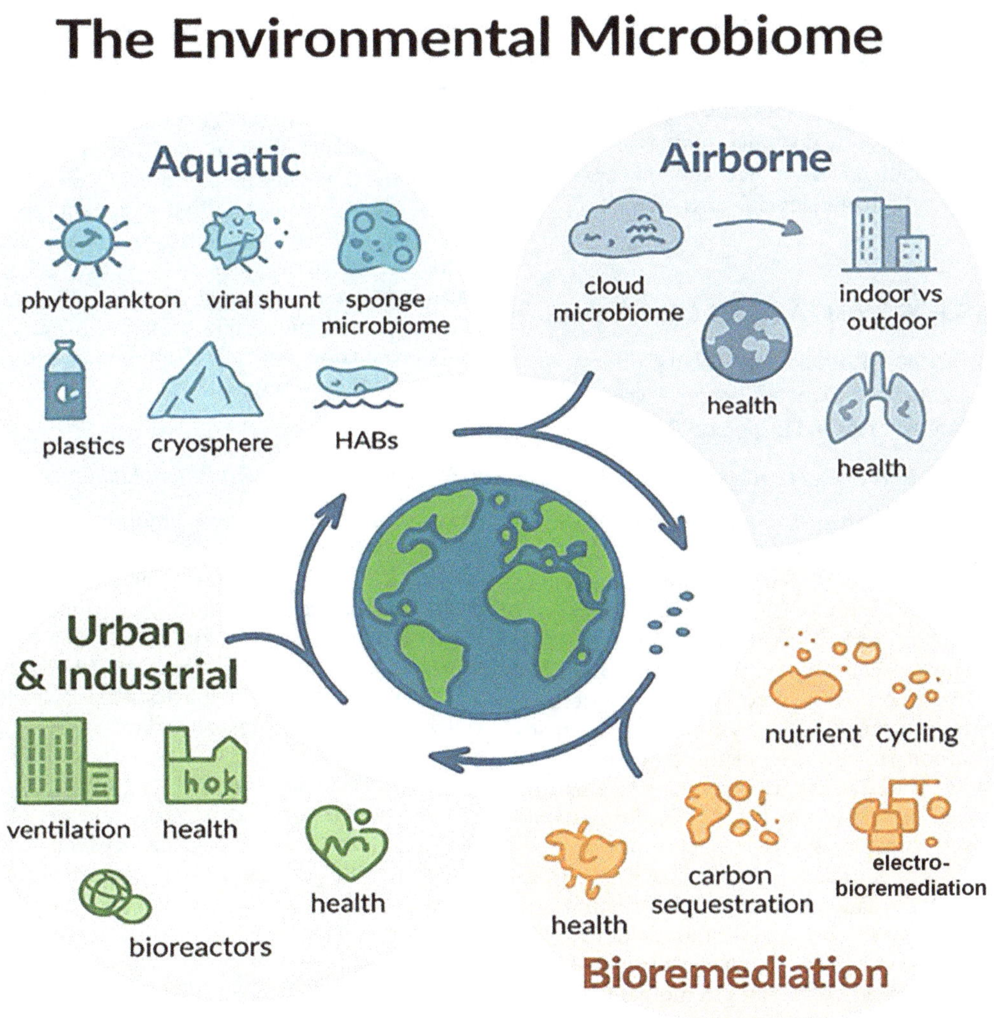


Figure 1.1 Schematic representation of the core components of the environmental microbiome. The illustration highlights microorganisms, their genetic material, metabolic functions, and ecological interactions that collectively sustain ecosystem structure and processes.

global health and sustainability policy are pressing frontiers (Reardon, 2017).

Taken together, these perspectives establish environmental microbiomes as fundamental to both planetary health and sustainable innovation. They operate as regulators of global processes, reservoirs of biotechnological potential, contributors to ecosystem services, and determinants of human and animal health (Cavicchioli et al., 2019). Their study bridges history and modern technological advances, integrates ecological theory with applied perspectives, and connects environmental science with ethics and policy (Al et al., 2024). This chapter builds on these foundations to explore microbiomes across diverse ecosystems, highlight their roles in ecological and biotechnological contexts, and consider future directions in predictive ecology, synthetic microbiomes, and global governance as demonstrated in Figure 1.1. By weaving together these threads, the chapter positions environmental microbiomes as a cornerstone of both the biosphere and the next generation of biotechnological discovery (L. Z. Liu et al., 2022).

## 1.2 MICROBIOME IN AQUATIC ECOSYSTEMS

### 1.2.1 Marine Microbiome in Global Carbon Cycling and Climate Change: Roles, Interactions, and Ecological Dynamics

#### *1.2.1.1 Role in Global Carbon Cycling*

Marine microorganisms drive several critical processes within the marine carbon cycle. Through photosynthesis, phytoplankton fixes atmospheric $CO_2$, converting it into organic carbon that enters the biological carbon pump (BCP) (Dang, 2020). This pump facilitates the transfer of particulate organic carbon (POC) from surface waters to the deep ocean, a sequestration pathway enabling carbon storage over centuries to millennia (Lampitt et al., 2023). Complementing this, the microbial carbon pump (MCP) transforms labile dissolved organic matter (DOM) into recalcitrant dissolved organic carbon (RDOC), a form of carbon resistant to microbial degradation and thus persistent in the ocean for thousands of years. Additionally, microbially induced carbonate precipitation (MICP) contributes to inorganic carbon sequestration through calcification processes often mediated by microbial communities (e.g., coccolithophores) (Jiao et al., 2024). The synergy among BCP, MCP, and MICP mechanisms forms a comprehensive framework for marine carbon storage, effectively mitigating atmospheric $CO_2$ and influencing the Earth's climate system. Marine macrophytes, sponges, and other macroorganisms further interact with microbiomes to enhance carbon transformation and sequestration dynamics through complex ecological networks (Fang & Achal, 2024; He et al., 2022).

#### *1.2.1.2 Climate Change Impact*

Climate change alters ocean temperature, acidification, nutrient dynamics, and stratification, profoundly impacting microbial community structure and functions. Viral ecology plays a pivotal role here, with marine viruses regulating microbial population dynamics via cell lysis, which is called the viral shunt (see Section 1.2.1.4); due to shunting microbial biomass into DOM and detritus, and thereby influencing carbon and nutrient fluxes (Locke et al., 2022). Climate-driven shifts in microbial biodiversity and viral-host interaction patterns alter these carbon cycling processes, potentially modifying the sequestration efficiency and feedback loops to climate change (Larkin et al., 2025). Such changes may affect the balance between microbial respiration (remineralization of organic carbon back to $CO_2$) and primary production, critically shaping marine carbon storage capacity (Dang, 2020).

#### *1.2.1.3 Interactions with Phytoplankton*

Phytoplankton, as primary producers, form the base of marine food webs and are central to carbon fixation. Their organic matter production supports various carbon transfer pathways, including sedimentation and aggregate formation promoted by transparent exopolymer particles (Richardson, 2019). Viral infection of phytoplankton leads to cell lysis, releasing organic carbon as DOM and fueling microbial loops, a key component of the viral shunt (see Section 1.2.1.4) (Kranzler et al., 2025). These interactions regulate nutrient recycling and influence the balance between carbon export to the deep ocean and carbon recycling within surface waters (Locke et al., 2022).

#### *1.2.1.4 Viral Shunt and Ecology of Viruses*

In the ocean, viruses reign supreme as the most numerous organisms, infecting marine microorganisms and altering organic carbon due to the death of microbes (Weitz & Wilhelm,

2012). The viral shunt disrupts the conventional food chain by transforming POC into dissolved organic carbon through lysis and feeding it to heterotrophic bacteria rather than to other bacteria (Shiah et al., 2022). This process enhances microbial respiration and recycling of nutrients but might lead to a reduced efficiency in the export of carbon to the deep ocean. Viral activities, therefore, critically mediate the carbon cycle, productivity of the ecosystem, and biogeochemical stability (Wei et al., 2022; Weitz & Wilhelm, 2012).

### 1.2.2 Freshwater Microbiome: Influence on Water Quality and Interactions with Pollutants Including Pharmaceuticals and Personal Care Products Pollution

#### *1.2.2.1 Influence on Water Quality*

The freshwater microbiome regulates elemental cycling, which affects the chemical and biological status of aquatic environments. Microbial communities use diverse metabolic pathways to regulate nitrogen, carbon, and phosphorus fluxes. They also drive organic matter degradation. These processes directly impact water clarity, oxygen levels, and nutrient availability (Muhie et al., 2025). Seasonal and spatial changes in microbial diversity shape ecosystem resilience and stability. Shifts in bacterial community structure are linked to changes in environmental conditions like temperature, pH, and nutrient inputs (Liu et al., 2024).

Proteobacteria and Actinobacteria are frequently associated with stable, high-quality freshwater bodies where they underpin critical biogeochemical processes. However, perturbations such as nutrient overload or physical disruptions can reduce microbial diversity and alter functional capacity, potentially undermining ecosystem services, including pollutant breakdown and self-purification (F. Zheng et al., 2022).

#### *1.2.2.2 Microbial Interactions with Pollutants Degradation Mechanisms*

Emerging pollutants, particularly pharmaceuticals and personal care products (PPCPs), are increasingly detected in freshwater environments at trace levels, where they can exert biologically active effects on resident microbial communities. The microbial degradation of PPCPs involves enzymatic processes, including oxidases, hydrolases, and reductases, which mediate the transformation of complex organic compounds into less harmful substances (Narayanan et al., 2023; X. Wang, Wang, et al., 2024). Recent studies demonstrate that certain bacterial taxa, such as *Stenotrophomonas maltophilia*, *Mycobacterium* spp., and various Proteobacteria, harbor genes that encode enzymes capable of degrading PPCPs and microplastics (MPs). However, the biodegradation rates vary widely depending on compound structure, microbial community composition, and environmental parameters including pH, temperature, and nutrient availability. For example, biodegradation efficiencies improve under neutral to alkaline pH and moderate temperature regimes, while acidic or highly polluted waters may inhibit microbial activity (Narayanan et al., 2023; X. Wang, Wang, et al., 2024). Community responses to PPCP pollution often include shifts toward antibiotic-resistant bacteria, with increased expression of antibiotic resistance genes (ARGs) and efflux pump systems such as SmeABC and regulators like *MtrA*. These adaptations provide resilience to contaminant stress but raise concerns for resistance gene propagation and pathogen emergence (Narayanan et al., 2023; X. Wang, et al. 2024; Wang, et al., 2024). Furthermore, combined exposure to multiple micropollutants can produce complex microbial responses, some compounds may inhibit degradation pathways for others, while some enhance microbial metabolism by serving as carbon sources. For instance, bisphenol A is more readily degraded by freshwater microbes compared to triclosan, with mixtures of these contaminants producing interactive effects on microbial degradation rates (Izabel-Shen et al., 2022; Narayanan et al., 2023).

#### *1.2.2.3 Implications for Freshwater Ecosystem Health and Management*

Microbial communities serve as both bioindicators and active agents in freshwater quality and contaminant mitigation. Leveraging advances in eDNA surveys and metagenomic analyses enhances the understanding of microbial community responses to pollution and enables predictive modeling of ecosystem health trajectories (Izabel-Shen et al., 2022; Narayanan et al., 2023). Effective freshwater management requires integrated approaches that address both nutrient and emerging pollutant inputs, promote microbial diversity, and prevent the spread of ARGs. Maintaining microbial functional potential for biodegradation is fundamental to sustaining the natural self-purification capacity of freshwater systems under growing anthropogenic stress (Izabel-Shen et al., 2022; Narayanan et al., 2023; X. Wang, Wang, et al., 2024).

### 1.2.3 Microbiome in Extreme Aquatic Environments: Hot Springs, Hypersaline Lakes, Deep-Sea Microbial Communities

Microbiomes in extreme aquatic environments like hot springs, hypersaline lakes, and deep-sea microbial communities exhibit unique adaptations and distinctive microbial compositions shaped by their challenging physicochemical conditions (Shu & Huang, 2022).

#### *1.2.3.1 Hot Springs Microbiome*

Hot springs harbor diverse microbiomes dominated by thermophilic and hyperthermophilic bacteria and archaea capable of surviving high temperatures, often exceeding 50°C. For example, microbial communities in hot springs with extremely high temperatures (around 89–90°C) are dominated by hyperthermophiles such as *Hydrogenobacter* and *Thermus*, while slightly cooler springs (around 52°C) host thermophiles like *Thermoanaerobaculum* and *Desulfurispora* (Debnath et al., 2019). Microbial diversity in hot springs is also influenced by pH and chemical composition. At neutral pH (5.7–8.5), diverse bacterial communities dominate, including phototrophic groups like Cyanobacteriota and Chloroflexota, whereas acidic springs (pH 1.5–3.6) harbor archaea like *Acidianus* and *Thermoplasma* (Karaseva et al., 2024; Roy et al., 2020). High-throughput DNA sequencing has enabled detailed metabarcoding studies revealing the link between physicochemical parameters and microbial composition in different springs. Bathing in hot springs also influences human gut microbiota by increasing specific bacterial populations, presumably due to the unique mineral properties of various kinds of springs (Çelik & Keskin, 2022; Takeda et al., 2024).

#### *1.2.3.2 Hypersaline Lakes Microbiome*

Hypersaline lakes, with salt concentrations often much higher than seawater, host microbial communities dominated by halophilic archaea and bacteria adapted to high salt and other stresses such as oxidative, metal, temperature fluctuations, and desiccation stresses. Microbes here are categorized as weak, moderate, or extreme halophiles, thriving in environments with 1–30% NaCl (Santos & Antón, 2019). As salinity increases in these hypersaline ecosystems, microbial diversity decreases; however, prokaryotes with extreme halophilic profiles, especially halophilic archaea such as Euryarchaeota, dominate in the most saline zones (Martínez et al., 2022b). These dominant communities also carry abundant stress response and ARGs, reflecting their survival strategies under harsh, nutrient-poor conditions. Studies of lakes like the Sambhar Salt Lake in India have shown these microbiomes to be enriched for metabolic features associated with energy harvesting and stress tolerance, enabling them to adapt and evolve in such extreme environments (Kheiri et al., 2023; Martínez et al., 2022a; Mehta et al., 2021).

#### *1.2.3.3 Deep-Sea Microbial Communities*

Deep-sea ecosystems are home to microorganisms known as deep-sea extremophiles. These microorganisms are specially adapted to harsh, high-pressure environments with low oxygen and low temperatures. Such deep-sea ecosystems are crucial to the decomposition and recycling of biogeochemical organic matter (Ge et al., 2025). Such ecosystems are characterized by the presence of Bacillota, Actinomycetota, and Pseudomonadota, which are known to live in deep-sea sediment and can metabolize different carbon sources (Ge et al., 2025). These and other deep-sea microorganisms are omnivorous and can metabolize different types of carbon nutrition. As an example of deep-sea extremophiles, some fungi, such as *Aspergillus*, are known to produce antimicrobials under deep-sea extensional pressures (Bao et al., 2025). They can also modify the composition of deep-sea bacterial communities. There is vertical linkage in surface and deep-sea microbes. Microbes from the surface and their proteins can integrate with deep-sea ecosystems. Some of them can reach the deep sea through sinking particles (S. Chen et al., 2024). Deep-sea wood-fall microbial communities illustrate great complexity in the interactions and functions between the extremophiles. The study of these extremophiles deepens the understanding of the role of microbes in nutrient cycles and deep ocean ecosystems (Bao et al., 2025; Giddings & Newman, 2022).

### 1.2.4 Cryosphere Microbiome: Microbiomes of Glaciers, Permafrost, and Sea Ice and Their Rapid Changes due to Climate Change

The cryosphere encompasses Earth's frozen water habitats, including glaciers, permafrost soils, snow, and sea ice. It constitutes approximately 20% of the planet's surface. These environments harbor unique microbial communities adapted to permanently or seasonally subzero conditions. The microorganisms dominating these ecosystems, including bacteria, archaea, fungi, and viruses, play crucial roles in biogeochemical cycling, ecosystem function, and climate feedbacks (Bourquin et al., 2022; Jungblut et al., 2022). Despite their vast extent and ecological importance, cryospheric microbiomes remain among the least understood on Earth. Rapid warming and accelerated melting of the cryosphere due to climate change threaten these specialized microbial communities, with implications for the global carbon cycle, microbial diversity, and potential pathogen release (Jungblut et al., 2022).

#### 1.2.4.1 Glacier and Ice Microbiomes

Glacial surfaces, ice sheets, and snowfields host cold-adapted microbial assemblages often dominated by taxa such as *Pseudomonas*, *Acinetobacter*, *Flavobacterium*, and psychrophilic lineages in Proteobacteria and Bacteroidota (Anesio & Laybourn-Parry, 2012; Ramasamy et al., 2023). These microbes thrive despite extreme physical stresses, including low temperatures, intense UV radiation, and nutrient limitation (Anesio & Laybourn-Parry, 2012; Ramasamy et al., 2023). They perform critical ecosystem processes, including primary production via photosynthetic microbes in cryoconite holes, organic matter recycling, and influencing albedo through pigmentation (e.g., red snow algae) (Anesio et al., 2017). Microbial diversity in snow and glacial ice tends to be lower than in downstream aquatic and soil environments, reflecting constrained environmental niches. However, glaciers serve as reservoirs of cold-adapted bacterial diversity that seed connected ecosystems downstream (Anesio et al., 2017; Wilhelm et al., 2013).

#### 1.2.4.2 Permafrost Microbiomes

Permafrost soils, permanently frozen for at least two consecutive years, store vast quantities of organic carbon and harbor diverse microbial communities adapted to subzero temperatures and anoxic conditions (Waldrop et al., 2025). These microbes include metabolically versatile bacteria and archaea, some capable of methanogenesis and other anaerobic pathways critical to greenhouse gas fluxes. With warming-induced thawing, permafrost microbiomes rapidly shift, increasing microbial activity and decomposition rates that release $CO_2$ and methane, potent greenhouse gases, thus creating positive climate feedback loops. Permafrost also harbors ancient microorganisms, some viable yet non-culturable, raising concerns about the release of potential pathogens (Waldrop et al., 2025; R. Wu et al., 2022).

#### 1.2.4.3 Sea Ice Microbiomes

The brine channels within sea ice support specialized microbial consortia, including psychrophilic bacteria, archaea, microalgae, and viruses, that are active during the polar winter and spring. These microbes contribute to polar primary production and nutrient cycling and serve as a base for polar food webs. Sea ice microbial communities are highly dynamic and sensitive to seasonal and climate-induced changes in sea ice extent and thickness (Bourquin et al., 2022; Jungblut et al., 2022).

#### 1.2.4.4 Impacts of Climate Change on the Cryosphere Microbiome

Climate change is driving rapid warming and melting across the cryosphere, profoundly shifting microbial communities and their roles in global biogeochemical cycles. As ice and permafrost environments thaw, conditions are altered, favoring opportunistic species over native, cold-adapted microorganisms (Yarzábal et al., 2021). This leads to reduced microbial diversity and disrupted ecosystem stability. Rising temperatures also accelerate the decomposition of organic matter, leading to enhanced greenhouse gas emissions from permafrost soils (Yarzábal et al., 2021; X. Zhu et al., 2024). This increased atmospheric burden directly contributes to further warming, which, in turn, accelerates the permafrost thaw. The melting and retreat of ice also have significant implications for microbial dispersal, releasing microbes into new freshwater and marine environments (Yarzábal et al., 2021). A more concerning aspect is the potential release of ancient, entrapped microorganisms, including pathogens and ARGs. This poses an unprecedented risk to human and animal health, introducing novel biological agents into environments that have no evolved resistance (Bourquin et al., 2022; Caruso & Rizzo, 2025; X. Zhu et al., 2024).

The resilience of cryospheric microbes is due to specific genomic and functional traits. Many exhibit an elevated guanine–cytosine (GC) content and a streamlined genome, and they produce specialized biomolecules like cold-active enzymes and antifreeze proteins. This metabolic versatility allows them to thrive in extreme, low-energy conditions (Bourquin et al., 2022; Jackson et al., 2024; Jungblut et al., 2022). These changes present both challenges and opportunities. There is a need to develop comprehensive methods to document global cryospheric microbiomes using advanced techniques like metagenomics and multi-omics. It is also essential to understand their functional roles and monitor public health risks. However, the unique adaptations of these microbes also provide positive opportunities for biotechnological exploration, including the discovery of cold-active enzymes and other novel biomolecules for industrial purposes (Jackson et al., 2024; Jungblut et al., 2022).

### 1.2.5 Harmful Algal Blooms: Microbiome Interactions Influencing and Influenced by Harmful Algal Blooms

Harmful algal blooms (HABs) are influenced by complex and dynamic interactions with microbial communities that shape and are shaped by these blooms. The microbiome interacts with HABs involving bacteria, viruses, fungi, and other microbes that can either promote or suppress algal bloom formation, impact toxin dynamics, and modulate nutrient cycling (Jackson et al., 2024).

#### *1.2.5.1 Microbial Interactions with HABs*

The microbiome associated with HABs, composed of diverse heterotrophic bacteria, viruses, fungi, and microalgae, engages in reciprocal interactions with bloom-forming algae. These interactions range from mutualistic to antagonistic, influencing bloom initiation, maintenance, and decline. Microbial consortia associated with HABs can facilitate algal growth by providing essential nutrients or growth factors or inhibit blooms through algicidal activities and competition (Bourquin et al., 2022; Jackson et al., 2024).

#### *1.2.5.2 Bacterial Heredity on HABs: A Molecular Perspective on Growth and Inhibition*

A web of bacteria continually evolves and interacts with HABs, contributing to their development. By colonizing the "phycosphere", some bacterial taxa promote the development of HAB species (Abate et al., 2024). Algae cells produce and release specific nutrients and vitamins that support the growth and proliferation of algal blooms (J. Zhu et al., 2024). In nitrogen-limited waters, some algal blooms are supported through the symbiotic relationship with diazotrophic bacteria, which provide nitrogen to cyanobacterial HABs. Other bacteria are termed algicidal bacteria and act as natural bloom terminators. These microorganisms produce compounds that lyse algal cells, either through physical attachment or through the release of bioactive molecules like prodigiosin or various amines (Coyne et al., 2022). Algicidal agents capable of producing ROS that induce oxidative stress are a promising area of research for using green methods to control blooms. The intricate and often species-specific nature of these bacterial–algal interactions directly affects bloom dynamics, influencing the succession and overall composition of the bloom community (Coyne et al., 2022).

#### *1.2.5.3 Viral and Fungal Role in HAB Dynamics*

In addition to bacterial interaction, fungi and viruses impose additional levels of regulation on HAB processes. Viruses that infect HAB algae and bacteria are involved in population regulation by executing lytic cycles that lead to the rapid killing of host cells. This widespread viral lysis brings about huge amounts of DOM into the water column, in turn subsidizing heterotrophic bacterial cycles and significantly modifying nutrient recycling in the bloom community (Coy et al., 2018; Gao et al., 2022). Parasitic fungi, such as chytrids, may also attack algal cells, contributing to the restriction of a HAB's size. The activity and presence of these fungi and viruses add a complex, by and large underappreciated, layer to the mechanisms governing the initiation, persistence, and collapse of toxic algal blooms (Rasconi et al., 2014).

#### *1.2.5.4 Microbial Community Shift during HABs*

The formation of a toxic algal bloom causes a significant temporal and spatial shift in the microbial community's structure and function. This shift results directly from changes in the physicochemical environment and the production of specific algal exudates during the bloom (Abate et al., 2024). Functional metagenomics has shown that co-occurring bacteria participate in nutrient cycling, detoxification, and secondary metabolite exchange with the HAB algae (Abate et al., 2024; Nwankwegu et al., 2019). These findings underscore that HABs are not single-species events but are instead driven by co-evolving microbial consortia. It is therefore necessary to monitor HAB ecology and develop predictive models by tracking the diversity and function of the entire microbial consortium, alongside the target HAB species (Abate et al., 2024; Nwankwegu et al., 2019).

#### *1.2.5.5 Role of Microbial Metabolites and Chemical Ecology*

Chemical signaling plays a central role in interactions within HAB communities. HAB species produce allelochemicals, specialized compounds that deter competitors or grazers as a means of defense. Their associated bacteria can metabolize or detoxify algal toxins. This process can critically influence the overall toxicity and fate of a bloom (Coyne et al., 2022). The initiation and collapse of a bloom are often mediated by complex chemical signaling networks. These include quorum sensing and nutrient sensing. Allelopathic interactions between algae and bacteria are a key component of the bloom's chemical ecology. These interactions dictate the relationships and communication that drive the entire process (Coolahan & Whalen, 2025; Coyne et al., 2022).

#### *1.2.5.6 Implications for HAB Management*

The growing understanding of these microbial interactions has significant implications for HAB management. The discovery of naturally occurring algicidal bacteria and their bioactive compounds presents a promising avenue for developing sustainable approaches to prevent the impacts of blooms (Coyne et al., 2022). However, the field effectiveness and ecological safety of these methods require further rigorous research before widespread adoption. Furthermore, a deeper comprehension of microbiome interactions is essential for developing more sophisticated biomonitoring tools and early warning systems based on detecting specific changes in microbial community composition that precede the formation of a bloom (Abate et al., 2024; Coyne et al., 2022).

### 1.2.6 Microplastics and Their Interactions with Aquatic Microbiomes

#### *1.2.6.1 Microplastics as Microbial Habitats (Plastisphere)*

MPs are plastic particles smaller than 5 mm that are pervasive pollutants in aquatic environments and serve as novel surfaces for microbial colonization. This newly created microbial habitat is collectively referred to as the "plastisphere" (Liang et al., 2025). Biofilms on MPs host diverse microbial communities that are distinct from those in the surrounding water or sediment (Liang et al., 2025). These communities frequently include bacteria capable of degrading plastics, such as those producing enzymes like PETase that break down polyethylene terephthalate, as well as other taxa enriched in plastic degradation pathways (Cai et al., 2023; Reddy et al., 2023). A concerning aspect of the plastisphere is its potential to concentrate ARGs. This can provide a vector for the dissemination of ARGs through aquatic food webs, particularly in urban or pollution-impacted waters, posing a potential risk to public health (Kim & Yoo, 2024; Siddique et al., 2025).

#### *1.2.6.2 Effects on Native Microbial Communities*

The presence of MPs can significantly alter the diversity and composition of native microbial communities. In freshwater systems, for instance, microbiomes associated with sentinel organisms like *Daphnia* show compositional shifts and functional gene enrichment in response to MPs exposure. These changes can include an increase in plastic-degrading enzymes and ARGs (Varg, Bergvall, et al., 2022; Varg & Svanbäck, 2023). By providing a new and unique habitat, MPs can disrupt microbial ecosystem functions by biasing microbial populations toward taxa that are well-adapted to plastisphere colonization. This shift may ultimately impact crucial processes such as nutrient cycling and organic matter decomposition within the aquatic environment (Yang, Liu, et al., 2020; Yin et al., 2023).

#### *1.2.6.3 Microplastic Degradation by Aquatic Microbes*

The biological degradation of MPs in aquatic environments is a process that is highly dependent on the type of polymer. Polymers like polyethylene terephthalate and polyamide are more susceptible to enzymatic degradation catalyzed by enzymes such as lipases, esterases, and proteases because they are hydrolysable (Pal et al., 2025). In contrast, non-hydrolyzable polymers like polyethylene and polypropylene are far more resistant to microbial breakdown due to their chemical stability, which results in a much slower and often incomplete degradation process (Pal et al., 2025; Tokiwa et al., 2009). A key factor that can significantly enhance this process is the formation of a biofilm on the surface of the MPs. This biofilm provides a stable surface for microbial adhesion and creates a microenvironment that is more conducive to enzymatic activity, thereby accelerating the rate of degradation (S. He et al., 2022).

#### 1.2.6.4 *Impacts on Aquatic Organisms and Microbiomes*

The presence of MPs can have a wide range of impacts on aquatic organisms and their associated microbiomes. MPs can be ingested by organisms across different trophic levels, leading to various physical and physiological impacts, including damage to the digestive tract, oxidative stress, and altered metabolism (Pal et al., 2025). The toxicity of MPs is further enhanced when they adsorb pollutants such as polychlorinated biphenyls (PCBs) and heavy metals (HMs). Additionally, MPs can influence the composition of the gut microbiota in fish and invertebrates, often causing a reduction in beneficial taxa and a corresponding increase in opportunistic or pathogenic microbes (X. Wu et al., 2025). These changes in the host's microbiota, induced by MPs exposure, may ultimately affect the host's health, immunity, and overall resilience to environmental stressors (Varg, Outomuro, et al., 2022; X. Wu et al., 2025).

#### 1.2.6.5 *Environmental and Ecological Consequences*

MPs may act as vectors for human-derived contaminants, including ARGs and antibiotics, which can then enter aquatic microbiomes. The altered microbiomes found on MPs surfaces and within host organisms have the potential to significantly impact aquatic biogeochemical cycles and the dynamics of the food web (Hongjin et al., 2025; Pal et al., 2025). Furthermore, urban aquatic environments, which tend to have higher concentrations of MPs, often show more pronounced microbial and functional alterations, underscoring the link between pollution levels and ecological impact (Jolaosho et al., 2025).

## 1.3 AIRBORNE MICROBIOME

The airborne microbiome comprises diverse communities of microorganisms suspended in the atmosphere, including bacteria, archaea, viruses, fungi, and microbial eukaryotes. These microorganisms are an integral but often understudied component of the Earth's biosphere, inhabiting a dynamic and complex environment influenced by multiple natural and anthropogenic factors (Prussin & Marr, 2015). The atmosphere acts as a vast and mobile habitat where microbial populations are transported over short and long distances, facilitating microbial dispersal and ecological connectivity between terrestrial, aquatic, and built environments (BEs). Airborne microbial communities are highly diverse, frequently dominated by bacterial phyla such as Proteobacteria, Actinobacteria, *Bacteroidetes*, and *Firmicutes* (Ruiz-Gil et al., 2020).

These communities vary spatially and temporally depending on geographic location, altitude, land use, weather conditions, and seasonal changes. Sources of airborne microbes include soil, plants, ocean surfaces, human and animal hosts, and BEs (Ruiz-Gil et al., 2020). The composition of the airborne microbiome is further shaped by physical processes such as wind, rain, and air mass movement. Importantly, the airborne microbiome influences ecosystem functioning and human health (X. Chen et al., 2020; Ruiz-Gil et al., 2020). Microbes transported through the air can participate in biogeochemical cycling, cloud formation, and atmospheric chemistry. At the same time, they can act as vectors of plant, animal, and human pathogens, contributing to disease transmission and allergy development (Šantl-Temkiv et al., 2022), as illustrated in Figure 1.2. Understanding the diversity, sources, dynamics, and ecological roles of the airborne microbiome is essential for elucidating its implications for environmental processes and public health (Ruiz-Gil et al., 2020; Šantl-Temkiv et al., 2022).

### 1.3.1 Composition and Diversity of the Airborne Microbiome

#### 1.3.1.1 *Taxonomic Composition*

The airborne microbiome represents a vast "pan-microbiome" comprising over 900 bacterial genera spanning approximately 70 phyla and also includes archaea, fungi, viruses, and various microbial eukaryotes. Dominant bacterial groups commonly found in both indoor and outdoor airborne communities typically include Proteobacteria, Actinobacteria, *Bacteroidetes, and Firmicutes* (Y. Chen et al., 2024). Interestingly, the core microbiome, defined as genera present in more than 90% of samples, is relatively small, constituting only about 8% of all detected genera. This indicates a high degree of heterogeneity and diversity in airborne microbial communities across different locations and temporal scales

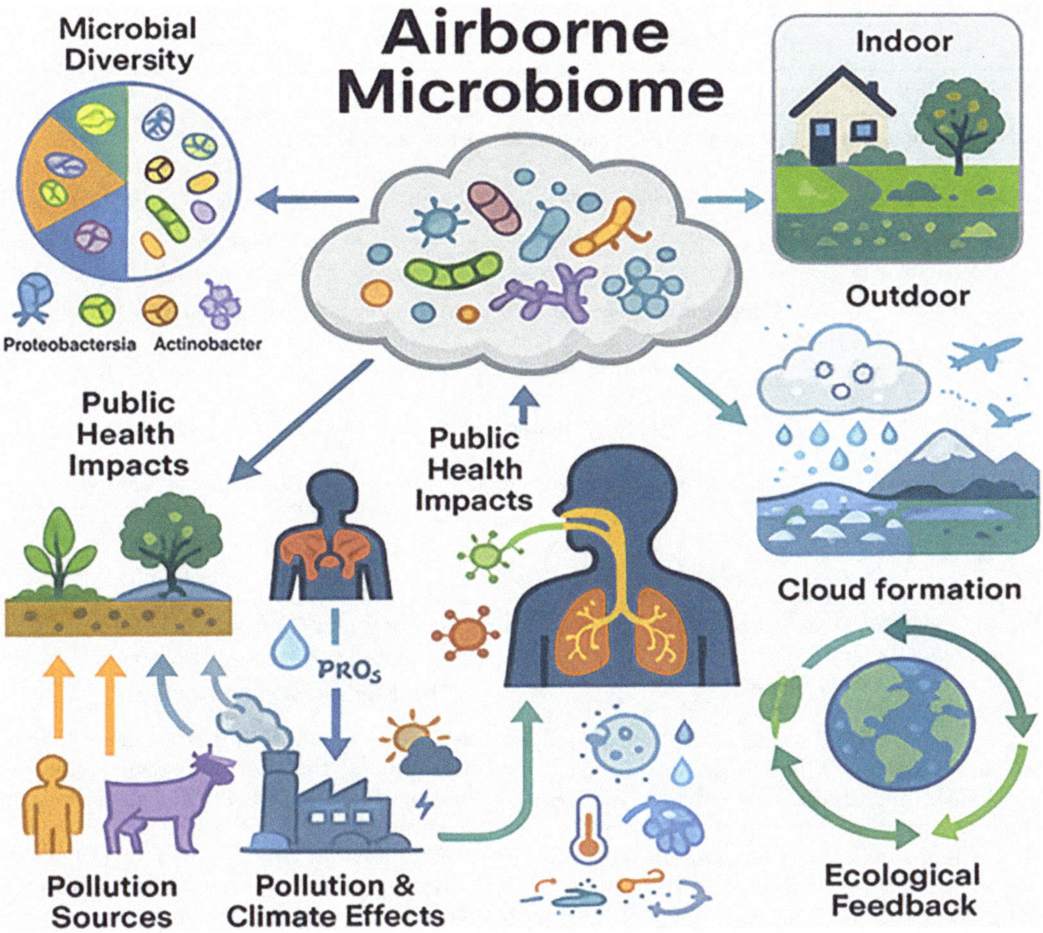


Figure 1.2 Schematic illustration of airborne microbiome at the nexus of pollution, climate, ecosystem function, and public health.

(Jiang et al., 2022). Anthropogenic environments, such as human settlements, livestock farms, and wastewater treatment plants (WWTPs), generally exhibit higher microbial richness and diversity compared to natural environments, often with a disproportionately higher presence of potential pathogens (Dodamani et al., 2022; Oluseyi Osunmakinde et al., 2019) (Table 1.2).

#### 1.3.1.2 Diversity Patterns

Alpha diversity metrics, such as the Chao1 richness index and Shannon diversity index, consistently demonstrate that airborne microbial richness and diversity are generally higher in anthropogenically influenced environments compared to natural settings. Seasonal, geographical, and environmental factors are key drivers of spatial and temporal variability in diversity. For instance, indoor air typically exhibits lower phylogenetic diversity than outdoor air. Climatic parameters like temperature, atmospheric pressure, and wind speed often correlate positively with airborne bacterial diversity, while the prevalence of pathogens can be significantly enhanced in urban and polluted areas. Urbanization, human activities, and air pollution collectively influence community composition by increasing microbial richness, but also by potentially increasing the presence of pathogens and antibiotic-resistant microbes (Jiang et al., 2022; Pollegioni et al., 2023).

### 1.3.2 Sources and Ecological Drivers

Source tracking analyses indicate that airborne microbes originate from diverse environments,

TABLE 1.2
*Key microbial features in the airborne microbiome*

| Microbial features | Details |
|---|---|
| Typical dominant phyla | Proteobacteria, Actinobacteria, Bacteroidetes, Firmicutes |
| Number of genera detected | ~944 (pan-microbiome) |
| Core microbiome | ~8% of genera shared across most samples |
| Diversity variation | Higher in anthropogenic environments, seasonal and climatic variation is significant |
| Main sources | Soil, plants, human body surfaces, animals, freshwater, saline water, WWTPs |
| Potential pathogens | Higher proportion in human settlements, livestock farms, and wastewater treatment |
| Ecological drivers | Temperature, atmospheric pressure, wind speed, and anthropogenic activity intensity |

including soil, plants, non-saline and saline waters, animals, humans, and waste treatment systems (J. Zhao et al., 2022). Microbial communities in the air represent a mixture of these sources, with outdoor air reflecting more soil and plant-associated microbes, while indoor air contains more human and animal-derived taxa. Weather and climatic conditions strongly influence microbial dispersal, survival, and community assembly in the atmosphere (Meadow et al., 2014b; Prussin & Marr, 2015).

### 1.3.3 Functional and Health Implications

The airborne microbiome also includes a small but significant fraction of potential pathogens, whose occurrence is greater in densely populated or industrial areas compared to natural environments (J. Zhao et al., 2022). Microbial community networks show complex patterns of co-occurrence, influenced by environmental conditions and anthropogenic impacts, affecting microbial interactions and possibly ecosystem and human health (Philippot et al., 2021a; J. Zhao et al., 2022).

This understanding is drawn from a global meta-analysis of air samples (3226 samples from 42 studies) that applied uniform bioinformatics processing to reduce technical biases, revealing a broad and complex airborne microbial ecosystem shaped by natural and human factors. Other studies corroborate these findings and expand on indoor/outdoor microbiome differences and seasonal dynamics (J. Zhao et al., 2022).

### 1.3.4 Indoor versus Outdoor Airborne Microbiomes: Distinct Characteristics and Implications

Indoor and outdoor airborne microbiomes exhibit distinct characteristics with significant implications for human health and environmental management (Y. Chen et al., 2024). Outdoor airborne microbiomes typically show higher microbial diversity and concentration, with bacterial levels often exceeding those found indoors by 1.26 to over four times, although indoor environments commonly harbor proportionally more fungi due to indoor moisture and growth conditions (Y. Chen et al., 2024; Prussin & Marr, 2015). The outdoor microbiome influences indoor air through ventilation and infiltration, causing the indoor airborne microbial community to partially reflect outdoor taxa, especially in spaces with natural ventilation (Meadow et al., 2014b; Prussin & Marr, 2015). However, indoor microbiomes are uniquely shaped by human occupants, pets, plants, and indoor surfaces, often dominated by human-associated bacteria and indoor fungi genera such as *Cladosporium*, *Penicillium*, and *Alternaria* (Leung et al., 2019; Rocchi et al., 2020; J. Zhao et al., 2022).

Indoor microbial exposure is affected by factors including occupancy density, ventilation type (natural versus mechanical), efficacy of air purifiers, and indoor dampness or mold growth, which can substantially elevate the presence of potentially pathogenic fungal spores linked to respiratory diseases (Y. Chen et al., 2024; Leung et al., 2019). The majority of indoor bioaerosols fall within the inhalable particle size range (<4.7 μm), increasing inhalation exposure risk. Seasonal cycles affect both indoor and outdoor communities, with microbial abundance peaking in warmer months (Ghosh et al., 2015). Crucially, indoor airborne microbiomes pose elevated health risks relative to outdoor air due to particle accumulation and closer human proximity, necessitating tailored indoor air quality interventions to mitigate allergy and infectious disease risks (Gaylarde et al., 2024). These differences underscore the complex interactions between outdoor environmental sources, indoor habitats, and occupants in shaping the airborne microbiome, pointing to the importance of integrated approaches in building design and public health strategies (Leung et al., 2019).

### 1.3.5 Effects of Pollution and Climate Change on Microbial Communities in the Air

#### *1.3.5.1 Impact of Air Pollution on Airborne Microbial Community Structure*

Air pollution, particularly the increase in particulate matter (PM2.5, PM10), gases like CO, $CO_2$, $O_3$, $NO_2$, and chemical components associated with haze and dust events, strongly influences the composition, diversity, and activity of airborne microbial communities (Aliu Olalekan Olatunji et al., 2024). Fine particulate matter (PM2.5) serves as a crucial physical substrate and transport vehicle for bacteria, fungi, viruses, and microbial fragments suspended in the atmosphere. Studies show a strong positive correlation between particulate matter concentration and the diversity and abundance of airborne microbes, with pollution episodes sometimes increasing microbial richness by providing energy and habitat niches for microbes on particles. Pollution alters bacterial and fungal community composition by selecting for certain taxa capable of surviving or thriving in polluted atmospheres (Aliu Olalekan Olatunji et al., 2024; K. Fu et al., 2024). For example, Proteobacteria (such as *Novosphingobium*) often dominate in highly polluted urban air, while some sensitive taxa decline. Seasonal and spatial studies reveal that air polluted by combustion (e.g., residential firewood use) leads to distinct microbial profiles compared to cleaner air (Jiang et al., 2022). Furthermore, polluted air, particularly during haze events, increases the abundance and diversity of ARGs associated with airborne bacteria, posing health risks by potentially disseminating antimicrobial resistance through the air (Xiu & Hu, 2025).

#### *1.3.5.2 Dysbiosis and Alteration of Human-Associated Airborne Microbiota*

Air pollution not only affects environmental microbes but also disrupts the human respiratory and skin microbiota. Inhaling fine particles and harmful gases can cause dysbiosis (an imbalance of microbial communities) in the respiratory tract (Mazumder & Hussain, 2024). This often leads to a decrease in beneficial bacteria, like *Fusobacteria* and *Bacteroidetes*, while increasing potentially pathogenic groups such as Proteobacteria and Firmicutes (Thomas & Nashwan, 2025). This imbalance is linked to a higher risk of respiratory infections, asthma, and chronic inflammation. Pollutant exposure also disrupts the pharyngeal microbial community, which is crucial for immune protection against airborne pathogens (Mazumder & Hussain, 2024). This imbalance further increases the risk of infection. Additionally, pollutants can enhance the production of reactive oxygen species, which damage epithelial barriers and kill off beneficial microbes, contributing to inflammation and chronic respiratory illness (Sierra-Vargas et al., 2023; Thomas & Nashwan, 2025).

#### *1.3.5.3 Influence of Climate Change on Airborne Microbial Communities*

Climate change significantly affects airborne microbiomes through altered atmospheric conditions. Rising temperatures, changes in humidity, and variations in wind speed collectively influence microbial dispersal, survival, and community assembly in the air (Ruiz-Gil et al., 2020). Warmer conditions may enhance microbial metabolism and increase the airborne survival of certain pathogens, while extreme climatic events can, in some cases, lead to a reduction in overall diversity. Furthermore, climate-driven changes in air pollution patterns, such as increased wildfire smoke and dust storms, directly alter the composition and function of the airborne microbiome, thereby impacting ecosystem processes and the risk of disease (Z. Fang et al., 2018; Ontiveros et al., 2021). Climate change also exacerbates the effects of existing pollution; for instance, by increasing the frequency and severity of haze events, which have been observed to significantly modify airborne bacterial community structures, often favoring certain resistant species (Ma et al., 2025).

#### *1.3.5.4 Ecological and Health Implications*

The impacts of climate change also include shifts in ecological functions, where changes in airborne microbial communities affect atmospheric biogeochemical cycles, cloud formation, and nutrient cycling; shifts toward pollution-tolerant and pathogenic taxa may disrupt these vital processes (Uwishema et al., 2023). Public health risks are heightened as airborne microbial alterations and increased ARGs exacerbate health concerns related to respiratory infections, allergy development, and chronic diseases, with vulnerable populations including children, the elderly, and immunocompromised individuals being most impacted (Ifedinezi et al., 2024). Consequently, understanding pollution and climate change impacts on airborne microbiomes is crucial for effective air quality management, disease surveillance, and targeted public health interventions (Liao et al., 2024).

### 1.3.6 Long-Range Transport of Airborne Microbes and Its Implications

#### *1.3.6.1 Mechanisms and Dynamics of Long-Range Transport*

Airborne microbes can remain suspended in the atmosphere for days to weeks, enabling their movement across thousands of kilometers, even across oceans and continents. The key mechanisms enabling such long-range transport include aerosolization and uplift, whereby microbes become airborne through various natural processes such as wind erosion of soil and dust, bubble bursting at ocean surfaces, evaporation, and emission from vegetation; dust storms and wildfires also entrain large quantities of microbes into the atmosphere (Mayol et al., 2017). Atmospheric suspension, where fine particles and aerosols (including PM2.5 and PM10) provide physical substrates that support microbial survival and transport, protecting microbes from desiccation, UV radiation, and temperature extremes, allows them to remain viable or at least genetically intact over long distances (X. Chen et al., 2020). Meteorological transport, involving wind currents, convection, and turbulence within the atmospheric boundary layer and free troposphere, facilitates horizontal and vertical movement of microbial-laden aerosols over intercontinental scales, with islands and landmasses potentially acting as "stepping stones" for microbial exchange over oceans (Manavi et al., 2025). Finally, residence time is a critical factor, with studies estimating that about 50% of airborne prokaryotes remain suspended for approximately 17 days, and eukaryotes for about 3.5 days, which is sufficient for intercontinental transport reaching distances of 6000 to over 20,000 km depending on factors such as size, weather, and deposition events (Bowers et al., 2013; Mayol et al., 2017).

#### *1.3.6.2 Sources and Composition of Microbial Communities in Long-Range Transport*

Airborne microbial communities transported over long distances comprise mixed-origin populations. These include a combination of marine-origin microbes, particularly over oceanic areas, representing roughly 25–35% of detected taxa, and terrestrial-origin microbes that dominate many regions, especially downwind from continental dust, urban aerosols, and vegetation. These taxa encompass bacteria, fungal spores, viruses, and microalgae; fungal spores, due to their aerodynamic properties, often dominate the eukaryotic fraction and can remain airborne for long periods (Al-Shaarani & Pecoraro, 2024; Mayol et al., 2017).

#### *1.3.6.3 Ecological and Biogeographical Implications*

These long-distance transported microbes also play critical roles in ecosystem connectivity, microbial biogeography, ocean–atmosphere exchange, and climate and atmospheric processes. Their movement links geographically disparate ecosystems, contributing to gene flow, microbial diversity, and the colonization of new habitats, thereby challenging traditional notions of microbial endemism by enabling widespread distribution of taxa (Mayol et al., 2017). Oceans act as both sources and sinks of these airborne microbes through exchange fluxes driven by sea spray and dry/wet deposition. Furthermore, airborne microbes influence cloud formation, precipitation processes, and atmospheric chemistry by acting as ice nuclei or participating in chemical transformations, demonstrating their far-reaching impact on Earth's systems (Griffin, 2007; Mayol et al., 2017).

#### *1.3.6.4 Public Health and Environmental Implications*

The long-range movement of pathogenic microbes increases the risk of disease transmission across regions, impacting plants, animals, and humans (Liao et al., 2024; Mayol et al., 2017). Airborne fungal spores and other allergens transported over long distances contribute to respiratory allergies and asthma outbreaks, while airborne transport may also facilitate the dissemination of ARGs and resistant bacteria between environments (Al-Shaarani & Pecoraro, 2024). Understanding these long-range transport patterns is crucial for environmental monitoring, assisting in modeling microbial exposure risks, and managing air quality effectively (F. Chen et al., 2024; Mayol et al., 2017).

#### *1.3.6.5 Deposition and Removal*

Microorganisms are removed from the atmosphere via two primary mechanisms: dry deposition and wet deposition. Dry deposition involves the settling of microbes onto surfaces, water bodies, and soils, a process mediated by gravity and turbulent fluxes (Farmer et al., 2020). Wet deposition, on the other hand, refers to the removal of microbes by precipitation events, which efficiently scavenge them from the air. This process often causes episodic decreases in airborne microbial abundance. The effects of wet deposition are variable, depending on factors such as rain frequency and intensity, but generally lead to localized and temporary reductions in airborne microbial loads (Farmer et al., 2020; Tignat-Perrier et al., 2019).

### 1.3.7 Cloud Microbiome and Its Role in Cloud Formation and Precipitation

The cloud microbiome encompasses diverse taxa of bacteria, fungi, archaea, viruses, and microeukaryotes suspended within cloud droplets, representing a dynamic and ecologically significant component of the atmosphere (Péguilhan et al., 2025). Clouds are extreme habitats characterized by low temperatures, which are often near freezing, osmotic stress, oxidants, e.g., reactive oxygen species, variable UV radiation exposure, and limited nutrient availability (Šantl-Temkiv et al., 2015). Despite these challenges, cloud-borne microorganisms exhibit adaptations that enable survival, metabolic activity, and participation in cloud microphysical and chemical processes (Joly et al., 2015; Šantl-Temkiv et al., 2015).

#### *1.3.7.1 Microbes as Cloud Condensation and Ice Nucleating Particles*

Microbial cells commonly range from 0.25 to 1 µm in size, overlapping with the size range of critical cloud condensation nuclei (CCN) (Lazaridis, 2019). The ability of bacterial and fungal cells to adsorb water vapor and nucleate liquid droplets contributes significantly to cloud droplet formation. Biological particles can act as supermicron CCN, influencing droplet size distributions and collision–coalescence processes, thus affecting precipitation efficiency (Lazaridis, 2019; Spracklen & Heald, 2014). Furthermore, certain bacteria, notably species of *Pseudomonas*, *Xanthomonas*, and *Erwinia*, produce specialized ice nucleation active proteins (INPs) embedded in their outer membranes (e.g., InaZ protein). These proteins promote ice crystal formation at temperatures as warm as −2°C, substantially higher than typical abiotic INPs (S. Huang et al., 2021; Roeters et al., 2021). These biological INPs strongly influence the glaciation of mixed-phase clouds, accelerating ice crystal formation, precipitation initiation, and modulating cloud lifetime and albedo effects (S. Huang et al., 2021). Fungal spores and pollen grains also contribute to ice nucleation, possessing surface proteins and exopolysaccharides that facilitate freezing under supercooled conditions (Duan et al., 2023; Iannone et al., 2011). The relative contribution of microbial INPs to total atmospheric INPs varies spatially and temporally; recent studies estimate that in some clouds, biological INPs can account for a substantial fraction of active ice nuclei (S. Huang et al., 2021).

#### *1.3.7.2 Microbial Survival, Metabolism, and Ecophysiology in Cloud Droplets*

Microorganisms suspended in cloud droplets face extreme stressors, including UV radiation, freeze–thaw cycles, nutrient scarcity, and oxidative damage. Despite these challenges, many remain metabolically active. Cloud-resident microbes have been shown to repair DNA damage, synthesize protective pigments, and scavenge reactive oxygen species, thereby enhancing their survival (Amato et al., 2007, 2017).

In addition, cloud microbes metabolize a range of organic substrates, including volatile organic compounds (VOCs) and carboxylic acids, altering aerosol chemistry and influencing the persistence of pollutants in the atmosphere. Some even participate in nitrogen and sulfur transformations, linking atmospheric chemistry with terrestrial and aquatic nutrient cycles (Péguilhan et al., 2025; Vaïtilingom et al., 2013). Their resilience enables long-range atmospheric transport. By surviving in droplets for days to weeks, cloud microorganisms can disperse across continents and oceans, seeding distant ecosystems and contributing to global microbial connectivity. This ecophysiological versatility highlights clouds as both transient microbial habitats and important vectors for microbial biogeography and ecosystem exchange (Sessitsch et al., 2023).

#### *1.3.7.3 Ecological and Climatic Implications*

Biological particles can significantly alter clouds, affecting their reflectivity and precipitation. By enhancing ice nucleation, microbes contribute to precipitation, which in turn changes regional rainfall and snow patterns and influences water cycles (S. Huang et al., 2021). Microbes also degrade organic compounds within clouds, which can alter aerosol composition and the atmospheric lifespan of pollutants like VOCs. Clouds also serve as a temporary home for microbes, enabling them to travel vast distances (Péguilhan et al., 2025). This microbial transport can lead to the colonization of new environments when microbes are deposited through precipitation, thereby impacting global biodiversity patterns. Ultimately, by modifying cloud properties and precipitation, microbes indirectly influence the Earth's radiation balance and climate systems, creating a biologically mediated climate feedback loop (Rosenfeld et al., 2014).

## 1.4 THE MICROBIOME IN URBAN AND INDUSTRIAL ENVIRONMENTS

### 1.4.1 Composition, Dynamics, and Implications for Health and Sustainability

Microbial communities and human populations coexist in cities, which are dynamic, complex ecosystems that are essential to preserving important ecosystem services in urban environments. The indoor microbiome is a complex and dynamic community of microorganisms found in the BE, which includes homes, workplaces, schools, and hospitals (Bruno et al., 2022; Gilbert & Stephens, 2018; National Academies of Sciences et al., 2017). Activities inside the building and interactions with the outside world continuously influence the indoor microbiome. In addition to being released by residents and their pets, microorganisms can enter buildings through open windows, doors, and ventilation systems (Gilbert & Stephens, 2018). They can re-enter the air through resuspension (e.g., stirring up dust) after settling on surfaces once indoors (Gilbert & Stephens, 2018).

Unprecedented population growth and the vast expansion of built-up areas have serious negative effects on human health even though they also provide advantages like better access to resources and sanitation (Bruno et al., 2022). Several classes of bacteria are common microbiological residents of BEs. Proteobacteria, Firmicutes, Actinobacteria, Cyanobacteria, and Bacteroidetes are among the phyla that are commonly found and are frequently dominant (Bruno et al., 2022; Leung et al., 2019). Fungi: They are common, especially in damp settings, and can include yeasts, molds, and mushrooms (Bruno et al., 2022; National Academies of Sciences et al., 2017). Viruses: Although less researched than bacteria and fungi, a variety of DNA and RNA viruses, including those linked to humans, bacteria, fungi, plants, and insects, have been found in BEs. Archaea: Because common DNA sequencing techniques do not adequately cover archaeal communities in BEs, there is a dearth of information on these organisms (Leung et al., 2019).

As indicated in Table 1.3 (Gilbert & Stephens, 2018; National Academies of Sciences et al., 2017), the surface (e.g., floors, walls, ventilation systems, and habitat), location within the building, and the types of materials present all have a significant impact on the composition of these communities. By offering surfaces for microbial colonization and influencing environmental factors like pH and humidity, building materials can have a substantial impact on the dynamics and composition of the indoor microbiome (Gilbert & Stephens, 2018; National Academies of Sciences et al., 2017). For example, compared to inorganic materials like metals and glass, organic materials like wood and paper are more susceptible to fungal growth (Kazemian et al., 2019). Therefore, the particular microbial communities that live in a built space can be influenced by the choice of building materials.

### 1.4.2 Factors Shaping the Indoor Microbiome

There are several factors shaping the indoor microbiome. These factors include the following:

a. **Human Occupancy**: The number and activity of people in a space is one of its most crucial features. The microbial makeup of indoor surfaces and air is directly impacted by the constant expulsion of microorganisms from human mouths,

TABLE 1.3
*Key microbial taxa and their environmental associations in urban settings*

| Habitat | Key microbial genera | Their implications |
|---|---|---|
| Built surfaces (transit systems) | *Pseudomonas*, *Brevundimonas*, *Stenotrophomonas* (skin- and soil-associated) | Critical for microbial dissemination; passengers acquire ARGs |
| Built surfaces (hospitals) | *Corynebacterium*, Staphylococcus, *Streptococcus*, *Acinetobacter* (human-associated, skin-associated); *Achromobacter*, *Elizabethkingia*, *Serratia* (aquatic/terrestrial in sinks) | Reservoirs for opportunistic pathogens and multi-drug-resistant organisms, human occupancy is a major driver; indiscriminate sterilization promotes resistome selection |
| Urban air (aerobiome) | Proteobacteria, Actinobacteria, Firmicutes, Bacteroidetes | Impacts public health (respiratory illnesses, allergies); diversity is generally protective; influenced by climate, urban design, and human activity; high air pollution can reduce diversity |
| Drinking water biofilms | Proteobacteria (most abundant phyla), *Pseudomonas* (most common genus) | Higher diversity at the consumer end, influenced by piping material and supply method; prevalence of ARGs (macrolides, fluoroquinolone) |
| Urban wastewater | *Acinetobacter*, *Aeromonas*, *Arcobacter*, *Pseudomonas*, *Trichococcus* (pipe-associated); human/animal gut microbiomes | Collectors and transporters of fecal pollution indicators; potential reservoirs for emerging human pathogens; altered nutrient flows; increased genetic exchange |
| Urban green spaces | *Hymenobacter*, *Pseudomonas*, *Beijerinckia* (typical phyllosphere bacteria); *Chroococcidiopsis*, *Flavisolibacter*, *Acinetobacter* (generalists in less green areas) | Influence plant health and ecosystem services; diversity shaped by green infrastructure and pollution; environmental selection favors specialists in greener areas and generalists in less vegetated urban areas |

stomachs, and skin (Iannone et al., 2011; Ruiz-Gil et al., 2020).

b. **Building Characteristics:** The design and structural elements of a building, such as its windows, ventilation systems, materials, and air filters, play a significant role in controlling the growth and spread of microorganisms (Lazaridis, 2019; Ruiz-Gil et al., 2020).

c. **Environmental Inputs**: Human-derived microbes can interact with environmental microbes that are brought in by outdoor air, soil, and water to form a distinct indoor ecosystem (Gilbert & Stephens, 2018; Meadow et al., 2014b; National Academies of Sciences et al., 2017).

d. **Pathogen Transmission**: The possibility of disease transmission through the air and surfaces, especially in places with lots of traffic (Gilbert & Stephens, 2018; Meadow et al., 2014b).

e. **Immune System Development**: According to the "biodiversity hypothesis", allergic and asthmatic diseases may become more common in overly sanitized urban settings because of a lack of exposure to a wide variety of environmental microbes (X. Fu et al., 2022; Fujimura & Lynch, 2015; Hanski et al., 2012; Kelly et al., 2022; M. Zhang et al., 2023). Immune system development is believed to benefit from exposure to a diverse microbiome, especially during early life (Kelly et al., 2022; M. Zhang et al., 2023).

### 1.4.3 Health Implications

Human health is significantly impacted by the indoor microbiome. Certain microorganisms can have negative effects, while others may be neutral or even helpful (Gilbert & Stephens, 2018). Numerous pathogens, such as influenza, norovirus, and MRSA, can spread more easily in the BE (Leung et al., 2019; National Academies of Sciences et al., 2017). Additionally, symptoms of sick building syndrome (SBS), such as headaches and respiratory irritation, are associated with poor indoor air quality and microbial contamination in HVAC systems (X. Fu et al., 2021). Asthma and Allergies: Variations in the microbiome of the BE are associated with an increased risk of asthma and allergies (Fujimura & Lynch, 2015; Hanski et al., 2012; Kelly et al., 2022). Children's asthma and allergy symptoms have been found to decrease when exposed to a diverse BE microbiome (X. Fu et al., 2022).

### 1.4.4 From the Hygiene Hypothesis to the Biodiversity Hypothesis

The early theory proposed that one of the main reasons for the increase in the prevalence of autoimmune and allergic diseases was that early childhood exposure to pathogens and common infections had decreased as a result of better sanitation and hygiene (X. Fu et al., 2022). This concept was developed in earlier studies, which contended that the problem is a wider loss of environmental biodiversity rather than merely a lack of exposure to a few particular pathogens. The biodiversity hypothesis, put forth by the authors, holds that the development of a robust and balanced immune system is hampered by a decreased diversity of microbial species in our immediate surroundings, especially in urban areas. This theory holds that exposure to dirt or infections is not the only factor that regulates the immune system, but to a diverse and rich community of non-pathogenic environmental microbes that our ancestors were constantly exposed to (X. Fu et al., 2022; Fujimura & Lynch, 2015; Hanski et al., 2012; Kelly et al., 2022).

### 1.4.5 Role of Ventilation Systems in Shaping the Indoor Microbiome

The indoor microbiome is significantly shaped by ventilation systems, especially HVAC systems (Meadow et al., 2014a; National Academies of Sciences et al., 2017). Although they are intended to keep people comfortable and supply fresh air, if they are not properly maintained, they may also serve as conduits for microbial contamination (Meadow et al., 2014a).

#### *1.4.5.1 Impact of Ventilation on Microbial Communities*

The introduction of outdoor air, which carries its own microbial communities, into indoor spaces by ventilation systems resulted in dilution of indoor microbes. Proper ventilation, particularly with outdoor air, can help reduce the levels of microorganisms and other indoor air pollutants (Meadow et al., 2014a; Sharpe et al., 2020).

- **Impact on Diversity**: Compared to naturally ventilated buildings, mechanically ventilated buildings may have less microbial diversity (Meadow et al., 2014a). Indoor bacterial communities are more likely to resemble outdoor communities in buildings with high outdoor air exchange rates (Meadow et al., 2014a; National Academies of Sciences et al., 2017).

Human-associated microbes: These microorganisms may proliferate in "closed mode" buildings with mechanical ventilation and little outdoor air (Cao et al., 2021; National Academies of Sciences et al., 2017).

- **Impact on Moisture and Growth**: HVAC parts such as drip pans and cooling coils have the potential to retain moisture, which can foster the growth of bacteria and fungi and result in the formation of biofilms (Bruno et al., 2022; Kazemian et al., 2019; National Academies of Sciences et al., 2017). The indoor air can then be infused with these (Kazemian et al., 2019; Meadow et al., 2014a).
- **Impact on Pathogen Transmission**: HVAC systems can help spread other viruses and bacteria through the air, particularly in crowded buildings, and have been linked to the spread of pathogens such as *Legionella pneumophila* (Leung et al., 2019; Meadow et al., 2014a; National Academies of Sciences et al., 2017).

**Microbial Communities in HVAC Systems:** A variety of microbial communities can be found in HVAC systems, especially in the air filters, cooling coils, fans, and water sinks (Sharpe et al., 2020; Watanabe et al., 2022). Certain bacterial genera have been found to be prevalent in various areas of air conditioners, such as *Methylobacterium–Methylorubrum* in water sinks, *Bacillus* in filter nets, and *Caproiciproducens* in air outlets (Watanabe et al., 2022). Potential hazards to public health have been highlighted by the discovery of potentially harmful microorganisms in HVAC systems, including *Mycobacterium*, *Bacillus*, and *Cupriavidus* (Sharpe et al., 2020; Watanabe et al., 2022).

### 1.4.6 Industrial Microbiome

While urban BEs largely shape human-associated microbial communities, industrial environments present unique selective pressures with broader ecological consequences. Different microbial communities with significant ecological and functional roles are also found in industrial settings, such as factories, wastewater treatment facilities, and contaminated sites (Q. Wang et al., 2020; Yang, Wang, et al., 2020). The microbial communities that break down pollutants and purify water can be profiled using the industrial microbiome. It includes important microbial taxa and functional genes that are necessary for environmental remediation and industrial process efficiency (Q. Wang et al., 2020). Prior studies have demonstrated the distinct microbial community and particular HMs, dyes, or organic chemicals that are present in industrial wastewater. Over time, the community adjusts to the toxic environment, and the pollutants can either inhibit or promote the growth of specific microbes (Q. Wang et al., 2020; Yang, Wang, et al., 2020; L. Zhang et al., 2020).

Operators can maximize plant conditions (such as aeration, temperature, and nutrient levels) to encourage the growth of advantageous microorganisms by knowing which microbes are necessary for effective treatment. Monitoring the microbial community has been proposed as a useful method for evaluating and enhancing a plant's performance (Q. Wang et al., 2020).

Other researchers looked at how the bacterial communities in urban river sediments were affected by long-term industrial contamination (Yang, Wang, et al., 2020). High-throughput sequencing was used in the study to demonstrate how contaminants such as organic compounds and HMs significantly change the diversity and structure of microbial communities. This study emphasizes the role of microbes as bioindicators of pollution and shows the substantial ecological impact of industrial activity on urban ecosystems (N. Ahmed et al., 2019; L. Zhang et al., 2020).

These pollutants act as a strong selective pressure, favoring microbes that can tolerate the toxic environment while eliminating more sensitive species (N. Ahmed et al., 2019; Maglione et al., 2024; Watanabe et al., 2022; L. Zhang et al., 2020).

## 1.5 ENVIRONMENTAL MICROBIOMES AS A SOURCE OF BIOACTIVE SECONDARY METABOLITES

Microbial secondary metabolites: what are they? Small molecules known as "microbial secondary metabolites" are produced by microorganisms mainly when they are in their stationary growth phase. These substances are crucial for ecological interactions but not for basic microbial survival like primary metabolites are. In different ecosystems, they have a variety of functions in communication, defense, symbiosis, and competition (Chater, 2016).

These substances are a wealth of advantageous bioactivities from a pharmaceutical standpoint. Among other things, these include hepatoprotective, antimicrobial, antitumor, anti-inflammatory, antivirulence, and hypoglycemic effects. As a result, microbial secondary metabolites constitute a substantial chemical entity reservoir with promising applications in biotechnology, agriculture, and drug discovery (Newman & Cragg, 2020).

**What are the genomic bases of secondary metabolite production?**
Microbial production of secondary metabolites typically relies on BGCs. BGCs are contiguous chromosomal regions harboring all necessary genes for the stepwise biosynthesis, modification, regulation, and export of a particular compound (Figure 1.3). BGCs often encode large modular enzyme systems such as polyketide synthases (PKSs) and nonribosomal peptide synthetases (NRPSs), which operate much like assembly lines to construct diverse molecular backbones from activated building blocks derived from primary metabolism. These clusters may also include hybrid PKS–NRPS systems, terpenoid synthases, or RiPP-type architectures, reflecting the remarkable

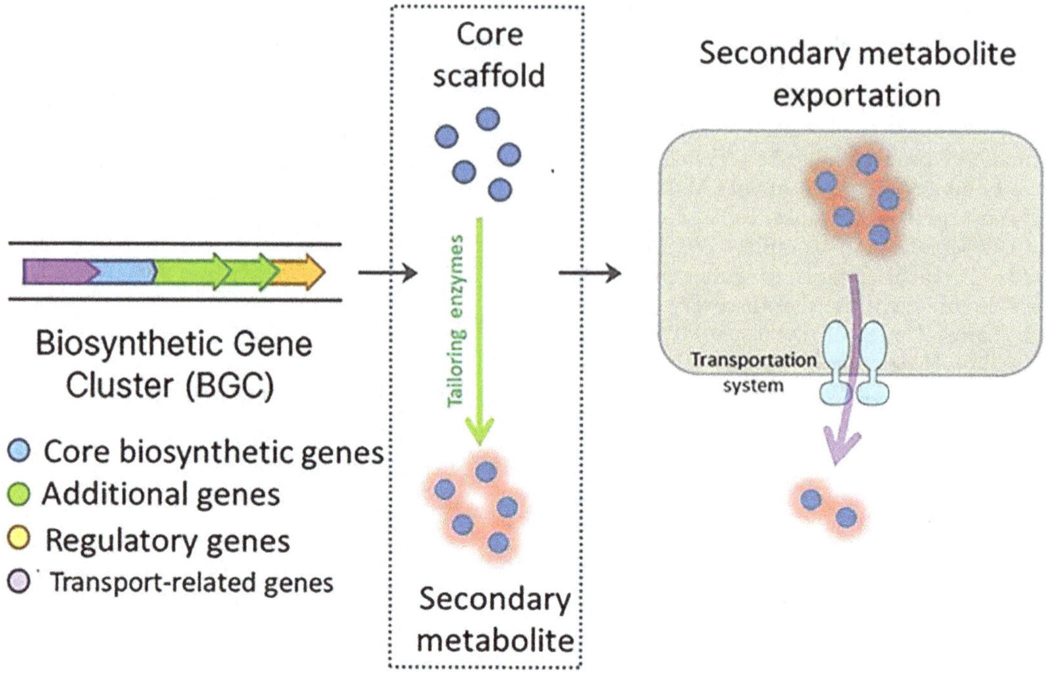


Figure 1.3 Schematic representation showing the bacterial production and export of secondary metabolites from biosynthetic gene clusters (BGCs).

biochemical diversity housed within microbial genomes (Cimermancic et al., 2014).

### 1.5.1 The Genomic Machinery of Microbial Bioactive Metabolites Production

Within each BGC, the core biosynthetic genes encode the modular enzymatic machinery that assembles the scaffold of the secondary metabolite. For PKS BGCs, each module typically contains a ketosynthase (KS), acyltransferase (AT), and acyl carrier protein (ACP). On the other hand NRPS modules include adenylation (A), thiolation (T), and condensation (C) domains that add specific amino acids in a defined sequence (Hwang et al., 2020). Hybrid clusters integrate both functionalities, as seen in natural products like tirandamycin, where PKS and NRPS modules combine to yield complex structures (Carlson et al., 2010). Adjacent to these scaffold-forming genes, tailoring enzyme genes often co-localize within the cluster. These genes encode enzymes such as methyltransferases, glycosyltransferases, halogenases, and cytochrome P450 monooxygenases, imparting further structural modifications that diversify the metabolite's bioactivity and ecological function (Hwang et al., 2020). The genomic architecture often also includes resistance and transporter genes, which allow the producer organism to export the metabolite or protect itself from its own bioactive product. Regulation of BGCs is mediated by both pathway-specific regulators (often encoded within the cluster) and global regulators that respond to environmental cues or cellular stress.

### 1.5.2 Shifting from the Traditional Methods of Microbial Bioprospecting

Traditional microbial bioprospecting relied primarily on culture-dependent isolation from readily accessible habitats (soil, plant surfaces, common fermentation sources), followed by activity-guided fractionation and chemical characterization. Classical workflows used broad selective media or enrichment cultures to obtain axenic isolates, followed by screening crude extracts or culture filtrates to detect antimicrobial, cytotoxic, or enzymatic activities. Positive leads advanced to iterative purification, structure elucidation (NMR, MS), and, if promising, biological profiling. These methods emphasize phenotypic screening and chemistry-first prioritization and underpinned natural-product drug discovery for much of the 20th century (Ribeiro da Cunha et al., 2019; Wohlleben et al., 2016).

The period from roughly the 1940s through the 1960s is often described as the golden age of antibiotic discovery. Many major antibiotic classes (penicillins, aminoglycosides, tetracyclines, macrolides, glycopeptides, and others) were discovered by systematic isolation of soil actinomycetes and fungi and rapid bioassay-guided chemical characterization. These discoveries transformed clinical medicine and established microbial secondary metabolites as a premier source of therapeutics. The success of this era demonstrates the high yield of culture-based, activity-guided exploration when novel taxa and diverse ecological sources were sampled intensively (Ribeiro da Cunha et al., 2019).

Despite early success, classical approaches have important limitations:

- **Cultivation Bias:** The majority of environmental microbes do not grow under standard laboratory conditions, so culture-dependent surveys sample only a small, non-representative fraction of microbial diversity.
- **Rediscovery Problem:** Repeated isolation from well-sampled habitats yields the same, previously described metabolites (high rediscovery rate), reducing returns from incremental sampling of the same ecological niches.
- **Silent/Cryptic Biosynthetic Potential:** Many cultured strains harbor BGCs that are not expressed under standard laboratory conditions; thus, chemical screens miss a large portion of encoded chemistry.
- **Scale and Throughput Limits:** Traditional fractionation and structure elucidation workflows are resource-intensive and low throughput relative to modern omics pipelines.

These constraints motivated the adoption of cultivation-independent and new cultivation approaches to access hidden diversity and cryptic chemistry.

### 1.5.3 Shifting toward Underexplored Niches

To overcome rediscovery and expand chemical novelty, bioprospecting has shifted to underexplored ecological niches and extreme habitats where microbial communities have unique selective pressures (and thus distinctive chemistries). Examples are described in the following sections.

#### *1.5.3.1 Marine Environments*

Marine microbes inhabit environments with extreme and fluctuating stressors including high pressure, variable temperatures, high salinity, and nutrient limitation, which drive the evolution of distinctive secondary metabolism. To survive, they produce specialized compounds such as siderophores for iron acquisition, biosurfactants and exopolymers for nutrient capture and biofilm formation, and unusual fatty acids or hopanoids to maintain membrane function under stress. Many biosynthetic pathways are tuned to marine-specific conditions, with some metabolites requiring seawater-derived nutrients or cofactors for expression. These selective pressures enrich marine microbes with biosynthetic capabilities that yield structurally and functionally novel compounds, rarely found in terrestrial counterparts (De Carvalho & Fernandes, 2010; Sedeek et al., 2025).

Marine sponges and tunicates host dense, diverse microbial consortia that can comprise up to 60% of their biomass, creating microenvironments rich in chemical signaling and competition. These symbiotic microbes are prolific producers of bioactive secondary metabolites. The host provides a stable, nutrient-rich, and chemically complex habitat, while the microbes contribute defensive and ecological fitness compounds that protect against pathogens, predators, and biofouling. This intimate symbiosis, coupled with the ecological pressures of marine benthic environments, makes sponge- and tunicate-associated microbiomes prolific sources of structurally unique and pharmacologically valuable metabolites (Gavriilidou et al., 2021; Kamel et al., 2022; Proksch et al., 2002; Samak et al., 2018; Sedeek et al., 2025; Taylor et al., 2007).

Furthermore, the deep-sea ecosystems, particularly at depths exceeding 100 m, remain ones of the planet's most underexplored environments. These environments harbor a remarkable diversity of microorganisms uniquely adapted to extreme conditions such as high pressure, low temperatures, and the perpetual absence of light. Recent studies have underscored the great potential of these microbes for drug discovery. In the last five years, a review highlighted the discovery of over 400 novel MNPs from deep-sea organisms collected from depths ranging from 100 to 11,000

m. A significant majority of these compounds, approximately four-fifths, exhibited bioactivity, with nearly half demonstrating potent cytotoxicity against various human cancer cell lines. Deep-sea fungi and actinomycetes have emerged as particularly significant sources of these compounds (Khan et al., 2025).

#### *1.5.3.2 Extreme Environments*

Extremophiles are those microorganisms that survive in conditions once thought incompatible with life. These microorganisms have evolved remarkable genomic and phenotypic adaptations to survive in environments with extreme temperatures, pressures, salinities, pH values, or chemical compositions. These adaptations are often underpinned by specialized enzymes and secondary metabolites that confer stability, functionality, and ecological competitiveness under such stresses. As a result, extreme habitats serve as reservoirs of unique biosynthetic pathways, making them valuable targets in the search for structurally novel and biologically potent natural products (Marzban & Tesei, 2025; Salwan & Sharma, 2022).

For example, hydrothermal vents, a prominent feature of the deep sea, release hot fluids rich in chemical energy sources like hydrogen and sulfide. These unique chemical environments support distinct microbial communities that rely on chemoautotrophy, the process of fixing inorganic carbon using energy derived from chemical reactions, suggesting the presence of unique metabolic pathways adapted to these conditions. Furthermore, hadal sediments, found in the deepest oceanic trenches, also harbor diverse and taxonomically novel microbial communities, indicating a promising potential to discover novel chemical entities (Schauberger et al., 2024; Wegener et al., 2024).

Similarly, polar ecosystems, particularly the Antarctic continent, are dominated by microbial life. These microorganisms play crucial roles in global biogeochemical cycles and serve as significant reservoirs of specialized metabolites. The long-term isolation and intense environmental pressures are characteristics of polar regions that have profoundly shaped the genomes of Antarctic bacteria, suggesting that they encode unique natural products not found elsewhere (Medeiros et al., 2025). Studies of Arctic Ocean metagenomes have revealed substantial biosynthetic diversity. For instance, 36% of the KS domain sequences recovered from Arctic Ocean samples could not be assigned to any known BGCs, highlighting a significant and novel biosynthetic potential (Rego et al., 2021).

#### *1.5.3.3 Host-Associated and Micro-Niche Habitats*

Microbial symbionts within host-associated and micro-niche environments, such as plant endophytes and insect-associated microbes, are shaped by intimate ecological interactions that foster unique chemical evolution. Plant endophytes, living asymptomatically within host tissues, play a crucial role in enhancing host defense against pathogens and herbivores. These organisms may even contribute metabolites typically attributed solely to plants, making them rich sources of novel natural products (Z. Li et al., 2023; Mousa & Raizada, 2013; Sedeek et al., 2023).

Similarly, insect-associated microbes are another prolific biosynthetic pool. For example, Actinobacteria living on the cuticle or within the nests of leaf-cutter ants produce antibiotics that protect the ants' cultivated fungal gardens from parasitic invaders. This defensive symbiosis exemplifies how microbial partners in tightly structured microhabitats generate potent bioactive compounds essential for host survival and ecosystem stability (Baranova et al., 2022).

## 1.6 ROLE OF THE ENVIRONMENTAL MICROBIOME IN BIOREMEDIATION AND ECOSYSTEM HEALTH

The environmental microbiome is a fundamental component of the biosphere, encompassing a vast diversity of bacteria, archaea, microbial eukaryotes, and their associated viruses. These microbial communities play pivotal roles in the global cycling of energy and elements, which is essential for maintaining a balanced and functional planet. For instance, environmental microbiomes are responsible for the detoxification of pollutants, the absorption of HMs, and the facilitation of carbon sequestration, which are all critical processes for ecosystem stability. The increasing recognition of microbial ecology and traits as fundamental drivers of global health is reflected in the unprecedented growth of the microbial-focused biotech industry, which is projected to reach a value of over $300 billion by 2032 (Ginnan et al., 2025; Soth et al., 2025).

For decades, the remediation of environmental pollutants has relied on conventional physical and chemical methods, which, despite being effective in some contexts, present significant drawbacks. These limitations include high costs, the need for specialized equipment, and the generation of secondary pollutants such as chemical wastes and sludge that require additional treatment. In response to these challenges, there has been a significant shift toward bioremediation, an alternative technology that uses biological agents like plants and microorganisms to remove or lessen the effects of environmental pollutants. This approach is widely recognized as a more efficient, eco-friendly, and cost-effective method for transforming contaminants (Ayilara & Babalola, 2023; Firincă et al., 2025).

Microbes are particularly favored over plants as agents of bioremediation due to their rapid growth rate and their capacity for genetic and metabolic manipulation. This allows for the enhancement of their function in pollutant degradation and adaptation to diverse environmental conditions. A deeper examination of microbial communities reveals that they play a foundational role in maintaining ecosystem health, influencing everything from soil structure to nutrient availability and the overall resilience of the environment against disturbances. The ultimate goal of understanding and utilizing these systems is to contribute to a sustainable future, ensuring environmental sustainability and mitigating global climate change. This interconnected web of microbial functions, from detoxification to nutrient cycling, forms the basis for a circular ecosystem model where waste is transformed into resources, and a healthy environment, in turn, nurtures human and animal well-being. The exploration of these connections is essential for developing comprehensive strategies that align with the holistic principles of global health and sustainability (Al Haj Chehadeh et al., 2025; Ayilara & Babalola, 2023; El-Sesy et al., 2025; Soth et al., 2025).

### 1.6.1 Microbial Communities in Pollutant Degradation (Oil Spills, Heavy Metals, Plastic Waste)

#### *1.6.1.1 Bioremediation of Hydrocarbon Contamination (Oil Spills)*

Petroleum hydrocarbon contamination, often resulting from oil spills, poses a severe threat to aquatic environments, leading to oxygen depletion, loss of habitats, and long-term biodiversity disruption. The degradation of these persistent pollutants is crucial for preserving the health of aquatic ecosystems. While various conventional technologies exist, biodegradation is chiefly responsible for the ultimate removal of petroleum tars that persist in the environment. Studies have shown that indigenous microbial communities in marine sediments possess an intrinsic capacity to mitigate such contamination. Key microbial genera involved in hydrocarbon degradation include *Pseudomonas*, *Erythrobacter*, *Bacillus*, *Alcanivorax*, *Marinobacter*, and *Acinetobacter* species, which metabolize these complex hydrocarbons into simpler, less toxic substances (Al Haj Chehadeh et al., 2025).

The efficiency of this natural biodegradation process is significantly influenced by environmental factors. For example, it was reported that biodegradation rates for n-alkanes and polycyclic aromatic hydrocarbons (PAHs) increase with higher temperatures. Emerging research is focusing on enhancing these natural processes through innovative approaches. A particularly promising strategy involves combining nanotechnology with degrading bacteria to create a synergistic nano-bio-hybrid system. For instance, green-synthesized magnetite nanoparticles (AL-MNPs) have been shown to enhance oil degradation when combined with oil-degrading bacteria. In this system, the nanoparticles act as promoters, improving the dispersion of hydrophobic contaminants and stimulating bacterial activity, while the microbes carry out the actual degradation. This hybrid system not only significantly reduces the degradation time and increases overall efficiency but also offers recyclability and a sustainable, cost-effective alternative to conventional methods (Al Haj Chehadeh et al., 2025; Debs et al., 2019; Feitkenhauer et al., 2003).

#### *1.6.1.2 Decontamination of Heavy Metals and Metalloids*

The bioaccumulation of HMs and metalloids such as cadmium, lead, and hexavalent chromium poses persistent risks to both ecosystems and human health. Microbes have evolved a diverse array of mechanisms to mitigate the toxic effects of these pollutants, which can be broadly categorized into mobilization and immobilization processes. These mechanisms range from passive physical processes to active, metabolism-driven transformations (Firincă et al., 2025).

A key metabolism-independent process is biosorption, where HMs are physically adsorbed onto the surface of a microbial cell through ion exchange and complexation with various functional groups. This process is passive and relies on the large surface area microbes provide, making it an excellent method for reducing the overall availability of HMs in a contaminated environment. A significant advantage of biosorption is that it can be a reversible process, allowing for desorption and subsequent recovery of economically valuable metals. In contrast, biotransformation is a metabolism-dependent process where microbes enzymatically convert toxic metal ions into less hazardous forms. For example, *Bacillus* species can biotransform toxic hexavalent chromium into the less soluble and less hazardous trivalent form. Another mobilization mechanism is bioleaching, where acidophilic microbes produce organic and inorganic acids that solubilize HMs from solid matrices, making them available for extraction. This method is considered an eco-friendly and energy-efficient alternative to conventional hydrometallurgical processes (Ayangbenro et al., 2018; Ayilara & Babalola, 2023; Y. Zhao et al., 2025).

Emerging approaches, such as genetic engineering and nanotechnology, are being explored to overcome the limitations of natural bioremediation processes, which can be slow and limited to small areas. Gene editing allows for the tailoring of specific metabolic traits, leading to increased pollutant tolerance, higher biodegradation efficiency, and improved enzymatic specificity. Similarly, nanotechnologies, particularly biogenic nanostructures and bio-based nanocomposites, can improve the efficiency and reduce the costs of bioremediation protocols by enhancing the sequestration of HMs. The overarching trend in this field is a shift from simply removing pollutants to integrating the remediation process into a circular economy model. This involves repurposing residual biomass for biogas and biochar production or recovering valuable metals through desorption, thereby transforming waste into a resource. This paradigm shift not only cleanses the environment but also creates sustainable value from pollutants (Firincă et al., 2025; Najam et al., 2025).

#### *1.6.1.3 Microbial Strategies for Plastic Waste Degradation*

The global accumulation of plastic waste poses a severe environmental challenge, with traditional methods like incineration and mechanical recycling having significant limitations, including the generation of secondary pollutants and a reduction in material quality. Microbial biodegradation offers an eco-friendly, sustainable, and cost-effective approach to addressing this pollution. Commendable biodegradative potentials have been observed in a variety of microorganisms, including bacteria and fungi. Fungi from the phylum Ascomycota, such as *Aspergillus niger* and *Aspergillus versicolor*, have demonstrated significant potential in degrading polymers like polyurethane (PUR) and low-density polyethylene (LDPE). Moreover, gut microorganisms from plastic-eating insects, such as *Tenebrio molitor* larvae, have been shown to assist in the degradation of PUR foam (M. Ahmed et al., 2025; Khatua et al., 2024; J. Liu et al., 2022; Najam et al., 2025; Orts et al., 2022; Rajan et al., 2024).

However, the primary obstacles to the widespread adoption of microbial biodegradation for plastic waste are its low efficiency and slow degradation rates. The susceptibility of plastics to degradation is highly dependent on their chemical structure, and the processes often require prolonged incubation periods. In response to these challenges, the field is moving beyond simple degradation toward a more sophisticated concept of "bio-upcycling". This approach involves using microbial metabolic pathways to convert plastic depolymerization products into valuable chemicals, effectively transforming waste plastics into feedstocks for microbial biosynthesis. This strategy represents a more sustainable solution than simple waste disposal, as it aligns with the principles of a circular economy and offers a path for creating new value from a pervasive environmental pollutant.

### 1.6.2 Role in Nutrient Cycling and Ecosystem Resilience

Microbiomes are essential for the biogeochemical cycling of essential elements, acting as the primary drivers of nutrient transport, mineralization, and solubilization in ecosystems. They are responsible for breaking down and transforming complex organic compounds into forms that can be assimilated by plants. The interaction of microbial communities with soil nutrients significantly alters their chemical reactivity and solubility, directly influencing their movement between the biosphere, atmosphere, soil, and water (Rousk & Bengtson, 2014; X. Wang, Chi, et al., 2024).

The contributions of specific microbial groups to these cycles are well-documented. For instance, mycorrhizal fungi, such as *Arbuscular mycorrhizal* fungi (AMF), facilitate the solubility and transport of insoluble soil nutrients, particularly phosphorus, to plant root systems, thereby increasing the effective volume of soil that roots can access. Similarly, bacterial families like Xanthobacteraceae and Bryobacteraceae possess genes related to phosphorus and nitrogen turnover, functioning as transporters and mineralizers. The inoculation of plants with bacterial strains such as *Pseudomonas*, *Arthrobacter*, and *Bacillus* has been shown to increase the nitrogen, phosphorus, and potassium content in seedlings, leading to substantial increases in crop yields under field conditions. Environmental microbiomes also play a direct role in carbon sequestration and the breakdown of organic compounds, contributing to soil health and long-term ecosystem function (Etesami et al., 2021; Jatana et al., 2021; Lee et al., 2008; Soth et al., 2025; Xie et al., 2022).

However, environmental stressors can profoundly disrupt these critical processes. A study on agricultural soils demonstrated that increasing pesticide diversity, particularly without nitrogen supplementation, drives substantial changes in microbial communities and nutrient cycling. This selective pressure can favor the enrichment of bacterial specialists and opportunists that are capable of degrading or resisting the pesticides, while reducing the proportion of generalist bacteria. This adaptive response can paradoxically lead to negative ecosystem-level consequences. The selection pressure may drive bacteria to streamline their average genome size to conserve energy, but this is correlated with an enhanced metabolic capacity for C, N, P, and S cycling, which can accelerate soil nutrient loss. This demonstrates a complex trade-off where a microbial community's short-term adaptation to a stressor can impair the long-term stability and health of the ecosystem by reducing functional redundancy and accelerating biogeochemical cycles (Ni et al., 2025).

**Microbiome resilience** is defined as a microbial community's capacity to either resist a perturbation or to recover and return to its original state of balance. The structural and functional characteristics of a community, including its diversity, the presence of keystone species, and the interplays between microbes, are all determinants of its resilience. Microorganisms achieve this resilience through various mechanisms, such as flexible metabolism, functional redundancy (where multiple species can perform the same function), and cooperative or competitive interactions. A widely accepted principle in ecology is that higher biodiversity positively influences ecosystem functioning, making more diverse communities more efficient, resistant, and resilient to disturbances (Eisenhauer et al., 2023; Philippot et al., 2021b; Xun et al., 2021; Zhou et al., 2025).

### 1.6.3 Potential Applications in Environmental Restoration and Biotechnology

Important players in environmental processes, soil microbial communities can be used as potential markers for ecological monitoring and restoration. It has been demonstrated that artificial or natural vegetation restoration techniques have a significant impact on degraded soil ecosystems, affecting water retention, nitrogen and carbon cycles, and other physical characteristics. Introducing functional microorganisms and inoculants, like the AMF, to improve soil health and promote vegetation growth is a crucial tactic in this field.

The intersection of microbiome research and advanced biotechnologies presents significant opportunities for advancing sustainable solutions. For instance, synthetic biology provides a transformative platform for engineering microbial systems with enhanced or novel traits for specific applications. Gene editing technologies have revolutionized this field, enabling the precise modification of genetic sequences in microbes to program cell functionalities. These engineered microbes can be used to create environmentally sustainable solutions, such as novel bioremediation platforms and waste-to-energy systems, by leveraging their unique metabolic capabilities (Aminian-Dehkordi et al., 2023; Ruan et al., 2024; Song & Prather, 2024; Tang & Song, 2024).

### 1.6.4 Biogeochemical Cycling of Trace Elements, Including Rare Earth Elements

Trace elements, which are released into the environment through both natural and anthropogenic activities, can become toxic at elevated levels and pose a significant public health concern. These elements can be taken up by crops and move up the food chain, eventually increasing human risks of disease. Microorganisms play a dominant role in the global biogeochemical cycling of these elements, influencing their speciation, mobility, and bioavailability. The behavior of toxic elements, in turn, can affect microbial diversity and activity, which can result in altered processes such as biomineralization, biosorption, methylation, and volatilization. Advanced analytical and molecular approaches, including metagenomics and metaproteomics, are helping to disclose the microbial community structures and functions, and to identify the novel catabolic genes and enzymes involved in the detoxification and transformation of these pollutants (Li et al., 2023; Pan et al., 2024; Romero-Crespo et al., 2023; Tang et al., 2024).

Rare earth elements (REE) are considered strategic metals due to their increasing importance in high-technology applications and renewable energy. The conventional hydrometallurgical extraction of REE is an energy-intensive process that often uses harsh acidic conditions and generates large amounts of toxic waste. This has prompted the development of more economically feasible and environmentally friendly solutions, including leveraging the roles of microorganisms in REE biogeochemical cycling. Microorganisms are crucial for the mobilization and enrichment of REE in both terrestrial and marine environments (Heilmann et al., 2021; Li et al., 2022).

Microorganisms mobilize REE from solid matrices via specific mechanisms. For example, *A. niger* produces organic acids like citric and oxalic acids, enabling acidolysis (proton-induced solubilization), while microbial secretion of chelating agents mediates complexolysis, both of which facilitate REE release from minerals. In marine environments, microorganisms contribute to REE cycling by promoting biosorption and bioaccumulation onto particulates, enhancing their transport, concentration, and eventual deposition in deep-sea sediments. Integrating microbial mobilization with hyperaccumulator plants in bioaugmentation-assisted phytoextraction offers an ecologically sound and cost-effective alternative to conventional REE recovery methods, combining microbial solubilization with plant uptake to form enriched bio-ores for harvesting (Jalali & Lebeau, 2021; Li et al., 2022; S. Liu & Deng, 2025).

## 1.7 FUTURE DIRECTIONS

By exporting the environmental microbiome component, microbial composition, and functional role, we need to understand not just which microbes are present, but what they are doing, how they are interacting, and how they respond to environmental stressors like pollution and climate change. This is driven by the developments in machine learning and multi-omics technologies (e.g., metagenomics, metatranscriptomics, metaproteomics, and metabolomics) to model and forecast microbiome behavior. Single-cell sequencing will offer unprecedented resolution, allowing researchers to characterize the unique genetic information of individual microbes within a complex community. For real-world uses, the field will shift toward microbiome engineering. This includes the development of air filtration systems that use microbes to eliminate airborne pathogens and allergens, the use of engineered microbial communities for improved bioremediation of pollutants in industrial wastewater and urban runoff, and the development of targeted microbial inoculants to repair damaged marine ecosystems. Utilizing environmental microbiomes as a sustainable tool for ecosystem health management and enhancement is the ultimate objective.

## REFERENCES


Abate, R., Oon, Y.-L., Oon, Y.-S., Bi, Y., Mi, W., Song, G., & Gao, Y. (2024). Diverse interactions between bacteria and microalgae: A review for enhancing harmful algal bloom mitigation and biomass processing efficiency. *Heliyon*, 10(17), e36503. https://doi.org/10.1016/j.heliyon.2024.e36503

Ahmed, M., Iram, S., Tabassum, N., Sajid, M., Paseutsakoun, K., Aleksza, L., & Székács, A. (2025). Biodegradation efficacy of Aspergillus niger and Trichoderma harzianum on Low-Density Polyethylene. *Polymers*, 17(10), 1303. https://doi.org/10.3390/polym17101303

Ahmed, N., Mahmoud, N. F., Solyman, S., & Hanora, A. (2019). Human nasal microbiome as characterized by metagenomics differs markedly between rural and industrial communities in Egypt. *OMICS: A Journal of Integrative Biology*, 23(11), 573–582. https://doi.org/10.1089/omi.2019.0144

Al Haj Chehadeh, B., Ali Ahmad, F., & Salam, D. A. (2025). Biodegradation of petroleum tar in contaminated sediments of the Eastern Mediterranean shores and associated microbial dynamics. *Applied and Environmental Microbiology*, 91(7). https://doi.org/10.1128/aem.00258-25

Al, M. A., Ren, K., Rensing, C., Hu, A., Tsyganov, A. N., Mazei, Y. A., & Mazei, N. G. (2024). Ecological and evolutionary processes involved in shaping microbial habitat generalists and specialists in urban park ecosystems. *mSystems*, 9(5), 00469–24. https://doi.org/10.1128/msystems.00469-24

Aliu Olalekan Olatunji, Janet Aderonke Olaboye, Chukwudi Cosmos Maha, Tolulope Olagoke Kolawole, & Samira Abdul. (2024). Environmental microbiology and public health: Advanced strategies for mitigating waterborne and airborne pathogens to prevent disease. *International Medical Science Research Journal*, 4(7), 756–770. https://doi.org/10.51594/imsrj.v4i7.1355

Al-Shaarani, A. A. Q. A., & Pecoraro, L. (2024). A review of pathogenic airborne fungi and bacteria: Unveiling occurrence, sources, and profound human health implication. *Frontiers in Microbiology*, 15. https://doi.org/10.3389/fmicb.2024.1428415

Amato, P., Joly, M., Besaury, L., Oudart, A., Taib, N., Moné, A. I., Deguillaume, L., Delort, A.-M., & Debroas, D. (2017). Active microorganisms thrive among extremely diverse communities in cloud water. *Plos One*, 12(8), e0182869. https://doi.org/10.1371/journal.pone.0182869

Amato, P., Parazols, M., Sancelme, M., Laj, P., Mailhot, G., & Delort, A.-M. (2007). Microorganisms isolated from the water phase of tropospheric clouds at the Puy de Dôme: Major groups and growth abilities at low temperatures: Microorganisms from the water phase of tropospheric clouds. *FEMS Microbiology Ecology*, 59(2), 242–254. https://doi.org/10.1111/j.1574-6941.2006.00199.x

Ameruoso, A., Villegas Kcam, M. C., Cohen, K. P., & Chappell, J. (2022). Activating natural product synthesis using CRISPR interference and activation systems in Streptomyces. *Nucleic Acids Research*, 50(13), 7751–7760. https://doi.org/10.1093/nar/gkac556

Aminian-Dehkordi, J., Rahimi, S., Golzar-Ahmadi, M., Singh, A., Lopez, J., Ledesma-Amaro, R., & Mijakovic, I. (2023). Synthetic biology tools for environmental protection. *Biotechnology Advances*, 68, 108239. https://doi.org/10.1016/j.biotechadv.2023.108239

Anesio, A. M., & Laybourn-Parry, J. (2012). Glaciers and ice sheets as a biome. *Trends in Ecology & Evolution*, 27(4), 219–225. https://doi.org/10.1016/j.tree.2011.09.012

Anesio, A. M., Lutz, S., Chrismas, N. A. M., & Benning, L. G. (2017). The microbiome of glaciers and ice sheets. *NPJ Biofilms and Microbiomes*, 3(1), 10. https://doi.org/10.1038/s41522-017-0019-0

Ayangbenro, A. S., Olanrewaju, O. S., & Babalola, O. O. (2018). Sulfate-reducing bacteria as an effective tool for sustainable acid mine bioremediation. *Frontiers in Microbiology*, 9. https://doi.org/10.3389/fmicb.2018.01986

Ayilara, M. S., & Babalola, O. O. (2023). Bioremediation of environmental wastes: The role of microorganisms. *Frontiers in Agronomy*, 5. https://doi.org/10.3389/fagro.2023.1183691

Bao, Z., Chen, B., Yu, K., Wei, Y., Liang, X., Yao, H., Liao, X., Xie, W., & Yin, K. (2025). Microbiome dynamics and functional profiles in deep-sea wood-fall micro-ecosystem: Insights into drive pattern of community assembly, biogeochemical processes, and lignocellulose degradation. *Applied and Environmental Microbiology*, 91(1), e02165–24. https://doi.org/10.1128/aem.02165-24

Baranova, A. A., Zakalyukina, Y. V., Ovcharenko, A. A., Korshun, V. A., & Tyurin, A. P. (2022). Antibiotics from insect-Aassociated actinobacteria. *Biology*, 11(11), 1676. https://doi.org/10.3390/biology11111676

Bardgett, R. D., & Putten, W. H. (2014). Belowground biodiversity and ecosystem functioning. *Nature*, 515(7528), 505–511. https://doi.org/10.1038/nature13855

Bengtsson-Palme, J., Kristiansson, E., & Larsson, D. G. J. (2018). Environmental factors influencing the development and spread of antibiotic resistance. *FEMS Microbiology Reviews*, 42(1), 68–80. https://doi.org/10.1093/femsre/fux053

Blainey, P. C. (2013). The future is now: Single-cell genomics of bacteria and archaea. *FEMS Microbiology Reviews*, 37(3), 407–427. https://doi.org/10.1111/1574-6976.12015

Blin, K., Shaw, S., Vader, L., Szenei, J., Reitz, Z. L., Augustijn, H. E., Cediel-Becerra, J. D. D., de Crécy-Lagard, V., Koetsier, R. A., Williams, S. E., Cruz-Morales, P., Wongwas, S., Segurado Luchsinger, A. E., Biermann, F., Korenskaia, A., Zdouc, M. M., Meijer, D., Terlouw, B. R., van der Hooft, J. J. J., … Weber, T. (2025). antiSMASH 8.0: Extended gene cluster detection capabilities and analyses of chemistry, enzymology, and regulation. *Nucleic Acids Research*. https://doi.org/10.1093/nar/gkaf334

Bourquin, M., Busi, S. B., Fodelianakis, S., Peter, H., Washburne, A., Kohler, T. J., Ezzat, L., Michoud, G., Wilmes, P., & Battin, T. J. (2022). The microbiome of cryospheric ecosystems. *Nature Communications*, 13(1), 3087. https://doi.org/10.1038/s41467-022-30816-4

Bowers, R. M., Clements, N., Emerson, J. B., Wiedinmyer, C., Hannigan, M. P., & Fierer, N. (2013). Seasonal variability in bacterial and fungal diversity of the near-surface atmosphere. *Environmental Science & Technology*, 47(21), 12097–12106. https://doi.org/10.1021/es402970s

Bruno, A., Fumagalli, S., Ghisleni, G., & Labra, M. (2022). The microbiome of the built environment: The nexus for urban regeneration for the cities of tomorrow. *Microorganisms*, 10(12), 2311. https://doi.org/10.3390/microorganisms10122311

Caesar, L. K., Montaser, R., Keller, N. P., & Kelleher, N. L. (2021). Metabolomics and genomics in natural products research: Complementary tools for targeting new chemical entities. *Natural Product Reports*, 38(11), 2041–2065. https://doi.org/10.1039/D1NP00036E

Cai, Z., Li, M., Zhu, Z., Wang, X., Huang, Y., Li, T., Gong, H., & Yan, M. (2023). Biological degradation of plastics and microplastics: A recent perspective on associated mechanisms and influencing factors. *Microorganisms*, 11(7). https://doi.org/10.3390/microorganisms11071661

Cao, L., Yang, L., Swanson, C. S., Li, S., & He, Q. (2021). Comparative analysis of impact of human occupancy on indoor microbiomes. *Frontiers of Environmental Science & Engineering*, 15(5), 89. https://doi.org/10.1007/s11783-020-1383-1

Carlson, J. C., Fortman, J. L., Anzai, Y., Li, S., Burr, D. A., & Sherman, D. H. (2010). Identification of the tirandamycin biosynthetic gene cluster from Streptomyces sp. 307-9. *ChemBioChem*, 11(4), 564–572. https://doi.org/10.1002/cbic.200900658

Caruso, G., & Rizzo, C. (2025). Cryosphere microbial communities as a reservoir of hidden risks to human and ecosystem's health. *Microbiological Research*, 299, 128261. https://doi.org/10.1016/j.micres.2025.128261

Cavicchioli, R., Ripple, W. J., Timmis, K. N., Azam, F., Bakken, L. R., Baylis, M., & Webster, N. S. (2019). Scientists' warning to humanity: Microorganisms and climate change. *Nature Reviews Microbiology*, 17(9), 569–586. https://doi.org/10.1038/s41579-019-0222-5

Çelik, I., & Keskin, E. (2022). Revealing the microbiome of four different thermal springs in Turkey with environmental DNA metabarcoding. *Biology*, 11(7). https://doi.org/10.3390/biology11070998

Chater, K. F. (2016). Recent advances in understanding Streptomyces. *F1000Research*, 5, 2795. https://doi.org/10.12688/f1000research.9534.1

Chen, F., Zhang, W., Mfarrej, M. F. B., Saleem, M. H., Khan, K. A., Ma, J., Raposo, A., & Han, H. (2024). Breathing in danger: Understanding the multi-faceted impact of air pollution on health impacts. *Ecotoxicology and Environmental Safety*, 280, 116532. https://doi.org/10.1016/j.ecoenv.2024.116532

Chen, S., Xie, Z.-X., Yan, K.-Q., Chen, J.-W., Li, D.-X., Wu, P.-F., Peng, L., Lin, L., Dong, C.-M., Zhao, Z., Fan, G.-Y., Liu, S.-Q., Herndl, G. J., & Wang, D.-Z. (2024). Functional vertical connectivity of microbial communities in the ocean. *Science Advances*, 10(21), eadj8184. https://doi.org/10.1126/sciadv.adj8184

Chen, X., Kumari, D., & Achal, V. (2020). A review on airborne microbes: The characteristics of sources, pathogenicity and geography. *Atmosphere*, 11(9). https://doi.org/10.3390/atmos11090919

Chen, Y., Liang, Z., Li, G., & An, T. (2024). Indoor/outdoor airborne microbiome characteristics in residential areas across four seasons and its indoor purification. *Environment International*, 190, 108857. https://doi.org/10.1016/j.envint.2024.108857

Cimermancic, P., Medema, M. H., Claesen, J., Kurita, K., Wieland Brown, L. C., Mavrommatis, K., Pati, A., Godfrey, P. A., Koehrsen, M., Clardy, J., Birren, B. W., Takano, E., Sali, A., Linington, R. G., & Fischbach, M. A. (2014). Insights into secondary metabolism from a global analysis of prokaryotic biosynthetic gene clusters. *Cell*, 158(2), 412–421. https://doi.org/10.1016/j.cell.2014.06.034

Coolahan, M., & Whalen, K. E. (2025). A review of quorum-sensing and its role in mediating interkingdom interactions in the ocean. *Communications Biology*, 8(1), 179. https://doi.org/10.1038/s42003-025-07608-9

Coy, S. R., Gann, E. R., Pound, H. L., Short, S. M., & Wilhelm, S. W. (2018). Viruses of eukaryotic algae: Diversity, methods for detection, and future directions. *Viruses*, 10(9). https://doi.org/10.3390/v10090487

Coyne, K. J., Wang, Y., & Johnson, G. (2022). Algicidal bacteria: A review of current knowledge and applications to control harmful algal blooms. *Frontiers in Microbiology*, 13, 871177. https://doi.org/10.3389/fmicb.2022.871177

Dang, H. (2020). Grand challenges in microbe-driven marine carbon cycling research. *Frontiers in Microbiology*, 11, 1039. https://doi.org/10.3389/fmicb.2020.01039

Darch, S. E., West, S. A., Winzer, K., & Diggle, S. P. (2012). Density-dependent fitness benefits in quorum-sensing bacterial populations. *Proceedings of the National Academy of Sciences*, 109(21), 8259–8263. https://doi.org/10.1073/pnas.1118131109

D'Costa, V. M., King, C. E., Kalan, L., Morar, M., Sung, W. W., Schwarz, C., & Wright, G. D. (2011). Antibiotic resistance is ancient. *Nature*, 477(7365), 457–461. https://doi.org/10.1038/nature10388

De Carvalho, C. C. C. R., & Fernandes, P. (2010). Production of metabolites as bacterial responses to the marine environment. *Marine Drugs*, 8(3), 705–727. https://doi.org/10.3390/md8030705

Debnath, T., Kujur, R. R. A., Mitra, R., & Das, S. K. (2019). Diversity of microbes in hot springs and their sustainable use. In T. Satyanarayana, B. N. Johri, & S. K. Das (Eds.), *Microbial Diversity in Ecosystem Sustainability and Biotechnological Applications: Volume 1. Microbial Diversity in Normal {\&} Extreme Environments* (pp. 159–186). Springer Singapore. https://doi.org/10.1007/978-981-13-8315-1_6

Debs, K. B., Cardona, D. S., da Silva, H. D. T., Nassar, N. N., Carrilho, E. N. V. M., Haddad, P. S., & Labuto, G. (2019). Oil spill cleanup employing magnetite nanoparticles and yeast-based magnetic bionanocomposite. *Journal of Environmental Management*, 230, 405–412. https://doi.org/10.1016/j.jenvman.2018.09.094

de Vries, F. T., Griffiths, R. I., Bailey, M., Craig, H., Girlanda, M., Gweon, H. S., ... and Hallin, S. (2018). Soil bacterial networks are less stable under drought than fungal networks. *Nature Communications*, 9(1), 3033. https://doi.org/10.1038/s41467-018-05516-7

Dodamani, S., Hattiholi, A., & Kurjogi, M. (2022). Chapter 11—Microbial communities: An effective tool for cleaning environment. In M. Shah, S. Rodriguez-Couto, C. B. V. De La Cruz, & J. Biswas (Eds.), *An Integration of Phycoremediation Processes in Wastewater Treatment* (pp. 231–248). Elsevier. https://doi.org/10.1016/B978-0-12-823499-0.00008-0

Duan, P., Hu, W., Wu, Z., Bi, K., Zhu, J., & Fu, P. (2023). Ice nucleation activity of airborne pollen: A short review of results from laboratory experiments. *Atmospheric Research*, 285, 106659. https://doi.org/10.1016/j.atmosres.2023.106659

Eisenhauer, N., Hines, J., Maestre, F. T., & Rillig, M. C. (2023). Reconsidering functional redundancy in biodiversity research. *NPJ Biodiversity*, 2(1), 9. https://doi.org/10.1038/s44185-023-00015-5

El-Sesy, M. E., Ibrahim, S. S., Yang, F., & Hegazy, A. (2025). Synergistic remediation of oil-water mixtures: Integration of nanoparticles and oil-degrading bacteria in a novel approach. *Frontiers in Microbiology*, 16. https://doi.org/10.3389/fmicb.2025.1590636

Etesami, H., Jeong, B. R., & Glick, B. R. (2021). Contribution of arbuscular mycorrhizal fungi, phosphate–solubilizing bacteria, and silicon to P uptake by plant. *Frontiers in Plant Science*, 12. https://doi.org/10.3389/fpls.2021.699618

Falkowski, P. G., Fenchel, T., & Delong, E. F. (2008). The microbial engines that drive Earth's biogeochemical cycles. *Science*, 320(5879), 1034–1039. https://doi.org/10.1126/science.1153213

Fang, C., & Achal, V. (2024). Enhancing carbon neutrality: A perspective on the role of microbially induced carbonate precipitation (MICP). *Biogeotechnics*, 2(2), 100083. https://doi.org/10.1016/j.bgtech.2024.100083

Fang, C., Sun, X., Fan, F., Zhang, X., Wang, O., Zheng, H., Peng, Z., Luo, X., Chen, A., Zhang, W., Drmanac, R., Peters, B. A., Song, Z., & Kristiansen, K. (2023). High-resolution single-molecule long-fragment rRNA gene amplicon sequencing of bacterial and eukaryotic microbial communities. *Cell Reports Methods*, 3(3), 100437. https://doi.org/10.1016/j.crmeth.2023.100437

Fang, Z., Guo, W., Zhang, J., & Lou, X. (2018). Influence of heat events on the composition of airborne bacterial communities in urban ecosystems. *International Journal of Environmental Research and Public Health*, 15(10), 2295. https://doi.org/10.3390/ijerph15102295

Farmer, D. K., Boedicker, E. K., & Debolt, H. M. (2020). Dry deposition of atmospheric aerosols: Approaches, observations, and mechanisms. *Annual Review of Physical Chemistry*, 72, 375–397. https://doi.org/10.1146/annurev-physchem-090519-034936

Feitkenhauer, H., Müller, R., & Märkl, H. (2003). Degradation of polycyclic aromatic hydrocarbons and long chain alkanes at 6070 °C by Thermus and Bacillus spp. *Biodegradation*, 14(6), 367–372. https://doi.org/10.1023/A:1027357615649

Ferrer, M., Martínez-Martínez, M., Bargiela, R., Streit, W. R., Golyshina, O. V., & Golyshin, P. N. (2016). Estimating the success of enzyme bioprospecting through metagenomics: Current status and future trends. *Microbial Biotechnology*, 9(1), 22–34. https://doi.org/10.1111/1751-7915.12309

Fierer, N. (2017). Embracing the unknown: Disentangling the complexities of the soil microbiome. *Nature Reviews Microbiology*, 15(10), 579–590. https://doi.org/10.1038/nrmicro.2017.87

Firincă, C., Zamfir, L.-G., Constantin, M., Răut, I., Jecu, M.-L., Doni, M., Gurban, A.-M., & Şesan, T. E. (2025). Innovative approaches and evolving strategies in heavy metal bioremediation: Current limitations and future opportunities. *Journal of Xenobiotics*, 15(3), 63. https://doi.org/10.3390/jox15030063

Flemming, H. C., & Wuertz, S. (2019). Bacteria and archaea on Earth and their abundance in biofilms. *Nature Reviews Microbiology*, 17(4), 247–260. https://doi.org/10.1038/s41579-019-0158-9

Franzosa, E. A., Hsu, T., Sirota-Madi, A., Shafquat, A., Abu-Ali, G., Morgan, X. C., & Huttenhower, C. (2015). Sequencing and beyond: Integrating molecular "omics" for microbial community profiling. *Nature Reviews Microbiology*, 13(6), 360–372. https://doi.org/10.1038/nrmicro3451

Fu, K., Zhou, Q., & Wang, H. (2024). Variability in microbial communities driven by particulate matter on human facial skin. *Toxics*, 12(7). https://doi.org/10.3390/toxics12070497

Fu, X., Norbäck, D., Yuan, Q., Li, Y., Zhu, X., Hashim, J. H., Hashim, Z., Ali, F., Hu, Q., Deng, Y., & Sun, Y. (2021). Association between indoor microbiome exposure and sick building syndrome (SBS) in junior high schools of Johor Bahru, Malaysia. *Science of the Total Environment*, 753, 141904. https://doi.org/10.1016/j.scitotenv.2020.141904

Fu, X., Ou, Z., & Sun, Y. (2022). Indoor microbiome and allergic diseases: From theoretical advances to prevention strategies. *Eco-Environment & Health*, 1(3), 133–146. https://doi.org/10.1016/j.eehl.2022.09.002

Fujimura, K. E., & Lynch, S. V. (2015). Microbiota in allergy and asthma and the emerging relationship with the gut microbiome. *Cell Host & Microbe*, 17(5), 592–602. https://doi.org/10.1016/j.chom.2015.04.007

Gao, Y., Lu, Y., Dungait, J. A. J., Liu, J., Lin, S., Jia, J., & Yu, G. (2022). The "regulator" function of viruses on ecosystem carbon cycling in the Anthropocene. *Frontiers in Public Health*, 10, 858615. https://doi.org/10.3389/fpubh.2022.858615

Gavriilidou, A., Mackenzie, T. A., Sánchez, P., Tormo, J. R., Ingham, C., Smidt, H., & Sipkema, D. (2021). Bioactivity screening and gene-trait matching across marine sponge-associated bacteria. *Marine Drugs*, 19(2). https://doi.org/10.3390/md19020075

Gaylarde, C. C., Baptista Neto, J. A., & da Fonseca, E. M. (2024). Indoor airborne microplastics: Human health importance and effects of air filtration and turbulence. *Microplastics*, 3(4), 653–670. https://doi.org/10.3390/microplastics3040040

Ge, Y., Wang, C., Etim, I.-I. N., Khan, S., Li, C., Yang, L., Liu, J., Yi, P., Liu, J., Sand, W., & Zhang, R. (2025). Deep-sea microbially influenced corrosion and biomineralization. *Frontiers in Microbiology*, 16, 1605909. https://doi.org/10.3389/fmicb.2025.1605909

Ghosh, B., Lal, H., & Srivastava, A. (2015). Review of bioaerosols in indoor environment with special reference to sampling, analysis and control mechanisms. *Environment International*, 85, 254–272. https://doi.org/10.1016/j.envint.2015.09.018

Giddings, L.-A., & Newman, D. J. (2022). Extremophilic fungi from marine environments: Underexplored sources of antitumor, anti-infective and other biologically active agents. *Marine Drugs*, 20(1). https://doi.org/10.3390/md20010062

Gilbert, J. A., Jansson, J. K., & Knight, R. (2014). The earth microbiome project: Successes and aspirations. *BMC Biology*, 12(1), 69. https://doi.org/10.1186/s12915-014-0069-1

Gilbert, J. A., & Stephens, B. (2018). Microbiology of the built environment. *Nature Reviews Microbiology*, 16(11), 661–670. https://doi.org/10.1038/s41579-018-0065-5

Ginnan, N., Crandall, S. G., Imchen, M., Dini-Andreote, F., Miyashiro, T. I., Singh, V., Ganda, E., & Bordenstein, S. R. (2025). Ecologically expanding the one health framework to unify the microbiome sciences. *mBio*, 16(6). https://doi.org/10.1128/mbio.03147-24

Gloor, G. B., Macklaim, J. M., Pawlowsky-Glahn, V., & Egozcue, J. J. (2017). Microbiome datasets are compositional: And this is not optional. *Frontiers in Microbiology*, 8, 2224. https://doi.org/10.3389/fmicb.2017.02224

Graves, J. L. (2022). Environment—Biogeochemical cycles, pollution. In *Principles and Applications of Antimicrobial Nanomaterials* (pp. 181–199). Elsevier. https://doi.org/10.1016/B978-0-12-822105-1.00007-X

Griffin, D. W. (2007). Atmospheric movement of microorganisms in clouds of desert dust and implications for human health. *Clinical Microbiology Reviews*, 20(3), 459–477. https://doi.org/10.1128/cmr.00039-06

Handelsman, J. (2004). Metagenomics: Application of genomics to uncultured microorganisms. *Microbiology and Molecular Biology Reviews*, 68(4), 669–685. https://doi.org/10.1128/MMBR.68.4.669-685.2004

Hanski, I., Von Hertzen, L., Fyhrquist, N., Koskinen, K., Torppa, K., Laatikainen, T., Karisola, P., Auvinen, P., Paulin, L., Mäkelä, M. J., Vartiainen, E., Kosunen, T. U., Alenius, H., & Haahtela, T. (2012). Environmental biodiversity, human microbiota, and allergy are interrelated. *Proceedings of the National Academy of Sciences*, 109(21), 8334–8339. https://doi.org/10.1073/pnas.1205624109

Hasin, Y., Seldin, M., & Lusis, A. (2017). Multi-omics approaches to disease. *Genome Biology*, 18(1), 83. https://doi.org/10.1186/s13059-017-1215-1

He, S., Jia, M., Xiang, Y., Song, B., Xiong, W., Cao, J., Peng, H., Yang, Y., Wang, W., Yang, Z., & Zeng, G. (2022). Biofilm on microplastics in aqueous environment: Physicochemical properties and environmental implications. *Journal of Hazardous Materials*, 424, 127286. https://doi.org/10.1016/j.jhazmat.2021.127286

He, Z., Lin, L., Wang, X., Qin, W., & Zhang, C. (2022). Editorial: Carbon storage by marine microorganisms for carbon neutrality. *Frontiers in Marine Science*, 9. https://doi.org/10.3389/fmars.2022.1018397

Heilmann, M., Breiter, R., & Becker, A. M. (2021). Towards rare earth element recovery from wastewaters: Biosorption using phototrophic organisms. *Applied Microbiology and Biotechnology*, 105(12), 5229–5239. https://doi.org/10.1007/s00253-021-11386-9

Hongjin, C., Ur Rahman, S., Rehman, A., Khan, A. A., & Khalid, M. (2025). Microplastics and antibiotic resistance genes as rising threats: Their interaction represents an urgent environmental concern. *Current Research in Microbial Sciences*, 9, 100447. https://doi.org/10.1016/j.crmicr.2025.100447

Huang, S., Hu, W., Chen, J., Wu, Z., Zhang, D., & Fu, P. (2021). Overview of biological ice nucleating particles in the atmosphere. *Environment International*, 146, 106197. https://doi.org/10.1016/j.envint.2020.106197

Huang, T., Hu, Q., Shen, Y., Anglés, A., & Fernández-Remolar, D. C. (2024). Biogeochemical cycles. In S. M. Scheiner (Ed.), *Encyclopedia of Biodiversity* (3rd ed., pp. 1–20). Elsevier. https://doi.org/10.1016/b978-0-12-822562-2.00347-9

Hwang, S., Lee, N., Cho, S., Palsson, B., & Cho, B.-K. (2020). Repurposing modular polyketide synthases and non-ribosomal peptide synthetases for novel chemical biosynthesis. *Frontiers in Molecular Biosciences*, 7. https://doi.org/10.3389/fmolb.2020.00087

Iannone, R., Chernoff, D. I., Pringle, A., Martin, S. T., & Bertram, A. K. (2011). The ice nucleation ability of one of the most abundant types of fungal spores found in the atmosphere. *Atmospheric Chemistry and Physics*, 11(3), 1191–1201. https://doi.org/10.5194/acp-11-1191-2011

Ifedinezi, O.V., Nnaji, N. D., Anumudu, C. K., Ekwueme, C. T., Uhegwu, C. C., Ihenetu, F. C., Obioha, P., Simon, B. O., Ezechukwu, P. S., & Onyeaka, H. (2024). Environmental antimicrobial resistance: Implications for food safety and public health. *Antibiotics*, 13(11). https://doi.org/10.3390/antibiotics13111087

Izabel-Shen, D., Li, S., Luo, T., Wang, J., Li, Y., Sun, Q., Yu, C.-P., & Hu, A. (2022). Repeated introduction of micropollutants enhances microbial succession despite stable degradation patterns. *ISME Communications*, 2(1), 48. https://doi.org/10.1038/s43705-022-00129-0

Jackson, V. L. N., Grevesse, T., Kilias, E. S., Onda, D. F. L., Young, K. F., Allen, M. J., Walsh, D. A., Lovejoy, C., & Monier, A. (2024). Vulnerability of Arctic Ocean microbial eukaryotes to sea ice loss. *Scientific Reports*, 14(1), 28896. https://doi.org/10.1038/s41598-024-77821-9

Jalali, J., & Lebeau, T. (2021). The role of microorganisms in mobilization and phytoextraction of rare Earth elements: A review. *Frontiers in Environmental Science*, 9. https://doi.org/10.3389/fenvs.2021.688430

Jatana, B. S., Kitchens, C., Ray, C., Gerard, P., & Tharayil, N. (2021). Dual inoculation with Arbuscular Mycorrhizal Fungi and phosphorus solubilizing fungi synergistically enhances the mobilization and plant uptake of phosphorus from meat and bone meal. *Frontiers in Soil Science*, 1. https://doi.org/10.3389/fsoil.2021.757839

Jiang, X., Wang, C., Guo, J., Hou, J., Guo, X., Zhang, H., Tan, J., Li, M., Li, X., & Zhu, H. (2022). Global meta-analysis of airborne bacterial communities and associations with anthropogenic activities. *Environmental Science & Technology*, 56(14), 9891–9902. https://doi.org/10.1021/acs.est.1c07923

Jiao, N., Luo, T., Chen, Q., Zhao, Z., Xiao, X., Liu, J., Jian, Z., Xie, S., Thomas, H., Herndl, G. J., Benner, R., Gonsior, M., Chen, F., Cai, W.-J., & Robinson, C. (2024). The microbial carbon pump and climate change. *Nature Reviews Microbiology*, 22(7), 408–419. https://doi.org/10.1038/s41579-024-01018-0

Jolaosho, T. L., Rasaq, M. F., Omotoye, E. V., Araomo, O. V., Adekoya, O. S., Abolaji, O. Y., & Hungbo, J. J. (2025). Microplastics in freshwater and marine ecosystems: Occurrence, characterization, sources, distribution dynamics, fate, transport processes, potential mitigation strategies, and policy interventions. *Ecotoxicology and Environmental Safety*, 294, 118036. https://doi.org/10.1016/j.ecoenv.2025.118036

Joly, M., Amato, P., Sancelme, M., Vinatier, V., Abrantes, M., Deguillaume, L., & Delort, A.-M. (2015). Survival of microbial isolates from clouds toward simulated atmospheric stress factors. *Atmospheric Environment*, 117, 92–98. https://doi.org/10.1016/j.atmosenv.2015.07.009

Jungblut, A. D., Velazquez, D., Cirés, S., Kleinteich, J., Kottekkatu Padinchati, K., Sattler, B., & Comte, J. (2022). Editorial: Digitizing frozen earth—Revealing microbial diversity and physiology in the cryobiosphere through "omics" tools, volume II. *Frontiers in Microbiology*, 13. https://doi.org/10.3389/fmicb.2022.1013398

Kamel, H. L., Hanora, A., & Solyman, S. M. (2022). Metataxanomic, bioactivity and microbiome analysis of Red Sea marine sponges from Egypt. *Marine Genomics*, 61(September 2021), 100920. https://doi.org/10.1016/j.margen.2021.100920

Karaseva, A. I., Elcheninov, A. G., Prokofeva, M. I., Klyukina, A. A., & Kochetkova, T. V. (2024). Microbial diversity of hot springs of the Kuril Islands. *BMC Microbiology*, 24(1), 547. https://doi.org/10.1186/s12866-024-03704-8

Karkaria, B. D., Fedorec, A. J. H., & Barnes, C. P. (2021). Automated design of synthetic microbial communities. *Nature Communications*, 12(1), 672. https://doi.org/10.1038/s41467-020-20756-2

Kayiranga, A., Isabwe, A., Yao, H., Shangguan, H., Coulibaly, J. L. K., Breed, M. F., & Sun, X. (2024). Distribution patterns of soil bacteria, fungi, and protists emerge from distinct assembly processes across subcommunities. *Ecology and Evolution*, 14(7), 11672. https://doi.org/10.1002/ece3.11672

Kazemian, N., Pakpour, S., Milani, A. S., & Klironomos, J. (2019). Environmental factors influencing fungal growth on gypsum boards and their structural biodeterioration: A university campus case study. *Plos One*, *14*(8), e0220556. https://doi.org/10.1371/journal.pone.0220556

Kelly, M. S., Bunyavanich, S., Phipatanakul, W., & Lai, P. S. (2022). The environmental microbiome, allergic disease, and asthma. *The Journal of Allergy and Clinical Immunology: In Practice*, *10*(9), 2206–2217.e1. https://doi.org/10.1016/j.jaip.2022.06.006

Kenshole, E., Herisse, M., Michael, M., & Pidot, S. J. (2021). Natural product discovery through microbial genome mining. *Current Opinion in Chemical Biology*, *60*, 47–54. https://doi.org/10.1016/j.cbpa.2020.07.010

Khan, S., Wang, T., Arifeen, M. Z. U., & Huang, S. (2025). Exploring the bioactive potential of deep-sea microorganisms: A review of recent discoveries. *Bioorganic Chemistry*, *161*, 108521. https://doi.org/10.1016/j.bioorg.2025.108521

Khatua, S., Simal-Gandara, J., & Acharya, K. (2024). Myco-remediation of plastic pollution: Current knowledge and future prospects. *Biodegradation*, *35*(3), 249–279. https://doi.org/10.1007/s10532-023-10053-2

Kheiri, R., Mehrshad, M., Pourbabaee, A. A., Ventosa, A., & Amoozegar, M. A. (2023). Hypersaline Lake Urmia: A potential hotspot for microbial genomic variation. *Scientific Reports*, *13*(1), 374. https://doi.org/10.1038/s41598-023-27429-2

Kim, H., & Yoo, K. (2024). Marine plastisphere selectively enriches microbial assemblages and antibiotic resistance genes during long-term cultivation periods. *Environmental Pollution (Barking, Essex: 1987)*, *344*, 123450. https://doi.org/10.1016/j.envpol.2024.123450

Knight, R., Vrbanac, A., Taylor, B. C., Aksenov, A., Callewaert, C., Debelius, J., & Dorrestein, P. C. (2018). Best practices for analysing microbiomes. *Nature Reviews Microbiology*, *16*(7), 410–422. https://doi.org/10.1038/s41579-018-0029-9

Kranzler, C. F., Busono, D. A., Walsh, G. J., Carrillo, A. C., Bidle, K. D., & Thamatrakoln, K. (2025). Taxonomically distinct diatom viruses differentially impact microbial processing of organic matter. *Science Advances*, *11*(18), eadq5439. https://doi.org/10.1126/sciadv.adq5439

Lampitt, R. S., Briggs, N., Cael, B. B., Espinola, B., Hélaouët, P., Henson, S. A., Norrbin, F., Pebody, C. A., & Smeed, D. (2023). Deep ocean particle flux in the Northeast Atlantic over the past 30 years: Carbon sequestration is controlled by ecosystem structure in the upper ocean. *Frontiers in Earth Science*, *11*. https://doi.org/10.3389/feart.2023.1176196

Larkin, A. A., Brock, M. L., Fagan, A. J., Moreno, A. R., Gerace, S. D., Lees, L. E., Suarez, S. A., Eloe-Fadrosh, E. A., & Martiny, A. C. (2025). Climate-driven succession in marine microbiome biodiversity and biogeochemical function. *Nature Communications*, *16*(1), 3926. https://doi.org/10.1038/s41467-025-59382-1

Larsson, D. G. J., & Flach, C. F. (2022). Antibiotic resistance in the environment. *Nature Reviews Microbiology*, *20*(5), 257–269. https://doi.org/10.1038/s41579-021-00649-x

Lazaridis, M. (2019). Bacteria as cloud condensation nuclei (CCN) in the atmosphere. *Atmosphere*, *10*(12). https://doi.org/10.3390/atmos10120786

Lee, K.-B., De Backer, P., Aono, T., Liu, C.-T., Suzuki, S., Suzuki, T., Kaneko, T., Yamada, M., Tabata, S., Kupfer, D. M., Najar, F. Z., Wiley, G. B., Roe, B., Binnewies, T. T., Ussery, D. W., D'Haeze, W., Den Herder, J., Gevers, D., Vereecke, D., … Oyaizu, H. (2008). The genome of the versatile nitrogen fixer Azorhizobium caulinodans ORS571. *BMC Genomics*, *9*(1), 271. https://doi.org/10.1186/1471-2164-9-271

Leung, M. H. Y., Tong, X., & Lee, P. K. H. (2019). Indoor microbiome and airborne pathogens. In *Comprehensive Biotechnology* (pp. 96–106). Elsevier. https://doi.org/10.1016/B978-0-444-64046-8.00477-8

Li, L., Meng, D., Yin, H., Zhang, T., & Liu, Y. (2023). Genome-resolved metagenomics provides insights into the ecological roles of the keystone taxa in heavy-metal-contaminated soils. *Frontiers in Microbiology*, *14*. https://doi.org/10.3389/fmicb.2023.1203164

Li, L., & Zhang, T. (2023). Roadmap to tackle antibiotic resistance in the environment under the one health framework. *mLife*, *2*(3), 1–10. https://doi.org/10.1002/mlf2.12078

Li, X., Liang, X., He, H., Li, J., Ma, L., Tan, W., Zhong, Y., Zhu, J., Zhou, M.-F., & Dong, H. (2022). Microorganisms accelerate REE mineralization in supergene environments. *Applied and Environmental Microbiology*, *88*(13). https://doi.org/10.1128/aem.00632-22

Li, Y.-Q., Xin, Y., Li, C., Liu, J., & Huang, T. (2023). Metagenomics-metabolomics analysis of microbial function and metabolism in petroleum-contaminated soil. *Brazilian Journal of Microbiology*, *54*(2), 935–947. https://doi.org/10.1007/s42770-023-01000-7

Li, Z., Xiong, K., Wen, W., Li, L., & Xu, D. (2023). Functional endophytes regulating plant secondary metabolism: Current status, prospects and applications. *International Journal of Molecular Sciences*, *24*(2), 1153. https://doi.org/10.3390/ijms24021153

Liang, X., Ma, Y., Li, J., Ye, Y., & Li, J. (2025). Impact of microplastics on microbial diversity and pathogen distribution in aquaculture ecosystems: A seasonal analysis. *Environmental Pollution*, *368*, 125796. https://doi.org/10.1016/j.envpol.2025.125796

Liao, H., Lyon, C. J., Ying, B., & Hu, T. (2024). Climate change, its impact on emerging infectious diseases and new technologies to combat the challenge. *Emerging Microbes & Infections*, 13(1), 2356143. https://doi.org/10.1080/22221751.2024.2356143

Liu, J., Liu, J., Xu, B., Xu, A., Cao, S., Wei, R., Zhou, J., Jiang, M., & Dong, W. (2022). Biodegradation of polyether-polyurethane foam in yellow mealworms (Tenebrio molitor) and effects on the gut microbiome. *Chemosphere*, 304, 135263. https://doi.org/10.1016/j.chemosphere.2022.135263

Liu, L. Z., Chen, X., Jin, G., Cai, M., & Qiu, Z. (2022). Microbial diversity in marine sediments of two hydrocarbon reservoir areas in the eastern Atlantic Ocean. *Aquatic Microbial Ecology*, 89, 1–15.

Liu, S., & Deng, Y. (2025). Marine rare earth elements: Distribution patterns, enrichment mechanisms and microbial interactions. *Journal of Marine Science and Engineering*, 13(8), 1471. https://doi.org/10.3390/jmse13081471

Liu, X., Zhang, L., Wang, Y., Hu, S., Zhang, J., Huang, X., Li, R., Hu, Y., Yao, H., & Wang, Z. (2024). Microbiome analysis in Asia's largest watershed reveals inconsistent biogeographic pattern and microbial assembly mechanisms in river and lake systems. *iScience*, 27(6), 110053. https://doi.org/10.1016/j.isci.2024.110053

Liu, Z., Zhao, Y., Huang, C., & Luo, Y. (2021). Recent advances in silent gene cluster activation in Streptomyces. *Frontiers in Bioengineering and Biotechnology*, 9. https://doi.org/10.3389/fbioe.2021.632230

Locey, K. J., & Lennon, J. T. (2016). Scaling laws predict global microbial diversity. *Proceedings of the National Academy of Sciences*, 113(21), 5970–5975. https://doi.org/10.1073/pnas.1521291113

Locke, H., Bidle, K. D., Thamatrakoln, K., Johns, C. T., Bonachela, J. A., Ferrell, B. D., & Wommack, K. E. (2022). Chapter two—Marine viruses and climate change: Virioplankton, the carbon cycle, and our future ocean. In M. J. Roossinck (Ed.), *Viruses and Climate Change* (Vol. 114, pp. 67–146). Academic Press. https://doi.org/10.1016/bs.aivir.2022.09.001

Ma, J., Yan, X., Wang, Y., Zhang, W., Ma, K., Li, X., Shen, F., & Han, Y. (2025). Insights into the effects of haze pollution on airborne bacterial communities and antibiotic resistance genes in fine particulate matter. *Environmental Pollution*, 378, 126494. https://doi.org/10.1016/j.envpol.2025.126494

Madsen, E. L. (2005). Identifying microorganisms responsible for ecologically significant biogeochemical processes. *Nature Reviews Microbiology*, 3(5), 439–446. https://doi.org/10.1038/nrmicro1151

Maestre-Carballa, L., Navarro-López, V., & Martinez-Garcia, M. (2022). A Resistome roadmap: From the human body to pristine environments. *Frontiers in Microbiology*, 13, 858831. https://doi.org/10.3389/fmicb.2022.858831

Maglione, G., Zinno, P., Tropea, A., Mussagy, C. U., Dufossé, L., Giuffrida, D., & Mondello, A. (2024). Microbes' role in environmental pollution and remediation: A bioeconomy focus approach. *AIMS Microbiology*, 10(3), 723–755. https://doi.org/10.3934/microbiol.2024033

Mallick, H., Rahnavard, A., McIver, L. J., Ma, S., Zhang, Y., Nguyen, L. H., & Huttenhower, C. (2021). Multivariable association discovery in population-scale meta-omics studies. *PLOS Computational Biology*, 17(11), 1009442. https://doi.org/10.1371/journal.pcbi.1009442

Manavi, S. E. I., Aktypis, A., Siouti, E., Skyllakou, K., Myriokefalitakis, S., Kanakidou, M., & Pandis, S. N. (2025). Atmospheric aerosol spatial variability: Impacts on air quality and climate change. *One Earth*, 8(3), 101237. https://doi.org/10.1016/j.oneear.2025.101237

Martínez, G. M., Pire, C., & Martínez-Espinosa, R. M. (2022a). Hypersaline environments as natural sources of microbes with potential applications in biotechnology: The case of solar evaporation systems to produce salt in Alicante County (Spain). *Current Research in Microbial Sciences*, 3, 100136. https://doi.org/10.1016/j.crmicr.2022.100136

Martínez, G. M., Pire, C., & Martínez-Espinosa, R. M. (2022b). Hypersaline environments as natural sources of microbes with potential applications in biotechnology: The case of solar evaporation systems to produce salt in Alicante County (Spain). *Current Research in Microbial Sciences*, 3, 100136. https://doi.org/10.1016/J.CRMICR.2022.100136

Marzban, G., & Tesei, D. (2025). The extremophiles: Adaptation mechanisms and biotechnological applications. *Biology*, 14(4), 412. https://doi.org/10.3390/biology14040412

Mayol, E., Arrieta, J. M., Jiménez, M. A., Martínez-Asensio, A., Garcias-Bonet, N., Dachs, J., González-Gaya, B., Royer, S.-J., Benítez-Barrios, V. M., Fraile-Nuez, E., & Duarte, C. M. (2017). Long-range transport of airborne microbes over the global tropical and subtropical ocean. *Nature Communications*, 8(1), 201. https://doi.org/10.1038/s41467-017-00110-9

Mazumder, M. H. H., & Hussain, S. (2024). Air-pollution-mediated microbial dysbiosis in health and disease: Lung-gut axis and beyond. *Journal of Xenobiotics*, 14(4), 1595–1612. https://doi.org/10.3390/jox14040086

McCarty, N. S., & Ledesma-Amaro, R. (2019). Synthetic biology tools to engineer microbial communities for biotechnology. *Trends in Biotechnology*, 37(2), 181–197. https://doi.org/10.1016/j.tibtech.2018.11.002

Meadow, J. F., Altrichter, A. E., Kembel, S. W., Kline, J., Mhuireach, G., Moriyama, M., Northcutt, D., O'Connor, T. K., Womack, A. M., Brown, G. Z., Green, J. L., & Bohannan, B. J. M. (2014a). Indoor airborne bacterial communities are influenced by ventilation, occupancy, and outdoor air source. *Indoor Air*, 24(1), 41–48. https://doi.org/10.1111/ina.12047

Meadow, J. F., Altrichter, A. E., Kembel, S. W., Kline, J., Mhuireach, G., Moriyama, M., Northcutt, D., O'Connor, T. K., Womack, A. M., Brown, G. Z., Green, J. L., & Bohannan, B. J. M. (2014b). Indoor airborne bacterial communities are influenced by ventilation, occupancy, and outdoor air source. *Indoor Air*, 24(1), 41–48. https://doi.org/10.1111/ina.12047

Medeiros, W., Kralova, S., Oliveira, V., Ziemert, N., & Sehnal, L. (2025). Antarctic bacterial natural products: From genomic insights to drug discovery. *Natural Product Reports*, 42(5), 774–787. https://doi.org/10.1039/D4NP00045E

Mehta, P., Yadav, M., Ahmed, V., Goyal, K., Pandey, R., & Chauhan, N. S. (2021). Culture-independent exploration of the hypersaline ecosystem indicates the environment-specific microbiome evolution. *Frontiers in Microbiology*, 12, 686549. https://doi.org/10.3389/fmicb.2021.686549

Morris, B. E., Henneberger, R., Huber, H., & Moissl-Eichinger, C. (2013). Microbial syntrophy: Interaction for the common good. *FEMS Microbiology Reviews*, 37(3), 384–406. https://doi.org/10.1111/1574-6976.12019

Mousa, W. K., & Raizada, M. N. (2013). The diversity of anti-microbial secondary metabolites produced by fungal endophytes: An interdisciplinary perspective. *Frontiers in Microbiology*, 4. https://doi.org/10.3389/fmicb.2013.00065

Muhie, S., Gautam, A., Mylroie, J., Sowe, B., Campbell, R., Perkins, E. J., Hammamieh, R., & Garcia-Reyero, N. (2025). Effects of environmental chemical pollutants on microbiome diversity: Insights from shotgun metagenomics. *Toxics*, 13(2). https://doi.org/10.3390/toxics13020142

Nagar, S., Bharti, M., & Negi, R. K. (2023). Genome-resolved metagenomics revealed metal-resistance, geochemical cycles in a Himalayan hot spring. *Applied Microbiology and Biotechnology*, 107(10), 3273–3289. https://doi.org/10.1007/s00253-023-12503-6

Najam, M., Javaid, S., Iram, S., Pasertsakoun, K., Oláh, M., Székács, A., & Aleksza, L. (2025). Microbial biodegradation of synthetic polyethylene and polyurethane polymers by pedospheric microbes: Towards sustainable environmental management. *Polymers*, 17(2), 169. https://doi.org/10.3390/polym17020169

Narayanan, M., Kandasamy, S., Lee, J., & Barathi, S. (2023). Microbial degradation and transformation of PPCPs in aquatic environment: A review. *Heliyon*, 9(8), e18426. https://doi.org/10.1016/j.heliyon.2023.e18426

Nathani, N. M., Rajyaguru, R. H., Prashanth, P. N. P., Mootapally, C., & Dave, B. P. (2020). Microbial omics: Role in ecological studies and environmental control measures. In K. M. Gothandam, S. Ranjan, N. Dasgupta, & E. Lichtfouse (Eds.), *Environmental Biotechnology Vol. 2* (Vol. 45, pp. 173–196). Springer International Publishing. https://doi.org/10.1007/978-3-030-38196-7_8

National Academies of Sciences, E., Engineering, N. A. of, Sciences, D. on E. and P., Division, H. and M., Studies, D. on E. and L., Environment, B. on I. and the C., Toxicology, B. on E. S. and, Sciences, B. on L., & Application, C. on M. of the B. E. F. R. to. (2017). Tools for characterizing microbiome–Built environment interactions. In *Microbiomes of the Built Environment: A Research Agenda for Indoor Microbiology, Human Health, and Buildings*. National Academies Press (US). www.ncbi.nlm.nih.gov/books/NBK458828/

Nemergut, D. R., Schmidt, S. K., Fukami, T., O'Neill, S. P., Bilinski, T. M., Stanish, L. F., & Ferrenberg, S. (2013). Patterns and processes of microbial community assembly. *Microbiology and Molecular Biology Reviews*, 77(3), 342–356. https://doi.org/10.1128/MMBR.00051-12

Neufeld, J. D., Dumont, M. G., Vohra, J., & Murrell, J. C. (2007). Methodological considerations for the use of stable isotope probing in microbial ecology. *Microbial Ecology*, 53(3), 435–442. https://doi.org/10.1007/s00248-006-9125-x

Newman, D. J., & Cragg, G. M. (2020). Natural products as sources of new drugs over the nearly four decades from 01/1981 to 09/2019. *Journal of Natural Products*, 83(3), 770–803. https://doi.org/10.1021/acs.jnatprod.9b01285

Ni, B., Xiao, L., Lin, D., Zhang, T.-L., Zhang, Q., Liu, Y., Chen, Q., Zhu, D., Qian, H., Rillig, M. C., & Zhu, Y.-G. (2025). Increasing pesticide diversity impairs soil microbial functions. *Proceedings of the National Academy of Sciences*, 122(2). https://doi.org/10.1073/pnas.2419917122

Nurk, S., Meleshko, D., Korobeynikov, A., & Pevzner, P. A. (2017). metaSPAdes: A new versatile metagenomic assembler. *Genome Research*, 27(5), 824–834. https://doi.org/10.1101/gr.213959.116

Nwankwegu, A. S., Li, Y., Huang, Y., Wei, J., Norgbey, E., Sarpong, L., Lai, Q., & Wang, K. (2019). Harmful algal blooms under changing climate and constantly increasing anthropogenic actions: The review of management implications. 3 *Biotech*, 9(12), 449. https://doi.org/10.1007/s13205-019-1976-1

Oluseyi Osunmakinde, C., Selvarajan, R., Mamba, B. B., & Msagati, T. A. M. (2019). Profiling bacterial diversity and potential pathogens in wastewater treatment plants using high-throughput sequencing analysis. *Microorganisms*, 7(11). https://doi.org/10.3390/microorganisms7110506

O'Malley, M. A. (2008). 'Everything is everywhere: But the environment selects': Ubiquitous distribution and ecological determinism in microbial biogeography. *Studies in History and Philosophy of Science Part C: Studies in History and Philosophy of Biological and Biomedical Sciences*, 39(3), 314–325. https://doi.org/10.1016/j.shpsc.2008.06.005

Ontiveros, V. J., Cáliz, J., Triadó-Margarit, X., Alonso, D., & Casamayor, E. O. (2021). General decline in the diversity of the airborne microbiota under future climatic scenarios. *Scientific Reports*, 11(1), 20223. https://doi.org/10.1038/s41598-021-99223-x

Orts, J. M., Parrado, J., Pascual, J. A., Orts, A., Cuartero, J., Tejada, M., & Ros, M. (2022). Polyurethane foam residue biodegradation through the Tenebrio molitor digestive tract: Microbial communities and enzymatic activity. *Polymers*, 15(1), 204. https://doi.org/10.3390/polym15010204

Pal, D., Prabhakar, R., Barua, V. B., Zekker, I., Burlakovs, J., Krauklis, A., Hogland, W., & Vincevica-Gaile, Z. (2025). Microplastics in aquatic systems: A comprehensive review of its distribution, environmental interactions, and health risks. *Environmental Science and Pollution Research*, 32(1), 56–88. https://doi.org/10.1007/s11356-024-35741-1

Pan, H., Wattiez, R., & Gillan, D. (2024). Soil metaproteomics for microbial community profiling: Methodologies and challenges. *Current Microbiology*, 81(8), 257. https://doi.org/10.1007/s00284-024-03781-y

Péguilhan, R., Rossi, F., Joly, M., Nasr, E., Batut, B., Enault, F., Ervens, B., & Amato, P. (2025). Clouds influence the functioning of airborne microorganisms. *Biogeosciences*, 22(5), 1257–1275. https://doi.org/10.5194/bg-22-1257-2025

Philippot, L., Griffiths, B. S., & Langenheder, S. (2021a). Microbial community resilience across ecosystems and multiple disturbances. *Microbiology and Molecular Biology Reviews*, 85(2),. https://doi.org/10.1128/mmbr.00026-20

Pieper, D. H., & Reineke, W. (2000). Engineering bacteria for bioremediation. *Current Opinion in Biotechnology*, 11(3), 262–270. https://doi.org/10.1016/S0958-1669(00)00094-X

Pollegioni, P., Cardoni, S., Mattioni, C., Piredda, R., Ristorini, M., Occhiuto, D., Canepari, S., Korneykova, M. V, Soshina, A. S., Calfapietra, C., & Gavrichkova, O. (2023). Variability of airborne microbiome at different urban sites across seasons: A case study in Rome. *Frontiers in Environmental Science*, 11. https://doi.org/10.3389/fenvs.2023.1213833

Proksch, P., Edrada, R. A., & Ebel, R. (2002). Drugs from the seas—Current status and microbiological implications. *Applied Microbiology and Biotechnology*, 59(2–3), 125–134. https://doi.org/10.1007/s00253-002-1006-8

Prussin, A. J. 2nd, & Marr, L. C. (2015). Sources of airborne microorganisms in the built environment. *Microbiome*, 3, 78. https://doi.org/10.1186/s40168-015-0144-z

Quince, C., Walker, A. W., Simpson, J. T., Loman, N. J., & Segata, N. (2017). Shotgun metagenomics, from sampling to analysis. *Nature Biotechnology*, 35(9), 833–844. https://doi.org/10.1038/nbt.3935

Raajaraam, L., & Raman, K. (2024). Modeling microbial communities: Perspective and challenges. *ACS Synthetic Biology*, 13(8), 2260–2270. https://doi.org/10.1021/acssynbio.4c00116

Ragozzino, C., Casella, V., Coppola, A., Scarpato, S., Buonocore, C., Consiglio, A., Palma Esposito, F., Galasso, C., Tedesco, P., Della Sala, G., de Pascale, D., Vitale, L., & Coppola, D. (2025). Last decade insights in exploiting marine microorganisms as sources of new bioactive natural products. *Marine Drugs*, 23(3), 116. https://doi.org/10.3390/md23030116

Rajan, A., Ameen, F., Jambulingam, R., & Shankar, V. (2024). Biodegradation of polyurethane by fungi isolated from industrial wastewater—A sustainable approach to plastic waste management. *Polymers*, 16(10), 1411. https://doi.org/10.3390/polym16101411

Ramasamy, K. P., Mahawar, L., Rajasabapathy, R., Rajeshwari, K., Miceli, C., & Pucciarelli, S. (2023). Comprehensive insights on environmental adaptation strategies in Antarctic bacteria and biotechnological applications of cold adapted molecules. *Frontiers in Microbiology*, 14. https://doi.org/10.3389/fmicb.2023.1197797

Rasconi, S., Grami, B., Niquil, N., Jobard, M., & Sime-Ngando, T. (2014). Parasitic chytrids sustain zooplankton growth during inedible algal bloom. *Frontiers in Microbiology*, 5. https://doi.org/10.3389/fmicb.2014.00229

Reardon, S. (2017). Indigenous groups fight for genome sovereignty. *Nature*, 548(7666), 146–147. https://doi.org/10.1038/nature.2017.22318

Reddy, C. N., Kallem, P., Mounika, K. V. S. S. N., Muqeet, A., Raj, J. C. J., Aishwarya, C. V. S., Gupta, R. K., Polisetti, V., Mishra, B., Yadavalli, R., Mandal, S. K., Hedenqvist, M. S., & Banat, F. (2023). Review of microplastic degradation: Understanding metagenomic approaches for microplastic degrading organisms. *Polymer Testing*, 128, 108223. https://doi.org/10.1016/j.polymertesting.2023.108223

Rego, A., Fernandez-Guerra, A., Duarte, P., Assmy, P., Leão, P. N., & Magalhães, C. (2021). Secondary metabolite biosynthetic diversity in Arctic Ocean metagenomes. *Microbial Genomics*, 7(12). https://doi.org/10.1099/mgen.0.000731

Ribeiro da Cunha, B., Fonseca, L. P., & Calado, C. R. C. (2019). Antibiotic discovery: Where have we come from, where do we go? *Antibiotics*, 8(2), 45. https://doi.org/10.3390/antibiotics8020045

Richardson, T. L. (2019). Mechanisms and pathways of small-phytoplankton export from the surface Ocean. *Annual Review of Marine Science*, 11(1), 57–74. https://doi.org/10.1146/annurev-marine-121916-063627

Rocchi, S., Reboux, G., Scherer, E., Laboissière, A., Zaros, C., Rouzet, A., Valot, B., Khan, S., Dufourg, M.-N., Leynaert, B., Raherison, C., & Millon, L. (2020). Indoor microbiome: Quantification of exposure and association with geographical location, meteorological factors, and land use in France. *Microorganisms*, 8(3). https://doi.org/10.3390/microorganisms8030341

Roeters, S. J., Golbek, T. W., Bregnhøj, M., Drace, T., Alamdari, S., Roseboom, W., Kramer, G., Šantl-Temkiv, T., Finster, K., Pfaendtner, J., Woutersen, S., Boesen, T., & Weidner, T. (2021). Ice-nucleating proteins are activated by low temperatures to control the structure of interfacial water. *Nature Communications*, 12(1), 1183. https://doi.org/10.1038/s41467-021-21349-3

Romero-Crespo, P., Jiménez-Oyola, S., Salgado-Almeida, B., Zambrano-Anchundia, J., Goyburo-Chávez, C., González-Valoys, A., & Higueras, P. (2023). Trace elements in farmland soils and crops, and probabilistic health risk assessment in areas influenced by mining activity in Ecuador. *Environmental Geochemistry and Health*, 45(7), 4549–4563. https://doi.org/10.1007/s10653-023-01514-x

Rosenfeld, D., Andreae, M. O., Asmi, A., Chin, M., De Leeuw, G., Donovan, D. P., Kahn, R., Kinne, S., Kivekäs, N., Kulmala, M., Lau, W., Schmidt, K. S., Suni, T., Wagner, T., Wild, M., & Quaas, J. (2014). Global observations of aerosol-cloud-precipitation-climate interactions: Aerosol-cloud-climate interactions. *Reviews of Geophysics*, 52(4), 750–808. https://doi.org/10.1002/2013RG000441

Rousk, J., & Bengtson, P. (2014). Microbial regulation of global biogeochemical cycles. *Frontiers in Microbiology*, 5. https://doi.org/10.3389/fmicb.2014.00103

Roy, C., Rameez, M. J., Haldar, P. K., Peketi, A., Mondal, N., Bakshi, U., Mapder, T., Pyne, P., Fernandes, S., Bhattacharya, S., Roy, R., Mandal, S., O'Neill, W. K., Mazumdar, A., Mukhopadhyay, S. K., Mukherjee, A., Chakraborty, R., Hallsworth, J. E., & Ghosh, W. (2020). Microbiome and ecology of a hot spring-microbialite system on the Trans-Himalayan Plateau. *Scientific Reports*, 10(1), 5917. https://doi.org/10.1038/s41598-020-62797-z

Ruan, Z., Chen, K., Cao, W., Meng, L., Yang, B., Xu, M., Xing, Y., Li, P., Freilich, S., Chen, C., Gao, Y., Jiang, J., & Xu, X. (2024). Engineering natural microbiomes toward enhanced bioremediation by microbiome modeling. *Nature Communications*, 15(1), 4694. https://doi.org/10.1038/s41467-024-49098-z

Ruiz-Gil, T., Acuña, J. J., Fujiyoshi, S., Tanaka, D., Noda, J., Maruyama, F., & Jorquera, M. A. (2020). Airborne bacterial communities of outdoor environments and their associated influencing factors. *Environment International*, 145, 106156. https://doi.org/10.1016/j.envint.2020.106156

Rusch, D. B., Halpern, A. L., Sutton, G., Heidelberg, K. B., Williamson, S., Yooseph, S., Wu, D., Eisen, J. A., Hoffman, J. M., Remington, K., Beeson, K., Tran, B., Smith, H., Baden-Tillson, H., Stewart, C., Thorpe, J., Freeman, J., Andrews-Pfannkoch, C., Venter, J. E., … Venter, J. C. (2007). The Sorcerer II Global Ocean Sampling Expedition: Northwest Atlantic through Eastern Tropical Pacific. *PLoS Biology*, 5(3), e77. https://doi.org/10.1371/journal.pbio.0050077

Salwan, R., & Sharma, V. (2022). Genomics of prokaryotic extremophiles to unfold the mystery of survival in extreme environments. *Microbiological Research*, 264, 127156. https://doi.org/10.1016/j.micres.2022.127156

Samak, M. El, Solyman, S. M., Hanora, A., & Sea, R. (2018). Antimicrobial activity of bacteria isolated from Red Sea marine invertebrates. *Biotechnology Reports*, 19, e00275. https://doi.org/10.1016/j.btre.2018.e00275

Šantl-Temkiv, T., Amato, P., Casamayor, E. O., Lee, P. K. H., & Pointing, S. B. (2022). Microbial ecology of the atmosphere. *FEMS Microbiology Reviews*, 46(4). https://doi.org/10.1093/femsre/fuac009

Šantl-Temkiv, T., Sahyoun, M., Finster, K., Hartmann, S., Augustin-Bauditz, S., Stratmann, F., Wex, H., Clauss, T., Nielsen, N. W., Sørensen, J. H., Korsholm, U. S., Wick, L. Y., & Karlson, U. G. (2015). Characterization of airborne ice-nucleation-active bacteria and bacterial fragments. *Atmospheric Environment*, 109, 105–117. https://doi.org/10.1016/j.atmosenv.2015.02.060

Santos, F., & Antón, J. (2019). Extremophiles: Hypersaline environments. *Encyclopedia of Microbiology*, 270–275. https://doi.org/10.1016/B978-0-12-809633-8.90673-3

Schauberger, C., Thamdrup, B., Lemonnier, C., Trouche, B., Poulain, J., Wincker, P., Arnaud-Haond, S., Glud, R. N., & Maignien, L. (2024). Metagenome-assembled genomes of deep-sea sediments: Changes in microbial functional potential lag behind redox transitions. *ISME Communications*, 4(1). https://doi.org/10.1093/ismeco/ycad005

Schloss, P. D., & Handelsman, J. (2003). Biotechnological prospects from metagenomics. *Current Opinion in Biotechnology*, 14(3), 303–310. https://doi.org/10.1016/S0958-1669(03)00067-3

Sedeek, A. M., Elfeky, H., Hanora, A. S., & Solyman, S. M. (2025). Genomic insights into biosynthesis and adaptation in the bioactive marine bacterium Streptomyces albidoflavus VIP-1 from the Red Sea. *BMC Microbiology*, 25(1), 372. https://doi.org/10.1186/s12866-025-04109-x

Sedeek, A. M., Ismail, M. M., Elsayed, T. R., & Ramadan, M. A. (2022). Recent methods for discovering novel bioactive metabolites, specifically antimicrobial agents, from marine-associated microorganisms. *Letters in Applied Microbiology*, 75(3). https://doi.org/10.1111/LAM.13728

Sedeek, A. M., Salah, I., Kamel, H. L., Soltan, M. A., Nour, E., Alshammari, A., Riaz Rajoka, M. S., & Elsayed, T. R. (2023). Genome-based analysis of the potential bioactivity of the terrestrial Streptomyces vinaceusdrappus strain AC-40. *Biology*, 12(3), 345. https://doi.org/10.3390/biology12030345

Sessitsch, A., Wakelin, S., Schloter, M., Maguin, E., Cernava, T., Champomier-Verges, M.-C., Charles, T. C., Cotter, P. D., Ferrocino, I., Kriaa, A., Lebre, P., Cowan, D., Lange, L., Kiran, S., Markiewicz, L., Meisner, A., Olivares, M., Sarand, I., Schelkle, B., … Kostic, T. (2023). Microbiome interconnectedness throughout environments with major consequences for healthy people and a healthy planet. *Microbiology and Molecular Biology Reviews*, 87(3), e00212–22. https://doi.org/10.1128/mmbr.00212-22

Sharpe, T., McGill, G., Dancer, S. J., King, M.-F., Fletcher, L., & Noakes, C. J. (2020). Influence of ventilation use and occupant behaviour on surface microorganisms in contemporary social housing. *Scientific Reports*, 10(1), 11841. https://doi.org/10.1038/s41598-020-68809-2

Shayanthan, A., Ordoñez, P. A. C., & Oresnik, I. J. (2022). The role of synthetic microbial communities (SynCom) in sustainable agriculture. *Frontiers in Agronomy*, 4. https://doi.org/10.3389/fagro.2022.896307

Shiah, F.-K., Lai, C.-C., Chen, T.-Y., Ko, C.-Y., Tai, J.-H., & Chang, C.-W. (2022). Viral shunt in tropical oligotrophic ocean. *Science Advances*, 8(41), eabo2829. https://doi.org/10.1126/sciadv.abo2829

Shu, W.-S., & Huang, L.-N. (2022). Microbial diversity in extreme environments. *Nature Reviews Microbiology*, 20(4), 219–235. https://doi.org/10.1038/s41579-021-00648-y

Siddique, A., Hubab, M., Rasheela, A. R. P., Samad, R., Al-Ghouti, M., Sayadi, S., & Zouari, N. (2025). Microplastics and their role in the emergence of antibiotic resistance in bacteria as a threat for the environment. *Environmental Chemistry and Ecotoxicology*, 7, 614–622. https://doi.org/10.1016/j.enceco.2025.03.006

Sierra-Vargas, M. P., Montero-Vargas, J. M., Debray-García, Y., Vizuet-de-Rueda, J. C., Loaeza-Román, A., & Terán, L. M. (2023). Oxidative stress and air pollution: Its impact on chronic respiratory diseases. *International Journal of Molecular Sciences*, 24(1). https://doi.org/10.3390/ijms24010853

Singh, R., Kumar, M., Mittal, A., & Mehta, P. K. (2016). Microbial enzymes: Industrial progress in 21st century. *3 Biotech*, 6(2), 174. https://doi.org/10.1007/s13205-016-0485-8

Song, Y., & Prather, K. L. J. (2024). Strategies in engineering sustainable biochemical synthesis through microbial systems. *Current Opinion in Chemical Biology*, 81, 102493. https://doi.org/10.1016/j.cbpa.2024.102493

Soth, S., Hampton, J. G., Alizadeh, H., Wakelin, S. A., & Mendoza-Mendoza, A. (2025). Microbiomes in action: Multifaceted benefits and challenges across academic disciplines. *Frontiers in Microbiology*, 16. https://doi.org/10.3389/fmicb.2025.1550749

Spracklen, D. V, & Heald, C. L. (2014). The contribution of fungal spores and bacteria to regional and global aerosol number and ice nucleation immersion freezing rates. *Atmospheric Chemistry and Physics*, 14(17), 9051–9059. https://doi.org/10.5194/acp-14-9051-2014

Stegen, J. C., Lin, X., Konopka, A. E., & Fredrickson, J. K. (2012). Stochastic and deterministic assembly processes in subsurface microbial communities. *ISME Journal*, 6(9), 1653–1664. https://doi.org/10.1038/ismej.2012.22

Sunagawa, S., Coelho, L. P., Chaffron, S., Kultima, J. R., Labadie, K., Salazar, G., Djahanschiri, B., Zeller, G., Mende, D. R., Alberti, A., Cornejo-Castillo, F. M., Costea, P. I., Cruaud, C., d'Ovidio, F., Engelen, S., Ferrera, I., Gasol, J. M., Guidi, L., Hildebrand, F., … Velayoudon, D. (2015). Structure and function of the global ocean microbiome. *Science*, 348(6237), 1261359. https://doi.org/10.1126/science.1261359

Takeda, M., Choi, J., Maeda, T., & Managi, S. (2024). Effects of bathing in different hot spring types on Japanese gut microbiota. *Scientific Reports*, 14(1), 2316. https://doi.org/10.1038/s41598-024-52895-7

Tang, H., & Song, M. (2024). Environment and synthetic biology. *Synthetic and Systems Biotechnology*, 9(3), 425. https://doi.org/10.1016/j.synbio.2024.03.016

Tang, H., Xiang, G., Xiao, W., Yang, Z., & Zhao, B. (2024). Microbial mediated remediation of heavy metals toxicity: Mechanisms and future prospects. *Frontiers in Plant Science*, 15. https://doi.org/10.3389/fpls.2024.1420408

Taylor, M.W., Radax, R., Steger, D., & Wagner, M. (2007). Sponge-associated microorganisms: Evolution, ecology, and biotechnological potential. *Microbiology and Molecular Biology Reviews*, 71(2). https://doi.org/10.1128/mmbr.00040-06

Thomas, T., & Nashwan, A. J. (2025). Air pollution and disrupted microbiomes: Tracing the impact on human health. *Cureus*, 17(8), e89267. https://doi.org/10.7759/cureus.89267

Thompson, L. R., Sanders, J. G., McDonald, D., Amir, A., Ladau, J., Locey, K. J., & Knight, R. (2017). A communal catalogue reveals Earth's multiscale microbial diversity. *Nature*, 551(7681), 457–463. https://doi.org/10.1038/nature24621

Thomsen, P. F., & Willerslev, E. (2015). Environmental DNA in ecology: A review. *Molecular Ecology*, 24(21), 5405–5419. https://doi.org/10.1111/mec.13405

Tignat-Perrier, R., Dommergue, A., Thollot, A., Keuschnig, C., Magand, O., Vogel, T. M., & Larose, C. (2019). Global airborne microbial communities controlled by surrounding landscapes and wind conditions. *Scientific Reports*, 9(1), 14441. https://doi.org/10.1038/s41598-019-51073-4

Timms, V. J., Hassan, K. A., Pearson, L. A., & Neilan, B. A. (2023). Cyanobacteria as a critical reservoir of the environmental antimicrobial resistome. *Environmental Microbiology*, 25(8), 1459–1473. https://doi.org/10.1111/1462-2920.16453

Tokiwa, Y., Calabia, B. P., Ugwu, C. U., & Aiba, S. (2009). Biodegradability of plastics. *International Journal of Molecular Sciences*, 10(9), 3722–3742. https://doi.org/10.3390/ijms10093722

Topçuoğlu, B. D., Lesniak, N. A., Ruffin, M. T., Wiens, J., & Schloss, P. D. (2020). A framework for effective application of machine learning to microbiome-based classification problems. *mBio*, 11(3), e00434–20. https://doi.org/10.1128/mBio.00434-20

Torres, L., Krüger, A., Csibra, E., Gianni, E., & Pinheiro, V. B. (2016). Synthetic biology approaches to biological containment: Pre-emptively tackling potential risks. *Essays in Biochemistry*, 60(4), 393–410. https://doi.org/10.1042/EBC20160013

Tyson, G. W., Chapman, J., Hugenholtz, P., Allen, E. E., Ram, R. J., Richardson, P. M., Solovyev, V. V., Rubin, E. M., Rokhsar, D. S., & Banfield, J. F. (2004). Community structure and metabolism through reconstruction of microbial genomes from the environment. *Nature*, 428(6978), 37–43. https://doi.org/10.1038/nature02340

Unit, B. (2025, July 30). *The Nagoya Protocol on Access and Benefit-sharing*. Secretariat of the Convention on Biological Diversity. https://www.cbd.int/abs/default.shtml

Uwishema, O., Masunga, D. S., Naisikye, K. M., Bhanji, F. G., Rapheal, A. J., Mbwana, R., Nazir, A., & Wellington, J. (2023). Impacts of environmental and climatic changes on future infectious diseases. *International Journal of Surgery (London, England)*, 109(2), 167–170. https://doi.org/10.1097/JS9.0000000000000160

Vaïtilingom, M., Deguillaume, L., Vinatier, V., Sancelme, M., Amato, P., Chaumerliac, N., & Delort, A.-M. (2013). Potential impact of microbial activity on the oxidant capacity and organic carbon budget in clouds. *Proceedings of the National Academy of Sciences*, 110(2), 559–564. https://doi.org/10.1073/pnas.1205743110

Varg, J. E., Bergvall, C., & Svanbäck, R. (2022). The stressful effects of microplastics associated with chromium (VI) on the microbiota of Daphnia Magna. *Frontiers in Environmental Science*, 10. https://doi.org/10.3389/fenvs.2022.875512

Varg, J. E., Outomuro, D., Kunce, W., Kuehrer, L., Svanbäck, R., & Johansson, F. (2022). Microplastic exposure across trophic levels: Effects on the host–microbiota of freshwater organisms. *Environmental Microbiome*, 17(1), 36. https://doi.org/10.1186/s40793-022-00429-x

Varg, J. E., & Svanbäck, R. (2023). Multi stress system: Microplastics in freshwater and their effects on host microbiota. *Science of the Total Environment*, 856(July 2022). https://doi.org/10.1016/j.scitotenv.2022.159106

Venter, J. C., Remington, K., Heidelberg, J. F., Halpern, A. L., Rusch, D., Eisen, J. A., & Smith, H. O. (2004). Environmental genome shotgun sequencing of the Sargasso Sea. *Science*, 304(5667), 66–74. https://doi.org/10.1126/science.1093857

Waldrop, M. P., Ernakovich, J. G., Vishnivetskaya, T. A., Schaefer, S. R., Mackelprang, R., Barta, J., O'Brien, J. M., Winkel, M., Barbato, R. A., Heffernan, L., Leewis, M.-C., Hewitt, R. E., Hultman, J., Sun, Y., Biasi, C., Bradley, J. A., Liebner, S., Ricketts, M. P., Muscarella, M. E., … Onstott, T. C. (2025). Microbial ecology of permafrost soils: Populations, processes, and perspectives. *Permafrost and Periglacial Processes*, 36(2), 245–258. https://doi.org/10.1002/ppp.2264

Walker, L. R., & Del Moral, R. (2003). *Primary Succession and Ecosystem Rehabilitation* (1st ed.). Cambridge University Press. https://doi.org/10.1017/CBO9780511615078

Wang, Q., Liang, J., Zhang, S., Yoza, B. A., Li, Q. X., Zhan, Y., Ye, H., Zhao, P., & Chen, C. (2020). Characteristics of bacterial populations in an industrial scale petrochemical wastewater treatment plant: Composition, function and their association with environmental factors. *Environmental Research*, 189, 109939. https://doi.org/10.1016/j.envres.2020.109939

Wang, X., Chi, Y., & Song, S. (2024). Important soil microbiota's effects on plants and soils: A comprehensive 30-year systematic literature review. *Frontiers in Microbiology*, 15. https://doi.org/10.3389/fmicb.2024.1347745

Wang, X., Wang, Y., Zhang, Z., Tian, L., Zhu, T., Zhao, Y., Tong, Y., Yang, Y., Sun, P., & Liu, Y. (2024). Effect, fate and remediation of pharmaceuticals and personal care products (PPCPs) during anaerobic sludge treatment: A review. *Environmental Science & Technology*, 58(43), 19095–19114. https://doi.org/10.1021/acs.est.4c06760

Watanabe, K., Yanagi, U., Shiraishi, Y., Harada, K., Ogino, F., & Asano, K. (2022). Bacterial communities in various parts of air-conditioning units in 17 Japanese houses. *Microorganisms*, 10(11), 2246. https://doi.org/10.3390/microorganisms10112246

Wegener, G., Molari, M., Purser, A., Diehl, A., Albers, E., Walter, M., Mertens, C., German, C. R., & Boetius, A. (2024). Hydrothermal vents supporting persistent plumes and microbial chemoautotrophy at Gakkel Ridge (Arctic Ocean). *Frontiers in Microbiology*, 15. https://doi.org/10.3389/fmicb.2024.1473822

Wei, W., Chen, X., Weinbauer, M. G., Jiao, N., & Zhang, R. (2022). Reduced bacterial mortality and enhanced viral productivity during sinking in the ocean. *The ISME Journal*, 16(6), 1668–1675. https://doi.org/10.1038/s41396-022-01224-9

Weitz, J. S., & Wilhelm, S. W. (2012). Ocean viruses and their effects on microbial communities and biogeochemical cycles. *F1000 Biology Reports*, 4, 17. https://doi.org/10.3410/B4-17

Wilhelm, L., Singer, G. A., Fasching, C., Battin, T. J., & Besemer, K. (2013). Microbial biodiversity in glacier-fed streams. *ISME Journal*, 7(8), 1651–1660. https://doi.org/10.1038/ismej.2013.44

Woese, C. R., & Fox, G. E. (1977). Phylogenetic structure of the prokaryotic domain: The primary kingdoms. *Proceedings of the National Academy of Sciences*, 74(11), 5088–5090. https://doi.org/10.1073/pnas.74.11.5088

Wohlleben, W., Mast, Y., Stegmann, E., & Ziemert, N. (2016). Antibiotic drug discovery. *Microbial Biotechnology*, 9(5), 541–548. https://doi.org/10.1111/1751-7915.12388

Wu, R., Trubl, G., Taş, N., & Jansson, J. K. (2022). Permafrost as a potential pathogen reservoir. *One Earth*, 5(4), 351–360. https://doi.org/10.1016/J.ONEEAR.2022.03.010

Wu, X., He, F., Xu, X., Wu, L., Rong, J., & Lin, S. (2025). Environmental health and safety implications of the interplay between microplastics and the residing biofilm. *Environment & Health (Washington, D.C.)*, 3(2), 118–132. https://doi.org/10.1021/envhealth.4c00148

Xie, K., Ren, Y., Chen, A., Yang, C., Zheng, Q., Chen, J., Wang, D., Li, Y., Hu, S., & Xu, G. (2022). Plant nitrogen nutrition: The roles of arbuscular mycorrhizal fungi. *Journal of Plant Physiology*, 269, 153591. https://doi.org/10.1016/j.jplph.2021.153591

Xiu, L., & Hu, S. (2025). Air pollution might contribute to antimicrobial resistance: A one health perspective. *Science in One Health*, 4, 100118. https://doi.org/10.1016/j.soh.2025.100118

Xu, M., & Wright, G. D. (2019). Heterologous expression-facilitated natural products' discovery in actinomycetes. *Journal of Industrial Microbiology and Biotechnology*, 46(3–4), 415–431. https://doi.org/10.1007/s10295-018-2097-2

Xun, W., Liu, Y., Li, W., Ren, Y., Xiong, W., Xu, Z., Zhang, N., Miao, Y., Shen, Q., & Zhang, R. (2021). Specialized metabolic functions of keystone taxa sustain soil microbiome stability. *Microbiome*, 9(1), 35. https://doi.org/10.1186/s40168-020-00985-9

Yang, Y., Liu, W., Zhang, Z., Grossart, H.-P., & Gadd, G. M. (2020). Microplastics provide new microbial niches in aquatic environments. *Applied Microbiology and Biotechnology*, 104(15), 6501–6511. https://doi.org/10.1007/s00253-020-10704-x

Yang, Y., Wang, L., Xiang, F., Zhao, L., & Qiao, Z. (2020). Activated sludge microbial community and treatment performance of wastewater treatment plants in industrial and municipal zones. *International Journal of Environmental Research and Public Health*, 17(2), 436. https://doi.org/10.3390/ijerph17020436

Yarzábal, L. A., Salazar, L. M. B., & Batista-García, R. A. (2021). Climate change, melting cryosphere and frozen pathogens: Should we worry…? *Environmental Sustainability (Singapore)*, 4(3), 489–501. https://doi.org/10.1007/s42398-021-00184-8

Yin, M., Yan, B., Wang, H., Wu, Y., Wang, X., Wang, J., Zhu, Z., Yan, X., Liu, Y., Liu, M., & Fu, C. (2023). Effects of microplastics on nitrogen and phosphorus cycles and microbial communities in sediments. *Environmental Pollution*, 318, 120852. https://doi.org/10.1016/j.envpol.2022.120852

Young, T. P., Petersen, D. A., & Clary, J. J. (2005). The ecology of restoration: Historical links, emerging issues and unexplored realms. *Ecology Letters*, 8(6), 662–673. https://doi.org/10.1111/j.1461-0248.2005.00764.x

Zhang, L., Tu, D., Li, X., Lu, W., & Li, J. (2020). Impact of long-term industrial contamination on the bacterial communities in urban river sediments. *BMC Microbiology*, 20(1), 254. https://doi.org/10.1186/s12866-020-01937-x

Zhang, M., Tang, H., Yuan, Y., Ou, Z., Chen, Z., Xu, Y., Fu, X., Zhao, Z., & Sun, Y. (2023). The role of indoor microbiome and metabolites in shaping children's nasal and oral microbiota: A pilot multi-omic analysis. *Metabolites*, 13(10), 1040. https://doi.org/10.3390/metabo13101040

Zhao, A., Lu, Y., Li, Q., Li, T., & Zhao, J. (2023). Metagenomics reveals the diversity and role of surface-water microbes in biogeochemical cycles in lakes at different terrain ladders. *Frontiers in Environmental Science*, 11, 1121775. https://doi.org/10.3389/fenvs.2023.1121775

Zhao, J., Jin, L., Wu, D., Xie, J., Li, J., Fu, X., Cong, Z., Fu, P., Zhang, Y., Luo, X., Feng, X., Zhang, G., Tiedje, J. M., & Li, X. (2022). Global airborne bacterial community—Interactions with Earth's microbiomes and anthropogenic activities. *Proceedings of the National Academy of Sciences*, 119(42), e2204465119. https://doi.org/10.1073/pnas.2204465119

Zhao, Y., Yuan, X., Ran, W., Zhao, Z., Su, D., & Song, Y. (2025). The ecological restoration strategies in terrestrial ecosystems were reviewed: A new trend based on soil microbiomics. *Ecology and Evolution*, 15(3). https://doi.org/10.1002/ece3.70994

Zheng, F., Zhang, T., Yin, S., Qin, G., Chen, J., Zhang, J., Zhao, D., Leng, X., An, S., & Xia, L. (2022). Comparison and interpretation of freshwater bacterial structure and interactions with organic to nutrient imbalances in restored wetlands. *Frontiers in Microbiology*, 13. https://doi.org/10.3389/fmicb.2022.946537

Zheng, J., Wang, H., Wang, X., Zeng, S., Yuan, S., & Yin, T. (2025). Genome mining and OSMAC strategies unveil diverse secondary metabolites from the endophytic fungus Diaporthe kyushuensis ZMU-48-1 with antifungal activity. *Frontiers in Microbiology*, 16. https://doi.org/10.3389/fmicb.2025.1604639

Zhou, H., Tang, L., Fenton, K. A., & Song, X. (2025). Exploring and evaluating microbiome resilience in the gut. *FEMS Microbiology Ecology*, 101(5). https://doi.org/10.1093/femsec/fiaf046

Zhou, P., Yang, X. L., Wang, X. G., Hu, B., Zhang, L., Zhang, W., & Shi, Z. L. (2020). A pneumonia outbreak associated with a new coronavirus of probable bat origin. *Nature*, 579(7798), 270–273. https://doi.org/10.1038/s41586-020-2012-7

Zhu, J., Yu, Z., He, L., Cao, X., Wang, W., & Song, X. (2024). Phycosphere bacterial composition and function in colony and solitary Phaeocystis globosa strains providing novel insights into the algal blooms. *Marine Pollution Bulletin*, 206, 116700. https://doi.org/10.1016/j.marpolbul.2024.116700

Zhu, X., Jia, G., & Xu, X. (2024). Accelerated rise in wildfire carbon emissions from Arctic continuous permafrost. *Science Bulletin*, 69(15), 2430–2438. https://doi.org/10.1016/j.scib.2024.05.022

CHAPTER 2

# Antibiotic Resistance Proliferation under the Influence of Environmental Determinants

## CHALLENGES AND MITIGATION

*Bhumika Maharana and Nishant A. Dafale*

## 2.1 INTRODUCTION

Antimicrobial resistance is a life-threatening challenge in the present scenario. Drug resistance in common gram-negative and -positive bacteria is an intimidating threat while dealing with bacterial infections (Frieri et al. 2017). Antibiotic-resistant bacteria (ARB) infect more than 2 million people annually, according to the Centers for Disease Control and Prevention (CDC). About 23,000 of them will pass away as a result of their sickness. In a time when there are few new medicines for bacterial illnesses, the difficult task of reducing antibiotic resistance (AR) falls to researchers, prescribers, and policymakers (Velez and Sloand 2016). Globally, AR poses a hazard to both human and animal health, and significant steps must be taken to lessen the dangers posed by environmental AR genes (ARGs). The adoption of technical solutions that can stop environmental contamination with bacteria and genes resistant to antibiotics, the identification of crucial points of control, and the creation of trustworthy surveillance and risk assessment protocols are some examples of these approaches. In order to overcome the growing resistance, correct measurement of potency and bioactivity is essential (Dafale et al. 2016). The dissemination of multidrug-resistant bacteria poses an alarming threat in the environment, due to improper antibiotic use and unmetabolized release into environmental matrices. The spread of multidrug AR can pass between hosts and across food webs (Parmar et al. 2018). Stagnation in the development of new antibiotics or antibacterial agents is one of the pressing challenges in combating AMR. This leaves the healthcare providers with limited treatment options when dealing with extensively drug-resistant strains. Velez and Sloand (2016) argue that prescribers, policymakers, and researchers now face the critical responsibility of mitigating resistance through strategies that extend the lifespan of existing antibiotics and enhance infection prevention and control. In the current environment, where the pharmaceutical pipeline is thin and resistance is rising, alternative approaches and innovative policies must be pursued aggressively.

Public awareness and education are also essential components of a long-term mitigation strategy. Informing healthcare providers, veterinarians, farmers, and the general public about the dangers of AMR and the importance of responsible ABU can reduce misuse of and overreliance on these drugs. Moreover, continued research into resistance mechanisms, transmission pathways, and ecological interactions is necessary to inform effective interventions and treatment alternatives. In conclusion, antimicrobial resistance is a multifaceted crisis deeply influenced by environmental determinants. Efforts to address it must extend beyond clinical settings and into the natural and

DOI: 10.1201/9781003637950-2

agricultural environments where resistance genes are maintained and propagated. This chapter proposes that sustained monitoring, evidence-based surveillance, policy reform, and innovation in both practice and treatment can provide viable solutions to reduce the environmental contribution to AR. Protecting human and animal health in the face of this growing threat will depend on our collective ability to implement coordinated, cross-sectoral strategies that recognize and act upon the full scope of the problem. The book chapter proposes that monitoring, surveillance of practice, policy, and new treatments provide solutions in regard to environmental variables contributing to AR in both the human and agricultural sectors (Figure 2.1).

## 2.2 EMERGENCE OF ANTIBIOTIC RESISTANCE: CONTRIBUTIONS OF ENVIRONMENTAL SELECTIVE PRESSURES

Bacteria developed AR long before the mass production of pharmaceuticals. Studies of ancient, uncontaminated environments—like caves and permafrost—reveal that resistance mechanisms existed in the pre-antibiotic era. This natural resistance likely arose from microbial competition and the production of antibiotic-like secondary metabolites (Larsson and Flach 2022). The discovery of antibiotics was a major milestone in 20th-century medicine, transforming healthcare. The development of antibiotics, including chloramphenicol, vancomycin, and cephalosporin, brought forward significant progress in healthcare, preventing fatal deaths and decreasing the mortality rate (Aminov 2010). The emergence of semi-synthetic antibiotics with broader antimicrobial spectra, like amoxicillin and quinolones, developed the foundation of the antibiotic era.

The era further grew with the development of new generations of cephalosporin, daptomycin, linezolid, vancomycin, which act against gram-negative bacteria (Iskandar et al. 2021). However, resistance inevitably limits their long-term effectiveness (Rehman et al. 2015). Since their introduction, antibiotics have been widely used in both human and animal health, often even in the absence of infection, including as growth promoters (Baquero et al. 2008). Persistent environmental stressors like residual antibiotics exert horizontal gene transfer (HGT) and genome plasticity, accelerating the spread of ARG. These conditions can suppress microbial defense systems like CRISPR-Cas, enabling mobile genetic elements to transfer resistance. Plasmids and prophages carrying resistance determinants are commonly found in environmental niches, aiding in the dissemination of resistance against common opportunistic pathogens (Das et al. 2022).

The overuse and misuse of antibiotics in clinical, veterinary, and agricultural settings have significantly accelerated the development and spread of resistant microorganisms (Ugoeze et al. 2024). This has led to the emergence of bacterial infections that are no longer treatable with first-line or even second-line antibiotics, thereby increasing patient morbidity, mortality, and healthcare costs. Common gram-positive and gram-negative bacteria, such as *Staphylococcus aureus*, *Klebsiella pneumoniae*, *Escherichia coli*, and *Pseudomonas aeruginosa*, are now frequently associated with multidrug resistance, making infections difficult to control and treat. Frieri et al. (2017) underscored the seriousness of the trend, highlighting resistance amongst gram-positive and gram-negative representing a difficult challenge in modern medicine. An often-underappreciated factor in the spread of AR is the role of environmental determinants. The natural environment has become a reservoir and transmission route for resistant bacteria and resistance genes (Lin et al. 2021). Wastewater from pharmaceutical production, agricultural fields, and hospitals contributes to the spread of resistance into soil and aquatic bodies. These environmental matrices become reservoirs for resistance genes, particularly via HGT amongst bacteria. Runoff from farms where antibiotics are heavily used in livestock is another major source of environmental contamination (Hall et al. 2023). In aquatic ecosystems, ARB and antibiotic residues interact with microbes, spreading AR and genetic exchange. A study by Bombaywala et al. (2021) examined ARGs in industrial and wetland sludge, along with pharmaceutical sediment, where pharma sites carried critical ARGs whereas ETP sludge showed efflux pump-like resistome. This requires robust addressing and surveillance to detect antibiotic residues and ARGs and to identify intervention points. Strengthening environmental regulations and stricter controls on industrial discharge are essential to reduce the environmental impact of antimicrobial agents. Addressing environmental contributors to resistance also requires a fundamental shift in policy and practice. There is a need for more integrated "One Health"

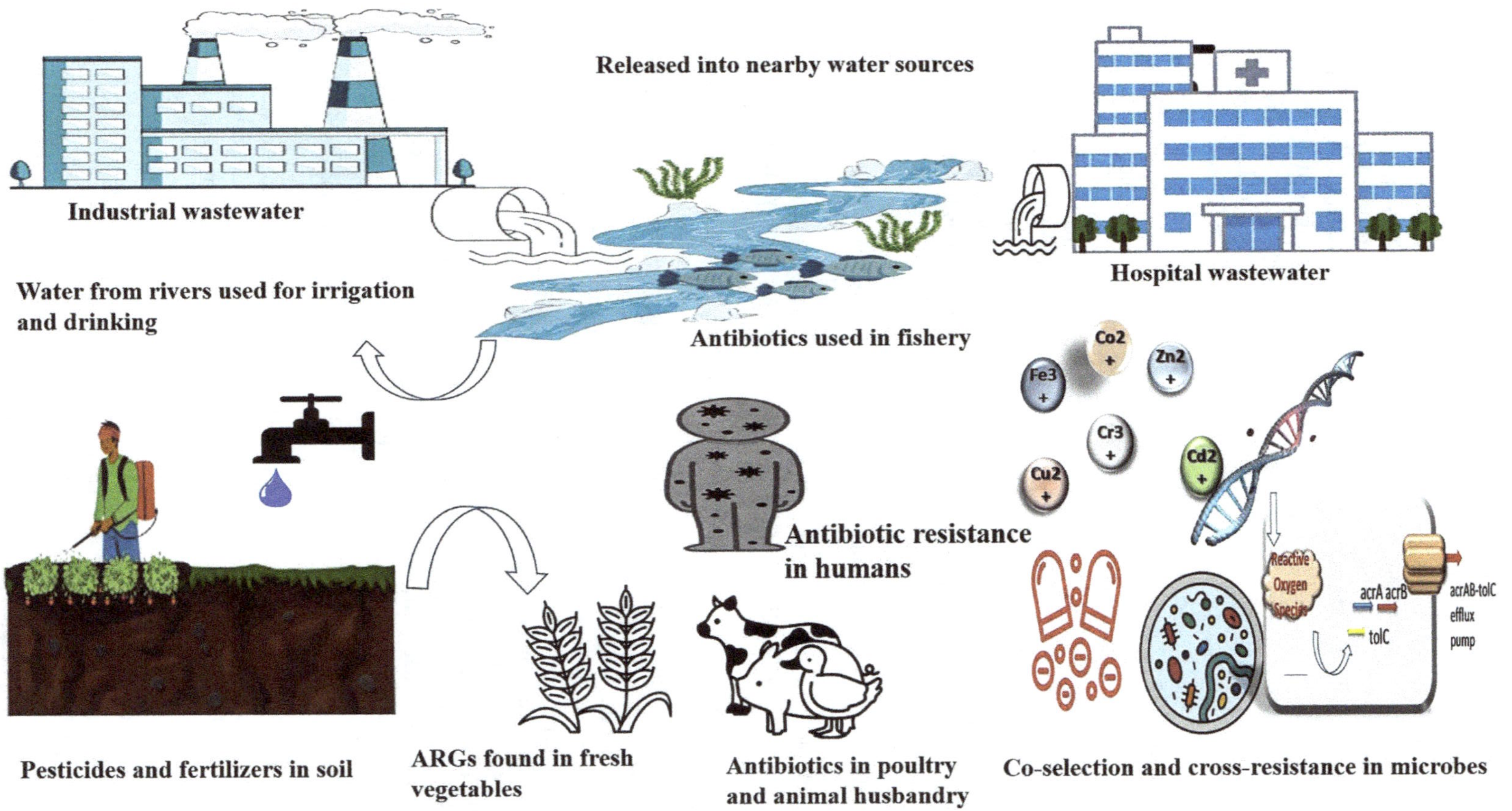


Figure 2.1 This diagram illustrates the major environmental drivers contributing to the proliferation and dissemination of antibiotic resistance genes (ARGs). Key determinants include pharmaceutical industry effluents, agricultural runoff, wastewater discharge, and untreated sewage, which introduce antibiotics and resistant bacteria into soil and aquatic environments. These pollutants create selective pressure that fosters the emergence and persistence of ARGs via mechanisms such as genetic mutation, efflux pump activation, and target protection.

approaches that treat human, animal, and environmental health as interconnected domains (Dafale et al. 2020). This overuse has increased selective pressure, driving the evolution and spread of AR. In order to understand the role of determinants in disseminating AR, it is essential to understand the mechanisms of AR and the effect of selective environment in dissemination.

## 2.3 EFFECT OF ENVIRONMENTAL DETERMINANTS OF AR DISSEMINATION PATHWAYS

AR can be both intrinsic and acquired, based on the genetic mechanisms of the bacteria to survive in a stressful environment. The gram-negative bacteria have adapted themselves to forming a complex barrier of membrane with flexible permeability. The bacteria, while taking up essential nutrients, can restrict the entry of antibiotics or other drugs (Whittle et al. 2021). The periplasm of gram-negative microbe is found to be the action site for beta-lactams and also their potential inactivation. Along with physical barriers, the bacterial envelope also hosts multiple drug efflux pumps, which solidifies the restriction of drug entry (Masi et al. 2017). The antibiotics get distributed to the inner membrane, cytoplasm, and periplasm after passing through the outer membrane protein. The outer membrane proteins (porins) are major contributors to resistance and drug permeability. The study on beta-lactams transport across the outer membrane shows an inverse relationship between hydrophobicity and permeation rate through porins amongst cephalosporins. The nature of OMP genes establishes the resistance on the basis of the types of antibiotics (Baslé et al. 2006). Hydrophobic antibiotics such as aminoglycosides, macrolides, rifamycins, and fusidic acid gain access to cells through the outer membrane bilayer, whereas tetracycline and quinolone use both porin- and lipid-mediated pathways. The core lipopolysaccharide region of cells plays a major role in restricting drug entry, and microbes expressing full-length LPS mostly possess intrinsic resistance (Armstrong et al. 2019).

It is a noteworthy fact that outer membrane proteins/genes are found to be affected by environmental selective pressure (Darby et al. 2023). A study by Pratt et al. (1994) on two-component system of ENVz–OmpR system shows the effect of changing the environment on the phosphorylation of OmpR. High levels of phosphorylation in OmpR lead to reduced ompF, and increase in ompC transcription. A high nutrient environment leads to smaller porin expression, and expression of smaller porins leads to more AR in microbes. sRNAs also play a role in the outer membrane structure of microorganisms. MicF, an sRNA, is recently known to have its expression regulated by environmental stress like high osmolarity. Similarly, the gene MicC is known to be activated in the presence of antibiotic stress and gets activated in the presence of beta-lactam (Dam et al. 2017).

Along with the prevention of the influx of antibiotics, bacteria have also developed efflux mechanisms to extrude antibiotics. This efflux process is carried out by proteins that help expel antibiotics from the inside of a cell for its survival. The first study by McMurry et al. (1983) showed the ability of *E. coli* in expelling tetracycline from the cytoplasm. Efflux pump mechanism is one of the major reasons for the development of multidrug AR. They can be both acquired and an inherent part of the resistance mechanism of bacteria. Innate resistance consists of multidrug efflux pumps present in chromosomes, whereas in some cases they are present in mobile genetic elements, which are mostly acquired through HGT. For example, Qac targets quaternary ammonium compounds, Tet pumps expel tetracyclines, and QepA and OqxAB confer quinolone resistance. Interestingly, narrow-spectrum pumps seem to spread more efficiently through HGT than broad-spectrum ones (Hernando-Amado et al. 2016). Efflux pumps have various families and groups, which include ATP-binding cassette, small multidrug resistance, multidrug and toxic compound extrusion, major facilitator superfamily, and resistance nodulation division. ATP-binding cassette is one of the largest membrane transporters, but unlike other efflux pumps, it utilizes ATP-driven pumps. These pumps consist of nucleotide-binding domains, which hydrolyze ATP and bring about conformational changes (Pal et al. 2014). The small multidrug resistance is composed of microbial membrane proteins, which carry out different transport functions like heavy hydrophobic drugs, antiseptics, as well as metabolite guanidium across the bimembrane layer. SMR transporter genes are increasingly spreading amongst bacterial populations, adapting to new physiological conditions and transport needs. This spread promotes the co-selection of nearby antimicrobial resistance genes in environmental reservoirs and shapes microbial communities in the human microbiome and human-influenced environments. SMR transporters thus play a key role in the ongoing natural selection at the human–microbe interface (Jang 2023). The major facilitator family (MFS) is the largest transporter of secondary active molecules. The transporters are responsible for transporting a broad spectrum of substrates, both uphill and downstream, with an electrochemical gradient (Drew et al. 2021). The family consists of three types of transporters, which are symporters, uniporters, and antiporters (Sauve et al. 2023). The resistance-nodulation-division superfamily is known for its ability to reduce the intracellular concentration of structurally diverse antimicrobials. They are mostly possessed by pathogenic gram-negative bacteria, which cause serious infections and disseminate resistance for multiple relevant antibiotics (Kavanaugh et al. 2024). These pumps utilize a proton motive force and efflux drugs through periplasm where the substrate binds through the outer membrane and other through the inner membrane of cytoplasm where the substrate binds through without losing substrate in the periplasm (Pal et al. 2014).

Growing AR is increasingly prevalent amongst pathogens through the modification of antibiotics. Glycopeptides and polymyxins get modified chemically in cell envelope, which leads to AR. Antibiotic targets undergo enzymatic modification and in some cases, resistance enzymes are mobilized readily in microbes (Schaenzer and Wright 2020). Quinolones inhibit topoisomerase by binding near the active site; however, amino acid substitutions in target proteins decrease binding efficiency and thus confer resistance. A mutation in penicillin binding protein 3 (PBP3) in *E. coli* was found to be resistant to aztreonam and avibactam (Bush et al. 2020; Periasamy et al. 2021).

## 2.4 HORIZONTAL GENE TRANSFER AND BIOFILM FORMATION IN ENVIRONMENTAL SELECTIVE PRESSURE

Most microbes form biofilms on surfaces, which facilitates AR in both the environment and clinical settings. These further promote the acquisition and dissemination of AR and are hotspots of AR. Amongst various polluted environments, water bodies are the most potential reservoirs where a wide range of clinical and terrestrial ARB thrive. Humans and animals get exposed to ARB more readily through water, where bacteria form multispecies communities through biofilms (Besemer 2015). The high density in these structures is formed due to a matrix of extracellular polymeric substances (EPS). EPS are composed of amyloid-like proteins, lipids, exopolysaccharides, and extracellular DNA (Fulaz et al. 2019). Biofilms get formed through attachment, microcolony formation, maturation, and detachment (Toyofuku et al. 2016). Genetic exchange between bacteria in biofilms is higher than that in the free-living state. The genes are based on mobile genetic elements like plasmids or ICEs and spread through HGT (Michaelis and Grohmann 2023). A study by Gold and Moellering (1997) showed a resistant bacillus strain and a sensitive staphylococcus strain in a biofilm created artificially. Later, isolates of *Staphylococcus* with resistance to tetracycline were recovered (Song et al. 2021). ARGs can spread through HGT, which mostly depends on environmental selective pressure, microbial population, and MGEs. Classically, the HGT can be divided into conjugation, transformation, and transduction.

Conjugation between bacterial communities occurs more frequently in biofilms owing to their dense structure and faster spread of mobile genetic elements. A high horizontal transfer rate makes the plasmids act as molecular parasites in biofilms. Biofilms offer an added advantage by providing spatial and structural stability while keeping conjugative pili intact (Madsen et al. 2012). Transformation is a process where bacteria take up exogenous DNA via transformation pilus and DNA transporter. DNA is integrated into chromosomes by homologous recombination or stays as an episome. Species including *Bacillus subtilis*, *Vibrio cholera*, and *Streptococcus pneumoniae* are known to be naturally transformable (Abe et al. 2021). Bacteria acquire AR through bacteriophages through the process of transduction. Temperate phages can integrate into the bacterial genome as prophages while carrying resistance factors which are expressed by the host. This process is referred to as the lysogenic cycle, which helps disseminate and maintain AR in bacteria through HGT without killing the bacteria (Mazaheri Nezhad Fard et al. 2011).

Although the mobility of resistance determinants occurs at a high frequency, these events usually do not lead to the formation of novel resistance in pathogens. New resistance usually occurs only in the presence of selective pressure for their maintenance and negligible survival costs. Therefore, it is crucial to look into the circumstances and degree of environmental selection for resistance in order to facilitate prophylactic interventions. Furthermore, knowing dispersal barriers is essential for assessing risks as well as preventing the spread of resistant diseases and novel resistance genes to people (Bengtsson-Palme et al. 2019). The natural evolution of resistance is usually driven by microbial competition for survival and production of secondary metabolites with antibiotic properties. Current use of antibiotics has magnified selection pressure in the environment, especially on human and environmental microbiota in the polluted environment. This pressure has led to the mobilization and horizontal transfer of ARGs in pathogenic bacteria, raising a challenge in infection treatment. As bacteria and genes move across environments and species, understanding the links between environmental variables and their effect on microbiota is essential (Larsson and Flach 2022). Environmental conditions not only shape microbial communities but also determine selective pressure, which helps the maintenance of the ARG pool. Major factors like pH, oxygen, and nutrient level, heavy metal concentration modulate ARG expression and gene transfer. In addition to that, antibiotic residues, industrial effluents, agricultural runoff act as selective agents that enrich resistant strains and activate mobile genetic elements. Elevated ROS levels and stress response genes were elevated in biomass exposed to antibiotics, indicating oxidative damage promoting AR enrichment suggesting the role of oxygen in spreading resistance (Bombaywala et al. 2024). These variables differ across environments (e.g., soil, freshwater, marine ecosystems, animal husbandry sites, and wastewater treatment plants), each presenting unique contexts for ARG evolution (Dafale et al. 2020) (Table 2.1).

## 2.5 ENVIRONMENTAL STRESS-INDUCED CO-SELECTION AND CROSS-RESISTANCE IN MICROBIAL COMMUNITIES

WHO also suggests that while predicting resistance emergence under environmental conditions, it is essential to account for co-selection mechanisms such as co- and cross-resistance. As per the WHO, co-resistance occurs when different resistance genes that confer resistance to different antibiotic classes are located together in DNA transfers simultaneously. In contrast, cross-resistance arises from mutations or genes that confer resistance to multiple drugs within the same class. Methicillin-resistant *S. aureus* (MRSA) exemplifies both co- and cross-resistance properties (Cantón and Ruiz-Garbajosa 2011). The environment is increasingly recognized as a key reservoir for the selection and dissemination of ARB and ARGs. Antibiotics and substances with antimicrobial properties can be well-established drivers of AR as a result of co-selection (Murray et al. 2024). Plant protection products, non-antibiotic drugs, metals, and biocides are often present within the nanogram to microgram per liter range but are enough to drive resistance amplification and persistence. Minimum inhibitory concentration, minimal selective concentration, and minimal increased persistence concentration indicate different effects on resistance dynamics. Pollutants from municipal, industrial, and agricultural sources contribute to a broad resistance selection window (Murray et al. 2024). Waste streams often cause the overlapping of metals and antibiotic waste, leading to co-contamination and microbial toxicity in water, soil, and sediments. Metals have a broader environmental impact compared to antibiotics due to their more frequent occurrence, release levels, and persistent nature. Cd, Cu, and Zn are found at concentrations far exceeding those of antibiotics. Additionally, some metals (e.g., zinc and copper) have lower sorption potential than antibiotics, increasing their bioavailability in soils and sediments (Gillieatt and Coleman 2024). Inefficiency of wastewater treatment plants could lead to further co-selection of metals and AR. These resistance determinants mostly co-localize with mobile genetic elements, facilitating their spread through HGT. Biofilm formation further enhances the persistence of multidrug and multi-metal resistance, reducing the microbe's sensitivity to antibiotics (Engin et al. 2023). A study by Fu et al. (2023) used Cd and sulfadiazine to explore the relationship of metal on antibiotic resistance and revealed the rise of ARGs level by up to 1.06-fold with Cd alone, up to 6.53-fold with SD alone, and up to 7.62-fold when combined. Proteobacteria and Bacteroidetes were the most dominant phyla with int1 playing a major role in AR propagation (Fu et al. 2023). Similarly, metal-polluted rivers are also the key hotspots of metal-induced AR. A study on metal-polluted rivers of India showed high levels of cobalt, nickel, zinc, and cadmium to be linked with ARGs of carbapenem and erythromycin (Gupta et al. 2022). A study of the lakes in Iran revealed the correlation between vanadium and tetracycline resistance and vanadium is a well-known marine

TABLE 2.1

*Environmental compartments and associated antibiotic resistance genes (ARGs)*

| Environmental compartment | Detected ARGs | Antibiotic-resistant bacteria | Persistence mechanisms | Monitoring and surveillance challenges | References |
|---|---|---|---|---|---|
| Fresh vegetables | *adef*, *tetA*, *ermA* | *Escherichia coli* | Transmitted via food web into human gut microbiome | Need for "good agricultural practices" to reduce fecal contamination | Holvoet et al. (2013) |
| Irrigation water | *tetW*, *blaTEM*, *sulI*, *qnrS*, *ermB* | *E. coli* | Fecal contamination from wastewater used for irrigation | Requires constant monitoring and improved wastewater treatment | Amato et al. (2021) |
| Irrigation water | *blaTEM*, *tetA*, *tetB*, *dfrA1-sat2-aadA1* | *E. coli* | Entry into crops via groundwater used for irrigation | Necessitates regular monitoring and ARG profiling in food products | Araújo et al. (2017) |
| Poultry farms | *Ampc*, *bla-SHV*, *bla-CTX-M* | *E. coli*, *Proteus mirabilis* | High ARG load due to antibiotic use in livestock | Lack of standard treatment for manure before environmental disposal | Dandachi et al. (2018) |
| Soil near livestock farms | *sul1*, *blaTEM*, *ermC*, *qnrS* | *Acinetobacter* spp., *E. coli*, *Bacillus* spp. | Uptake of manure-treated soils | Variability in soil type and ARG persistence under different conditions | Zhu et al. (2013) |
| Municipal wastewater | *blaNDM-1*, *blaOXA-48*, *qnrB*, *tetA*, *ermF* | *Klebsiella pneumoniae*, *E. coli*, *Pseudomonas aeruginosa* | Resistance genes on plasmids and integrons; frequent HGT | Limited surveillance in many low- and middle-income countries | Riquelme et al. (2022) |
| Hospital wastewater | *blaKPC*, *blaNDM*, *blaOXA-1*, *tetM*, *qnrS* | *Acinetobacter baumannii*, *Enterobacter cloacae* | Enrichment by pharmaceutical residues; MDR gene clusters | Often untreated or mixed with domestic sewage; low oversight | Kümmerer (2003) |

(Continued)

Table 2.1 (Continued)

| Environmental compartment | Detected ARGs | Antibiotic-resistant bacteria | Persistence mechanisms | Monitoring and surveillance challenges | References |
|---|---|---|---|---|---|
| Surface water (rivers/lakes) | *sul1*, *qnrA*, *tetX*, *blaSHV* | *E. coli*, *Enterococcus* spp., *Aeromonas* spp. | Runoff from urban and agricultural sources; natural biofilm reservoirs | Infrequent ARG monitoring in aquatic ecosystems | Berendonk et al. (2015) |
| Marine water (estuaries/ coasts) | *tetM*, *qnrS*, *ermB*, *blaTEM* | *Vibrio* spp., *Pseudomonas* spp. | Coastal discharge of untreated waste; ARGs absorbed in sediment | ARGs persist in sediment and can re-enter human systems via seafood | Su et al. (2020) |
| Treated water | *sul1*, *tetW*, *blaTEM* (*at low levels*) | *Rare detection: E. coli*, *Pseudomonas* spp. | Resistant genes survive filtration or originate from biofilm in pipes | Emerging concern; few countries screen treated water for ARGs | Rodriguez-Mozaz et al. (2020) |
| Biofilms in pipes | *Bla* and *tet* | *Pseudomonas* spp., *Acinetobacter* spp. | Biofilms shield bacteria and ARGs from disinfectants | Difficult to sample and quantify in real time | Blaustein et al. (2016) |
| Landfill leachate | *tetM*, *sul2*, *blaCTX-M*, *qnrB* | *E. coli*, *Stenotrophomonas* spp. | Leachate promotes selective pressure and mobility of ARGs | Few guidelines for leachate monitoring | Wang et al. (2015) |
| Compost from organic waste | *tetW*, *ermB*, *sul1* | *Enterococcus* spp., *Bacillus* spp. | Composting can reduce but not eliminate ARGs | Composting processes vary; inconsistent results | Qian et al. (2019) |
| Sediments near aquaculture farms | *floR*, *sul1*, *tetM*, *qnrA* | *Vibrio* spp., *Shewanella* spp. | Accumulation due to medicated feed and fecal residues | Rarely monitored for resistance genes | Cabello et al. (2013) |
| Urban stormwater runoff | *sul1*, *blaTEM*, *qnrS* | *E. coli*, *Aeromonas* spp. | Runoff collects ARGs from multiple sources (roads, sewers) | ARG source tracking is complex | Liguori et al. (2021) |

| Recreational waters | *qnrB*, *tetX*, *sul1* | *E. coli*, *Pseudomonas* spp. | ARGs persist due to fecal contamination and low chlorine | Health risks underestimated; inadequate routine surveillance | Nappier et al. (2020) |
|---|---|---|---|---|---|
| Septic tanks | *blaTEM*, *tetW*, *sul1* | *E. coli*, *Clostridium* spp. | ARGs accumulate due to poor sanitation infrastructure | Often excluded from formal surveillance systems | Tan et al. (2021) |
| Airborne particulates near farms | *tetX*, *sul2*, *blaSHV* | *E. coli*, *Staphylococcus aureus* | Dissemination through dust and aerosols from animal facilities | ARG presence in air rarely assessed; underregulated | McEachran et al. (2015) |
| River sediments downstream of pharmaceutical plants | *blaNDM*, *blaCTX-M*, *qnrS*, *sul1* | *E. coli*, *Klebsiella* spp. | High selective pressure from antibiotic residues | Poor oversight and limited transparency in emissions | Larsson (2008) |
| Industrial effluents (non-healthcare) | *blaTEM*, *tetM*, *sul2* | *Pseudomonas* spp., *Bacillus* spp. | Occasional antibiotic use in dye/textile/paper industries | ARG impact underestimated in industrial zones | Milaković et al. (2020) |
| Sludge from wastewater treatment plants | *tetA*, *sul1*, *blaOXA*, *ermB* | *Acinetobacter* spp., *E. coli*, *Enterococcus* spp. | ARGs retained and concentrated in biosolids | Biosolid reuse in agriculture raises exposure risk | Munir et al.(2011) |

Overview of key environmental sources of ARG dissemination, commonly detected ARGs and resistant bacteria, persistence mechanisms contributing to ARG survival, and challenges in monitoring and surveillance efforts.

fuel additive and could play a key role in microbial efflux mechanisms. These lakes were located near big cities, with lots of effluent discharge from industries, urban wastewater, and agricultural and poultry runoff, and were rich in *tetC*, *sul1*, *sul2* (Komijani et al. 2021). This co-resistance phenomenon also exists in terrestrial and surface soil. A study confirmed the promotion of co-selection of ARGs and MRGs in metal-polluted soil through metagenomics. Co-occurrence of ARGs and MRGs, such as aminoglycoside and copper resistance and multi-metal and multidrug resistance determinants, was abundant under high heavy metal pressure. They also increased the diversity of resistance determinants and reshaped the microbial populations while enhancing HGT (Wang et al. 2021b). The above studies support and highlight that metals play a crucial role in ARG drive in the environment, and metal pollution should be intervened and mitigated in the near future. Apart from heavy metals, pesticides are also one of the emerging pollutants to have the same alarming impact on growing AR. Selective pesticide pressure promotes the development of AR in bacteria through multiple mechanisms, like activation of efflux pumps, reduced permeability, and induction of gene mutations. A key pathway is HGT as pesticides enhance conjugative transfer of ARGs by boosting MGE, which facilitates AR dissemination (Qiu et al. 2022). A study of lakes and wetlands measured the elements and pesticide levels and showed a correlation of beta-lactamase with boron, iron, lithium, sodium, and phosphorus. It also revealed a correlation of tetA and tetB with parathion levels. Overall, organophosphates and heavy metals appear to drive multidrug resistance by inducing mutations, altering gene expression, and enhancing HGT (Yasemi et al. 2025). Similarly, the combined effect of pesticides and antibiotics elevates virulence factors, posing a global health crisis. In a microcosm study, zinc thiazole and oxytetracycline altered microbial communities in soil while increasing ARGs and virulence factors in soil, earthworm gut, and phyllosphere. Zinc thiazole alone disrupted nutrient metabolism and, while combined with oxytetracycline, reduced phyllosphere diversity and shifted resistance profiles. *Citrobacter* and *Aeromonas* were key ARG/VFG hosts, with the co-occurrence of ARGs and MGEs detected. Notably, ZT boosted drug-resistant *Acinetobacter* spp. in the phyllosphere. These findings highlight pollutant-driven resistance spread and related health concerns (Wang et al. 2024).

## 2.6 ENVIRONMENTAL PATHWAYS OF ANTIBIOTIC RESISTANCE FROM WASTEWATER TO ECOSYSTEMS

Residues of industrial chemicals, specifically from the pharmaceutical industry, can contaminate the nearby water sources through waste discharge and accidental leaks. While the source could be viewed as minimal at a global scale, this creates a localized hotspot for ARG breeding (Thomas et al. 2007). The exponential rise in the number of pharmaceutical companies has led to antibiotic production and released waste contributing to pollution. Contamination of natural sources leads to the selection of resistance genes via mechanisms like mutations and multidrug efflux pumps (Ajekiigbe et al. 2025).

According to papers published on high emissions of pharmaceuticals in Hyderabad, India, the concentrations of drugs in effluent from wastewater were higher than that of patients taking medicines. The concentration of ciprofloxacin, a broad-spectrum antibiotic, was as high as 31 mg/L which is approximately 1 million times greater than the levels that are regularly found in treated municipal sewage effluents (Larsson 2014). Environment plays a major role in the emergence of novel phenotypes, especially in antibiotic selective pressure. Pharmaceutical wastewater could become a major hub of antibiotic selective pressure, as they can reach highly resistive-selective concentration in Asian countries. Similarly, Croatia also reported macrolide resistance in wastewater and sediments (Bengtsson-Palme et al. 2018). Studies show that wastewater exhibits ARGs even with lower antibiotic concentrations. Industries release heavy metals and pesticides which promote gene exchange linked to AR, creating a selective pressure. Co-resistance and co-selection can induce resistance in microbes even in the absence of antibiotics (Hubeny et al. 2021). Textile, paper, and plastic industries usually release tons of colored dyes and heavy metals and are more hazardous to animal and human health. Toxic heavy metals like mercury, chromium, lead, and cadmium have no biological role and are known for their toxic, mutagenic, and carcinogenic effects on human and aquatic life. Discharging untreated wastewater into nearby aquatic sources can lead to groundwater contamination (Ahmed et al. 2020).

## 2.7 ENVIRONMENTAL CONTROLS ON THE EMERGENCE AND SPREAD OF SOIL RESISTOME

Soil is one of the major reservoirs and hotspots of AR, and its resistome is influenced by phytochemical properties, changing microbiome. Soil is known to harbor 30% of ARGs along with a wide range of microbial population. ARGs clinically relevant to antibiotics like beta-lactam and tetracycline are found in soil, including ancient permafrost. Exogenous ARGs can enter soil through wastewater, manure, making soil a significant pathway for ARGs and posing risk to the environment (Han et al. 2022). Fertilization, irrigation, and cropping system management can alter both intrinsic and acquired resistance in the soil resistome. Heavy metals and antibiotics in manure can act as co-selecting agents and enhance the dissemination of AR (Pan and Chu 2018). Pesticides can also lead to the growing co-resistance of antibiotics and heavy metals in irrigated fields (Maharana et al. 2024). Similarly, irrigation of fields with sewage effluent carrying AR can potentially facilitate the global dissemination of clinically relevant ARGs. Domestic wastewater/sewage, whether treated or untreated, is often used for irrigation due to its easy availability despite residual organic matter and nutrients (Saxena et al. 2021). This could serve as another major hotspot for directly introducing ARG-inducing substances into the soil. Although environmental pollutants can trigger co-selection, they also improve microbial tolerance to antibiotics. Resistant microbes can become dominant in soil communities, increasing the potential for ARG transfer via HGT (Wang et al. 2021a).

## 2.8 ENVIRONMENTAL STRESSORS SHAPING AR IN NATURAL AND RECREATIONAL WATER SOURCES

Aquatic ecosystems consisting of rivers, lakes, receive runoff from urban, industrial, and agricultural areas. These water bodies also support biodiversity and recreational activities, including swimming and fishing. Recent studies have identified ARGs like *bla*TEM, *sul1*, *tetX* in urban water sources (Berendonk et al. 2015). Along with recreational activities, these water bodies also provide the majority of drinking water to the population. Effluents from industries also affect groundwater which is a major source of drinking water, thus making it a crucial hotspot for AR. Biofilm formations in drinking water systems further facilitate HGT, accelerating AR dissemination (Li et al. 2019). Effluents from treatment plants, hospitals, and animal farms discharge into water bodies. Rivers continuously transport the contamination downstream whereas lakes tend to accumulate ARG owing to their stagnant nature. Since lakes constitute 90% of the world's surface water, compared to 2% of rivers, their role is higher in storing resistance elements and posing a greater risk to water quality and public health 202 (Chen et al. 2019). Since lakes hold 90% of surface freshwater, they are most vulnerable to pollutant accumulation, and AR is one of such emerging pollutants in lakes. A meta-analysis of several lakes detected AR of 57 antibiotics, including sulfamethoxazole, tetracycline, and erythromycin. While sulfonamide levels showed no regional difference, quinolone concentrations varied significantly. Erythromycin posed the highest ecological risk. Despite this, research on the role of aquatic organisms in ARG propagation in lakes remains limited (Yang et al. 2019). In addition to effluents and discharge, water bodies have also been contaminated with plastics from anthropogenic activities. Plastics have been accumulated in various water bodies and oceans and have served as a reservoir for colonization for microbial communities. A study by Andrzejak et al. (2023) studied water quality, AR, and microbial ecology of biofilms on plastic in public boats. The biofilms were studied for nutrients, coliforms, and resistant bacteria, where 86% showed multidrug resistance, and *sul2*, *bacA*, and *ampA* were detected (Andrzejak et al. 2023). A metagenomic study on North Pacific Gyre revealed higher diversity and abundance of ARGs and MRGs on plastic particles compared to seawater. Multidrug and metal resistance genes were most prevalent but no significant difference was found in macro- and microplastic effects in AR dissemination. Co-occurrence patterns suggest co-selection by antibiotics and metals, which drives resistance on plastic (Yang et al. 2019).

## 2.9 THE AIRBORNE DIMENSION OF AR: THE ROLE OF AIR IN MICROBIAL GENE DISSEMINATION

Although various dispersal mechanisms for ARGs exist, atmospheric transport has been underestimated. Fresh snow can disseminate ARGs over larger areas; however, their link to air pollution is not well understood. In a study by Zhu et al. (2021), 44 snow samples from North America, Europe, and China were analyzed through qPCR. Results showed air pollution significantly contributes to air pollution in snow, highlighting snowfall as a vector of ARG spread (Zhu et al. 2021). Similarly, a global study profiled the relative abundance of 30 ARGs across 19 cities where ARG levels varied by 100-fold, with beta-lactam and quinolone being the most abundant. Network analysis highlighted the bacterial contributors to airborne ARGs (Li et al. 2018). ARG levels in the air are higher in urban areas in summer and in rural areas in winter. Indoor human activities also increase airborne ARG and could possibly be transferred through inhalation. Sampling and cross-study comparisons need to be more developed, and methodologies need to be standardized in order to assess atmospheric role in AR spread (Kormos et al. 2022). Hospitals and clinical settings are key hotspots of airborne microbial contamination. Frequent use of antibiotics and disinfection practices drives the emergence of ARBs. Hospital environments often harbor resistant strains, and environmental resistome have been identified as reservoirs of ARGs that can transfer to clinical pathogens causing nosocomial infections. Overall, 95% of airborne bacterial isolates from hospital-operating rooms were resistant to antibiotics, with many showing multidrug resistance, highlighting the abundance of ARB in hospital air (Gwenzi et al. 2022). A global study by Zhou et al. (2023) showed strong and increasing correlations between PM 2.5 levels and AR. In 2018, PM 2.5-related AR was linked to 480,000 pre-mature deaths and welfare loss (Zhou et al. 2023). Meeting the WHO's 2050 PM 2.5 target of 5 $\mu g/m^3$ could reduce AR by 16.8% and prevent 23.4% of related deaths, saving approximately $640 billion. This is crucial in order to navigate the less explored areas of airborne transmission of AR (Zhou et al. 2023).

## 2.10 SURVEILLANCE AND MONITORING OF ENVIRONMENTAL ANTIBIOTIC RESISTANCE

The widespread distribution of AR, across various environmental compartments, highlights the interconnected nature of resistance dissemination. From soil to marine environments, to vegetables and urban infrastructure, each compartment plays an important role in the ecological resistome. The ARG spread is no longer limited to clinical settings; rather, it has spread to the environment and requires a holistic One Health approach. The continued proliferation of ARGs in the environment poses a multifaceted and escalating threat to global health, food safety, and ecosystem stability. AR is a widespread human-driven issue, and addressing it would require an interdisciplinary approach combining environmental science, public health, microbiology, and mitigation policies. There is a need for a paradigm shift toward upstream interventions that prevent the release of ARGs into the environment in the first place, including genes in soil, wastewater, and biofilms, which enhance long-term survival and gene transfer. Even advanced treatments can't fully eliminate the genes, highlighting the need for upstream solutions and their release at source (Manaia et al. 2016). Regulating the use of antibiotics in animal husbandry, reducing unnecessary prescription in both human and veterinary medicine, and enhancing biosecurity are vital steps. Technologies like advanced filtration systems, oxidation processes, and biological treatments for the deactivation of ARGs should be included before discharging waste. The implementation of stricter controls on pharmaceutical waste discharge and better enforcement of waste treatment protocols can significantly reduce the environmental burden of antimicrobial agents.

AR varies by region, influenced by local disease prevalence, antibiotic usage, and testing practices. Effective surveillance systems must consider these differences when selecting indicators, antibiotics monitored, and testing methods. AR surveillance is a structured process for tracking resistance with consistent methodologies and proper time intervals (Diallo et al. 2020). Addressing AMR in food production requires targeting high-risk areas like farm soil, manure, and wastewater where ARB and ARGs accumulate. Reducing antibiotic use through stricter veterinary regulations and improving sanitation practices can help limit resistance spread. Integrated surveillance using tools like PCR and genome sequencing is essential to track ARB/ARGs across the food chain and provide early warnings for outbreaks. These actions support a One Health approach from farm to consumer (Tiedje et al. 2023). A study by Huijbers et al. (2019) outlined five objectives for surveillance of AR in the environment. The objectives included assessing the risk of human exposure to resistant bacteria via the environment, evaluating the way pollution drives AR evolution. This is followed by examining ecological risks to ecosystems and their functions, estimating population-level resistance and antibiotic use, particularly in settings lacking robust clinical data (Huijbers et al. 2019). The WHO global strategy identified laboratory-based surveillance as essential for managing AR and evaluating interventions. Individual countries have developed regional and national surveillance networks, but a global framework is absent, leading to failure in comparison of national efforts (Grundmann et al. 2011). Advances in rapid, accurate, and cost-effective molecular tools for detecting AMR and monitoring environmental pollution are essential for managing resistance. Holistic implementation of antibiotic stewardship through national policies is equally crucial to curb environmental and clinical resistance (Bobate et al. 2023). Most ARG-predicting models often overlook the complexities of HGT and the persistence of ARGs. High-throughput sequencing is a preferred method for analyzing multiple ARG classes, including fluroquinolones, beta-lactams, tetracycline, vancomycin, and multidrug efflux pumps. Integration of mobile genetic elements as a marker to estimate horizontal transfer needs to be incorporated into models. Monitoring antibiotic concentrations in organic waste remains challenging, typically relying on techniques such as high-performance liquid chromatography, gas chromatography, and mass spectrometry (Nava et al. 2022).

WHO launched the Global Action Plan in 2015, promoting antimicrobial use leading to national action plan. Low and middle-income countries face financial and staffing challenges. To support GAP goals, WHO introduced AWaRe classification grouping antibiotics as access (narrow spectrum, low resistance risk), watch (broad-spectrum, higher resistance risk), and reserve (last-resort options for infection) to guide responsible antibiotic use and preserve efficacy (Saleem et al. 2025). Similarly, the World Organization for Animal Health and FAO jointly adopted a plan promoting integrated AMR surveillance. Europe followed the 2017 One Health AMR Action Plan, calling for unified monitoring of AR and residue sectors. European countries had established mandatory voluntary AR surveillance for animals, humans, and food by 2021. In France, numerous programmers monitor ABR, ABU, and antibiotic residues across human, animal, food, and environmental sectors. The 2016 inter-ministerial roadmap aimed to improve data sharing and cross-sector collaboration. However, the diversity of these programmers complicates system-wide assessment, making comprehensive mapping crucial for better coordination and evaluation (Collineau et al. 2023). Wastewater surveillance of ARBs and ARGs has limited utility for informing public health actions such as outbreak investigations due to the intrinsic biology of pathogens and the dynamics of ARG exchange. While emerging ARGs can be detected in wastewater, it remains unclear whether this leads to earlier or enhanced public health responses beyond existing surveillance systems. Long-read sequencing of pathogens from wastewater could help link ARGs to specific species (Foxman et al. 2025).

## 2.11 CONCLUSION

The environmental dimension of AR presents a significant and often underestimated challenge to global health. The diverse environmental determinants of contaminated wastewater, agricultural runoff, air particles, rivers, lakes, and artificial aquatic systems play a vital role as reservoirs of the alarming threat. They also play a major role in co-selection and cross-resistance along with the horizontal transfer of ARGs across the microbial populations, including common opportunistic pathogens. These complicate efforts to track and control resistance at the source level. The technical limitation of current surveillance methods, like qPCR and DNA sequencing, to distinguish non-viable microbes could hinder proper assessment. There is also a lack of global framework or standard of monitoring protocols across matrices and geographic regions, which restricts modeling efforts. A multidisciplinary approach integrating advances in molecular diagnostics along with the surveillance of mobile genetic elements is required to address the above problems. Environmental mitigation strategies should include improved antibiotic stewardship, stricter regulation of pharmaceutical and agricultural discharges. Additionally, the implementation of effective engineered systems of wastewater treatment is capable of reducing both residual antibiotics and ARGs. An advanced One Health approach underpinned by coordinated databases, standardized monitoring protocols, and cross-sectoral policy alignment is essential to mitigate environmental contributions to AR and support sustainable environmental health.

## REFERENCES

Abe K, Nomura N, Suzuki S (2021) Biofilms: Hot spots of horizontal gene transfer (HGT) in aquatic environments, with a focus on a new HGT mechanism. *FEMS Microbiol Ecol* 96:1–12. https://doi.org/10.1093/FEMSEC/FIAA031

Ahmed S, Aktar S, Zaman S, et al (2020) Use of natural bio-sorbent in removing dye, heavy metal and antibiotic-resistant bacteria from industrial wastewater. *Appl Water Sci* 10:1–10. https://doi.org/10.1007/s13201-020-01200-8

Ajekiigbe VO, Agbo CE, Ogieuhi IJ, et al (2025) The increasing burden of global environmental threats: Role of antibiotic pollution from pharmaceutical wastes in the rise of antibiotic resistance. *Discov Public Health* 22. https://doi.org/10.1186/s12982-025-00506-9

Amato M, Dasí D, González A, et al (2021) Occurrence of antibiotic resistant bacteria and resistance genes in agricultural irrigation waters from Valencia city (Spain). *Agric Water Manag* 256. https://doi.org/10.1016/j.agwat.2021.107097

Aminov RI (2010) A brief history of the antibiotic era: Lessons learned and challenges for the future. *Front Microbiol* 1:1–7. https://doi.org/10.3389/fmicb.2010.00134

Andrzejak T, Raje H, LaFleur G, et al (2023) Water quality and antibiotic resistance in the recreational waters. *Bioresour Technol* 370:128546. https://doi.org/10.1016/j.biortech.2022.128546

Araújo S, Silva IAT, Tacão M, et al (2017) Characterization of antibiotic resistant and pathogenic Escherichia coli in irrigation water and vegetables in household farms. *Int J Food Microbiol* 257:192–200. https://doi.org/10.1016/j.ijfoodmicro.2017.06.020

Armstrong AE, Rossoff J, Hollemon D, et al (2019) Cell-free DNA next-generation sequencing successfully detects infectious pathogens in pediatric oncology and hematopoietic stem cell transplant patients at risk for invasive fungal disease. *Pediatr Blood Cancer* 66. https://doi.org/10.1002/pbc.27734

Baquero F, Martínez JL, Cantón R (2008) Antibiotics and antibiotic resistance in water environments. *Curr Opin Biotechnol* 19:260–265. https://doi.org/10.1016/j.copbio.2008.05.006

Baslé A, Rummel G, Storici P, et al (2006) Crystal structure of osmoporin OmpC from E. coli at 2.0 Å. *J Mol Biol* 362:933–942. https://doi.org/10.1016/j.jmb.2006.08.002

Bengtsson-Palme J, Kristiansson E, Larsson DGJ (2018) Environmental factors influencing the development and spread of antibiotic resistance. *FEMS Microbiol Rev* 42:68–80. https://doi.org/10.1093/femsre/fux053

Bengtsson-Palme J, Milakovic M, Švecová H, et al (2019) Industrial wastewater treatment plant enriches antibiotic resistance genes and alters the structure of microbial communities. *Water Res* 162:437–445. https://doi.org/10.1016/j.watres.2019.06.073

Berendonk TU, Manaia CM, Merlin C, et al (2015) Tackling antibiotic resistance: The environmental framework. *Nat Rev Microbiol* 13:310–317. https://doi.org/10.1038/nrmicro3439

Besemer K (2015) Biodiversity, community structure and function of biofilms in stream ecosystems. *Res Microbiol* 166:774–781. https://doi.org/10.1016/j.resmic.2015.05.006

Blaustein RA, Shelton DR, Van Kessel JAS, et al (2016) Irrigation waters and pipe-based biofilms as sources for antibiotic-resistant bacteria. *Environ Monit Assess* 188:1–12. https://doi.org/10.1007/s10661-015-5067-4

Bobate S, Mahalle S, Dafale NA, Bajaj A (2023) Emergence of environmental antibiotic resistance: Mechanism, monitoring and management. *Environ Adv* 13:100409. https://doi.org/10.1016/j.envadv.2023.100409

Bombaywala S, Bajaj A, Dafale NA (2024) Deterministic effect of oxygen level variation on shaping antibiotic resistome. *J Hazard Mater* 465. https://doi.org/10.1016/j.jhazmat.2023.133047

Bombaywala S, Dafale NA, Jha V, et al (2021) Study of indiscriminate distribution of restrained antimicrobial resistome of different environmental niches. *Environ Sci Pollut Res* 28:10780–10790. https://doi.org/10.1007/s11356-020-11318-6

Bush NG, Diez-Santos I, Abbott LR, Maxwell A (2020) Contributions to antibiotic resistance. *Molecules* 25:5662.

Cabello FC, Godfrey HP, Tomova A, et al (2013) Antimicrobial use in aquaculture re-examined: Its relevance to antimicrobial resistance and to animal and human health. *Environ Microbiol* 15:1917–1942. https://doi.org/10.1111/1462-2920.12134

Cantón R, Ruiz-Garbajosa P (2011) Co-resistance: An opportunity for the bacteria and resistance genes. *Curr Opin Pharmacol* 11:477–485. https://doi.org/10.1016/j.coph.2011.07.007

Chen H, Jing L, Yao Z, et al (2019) Prevalence, source and risk of antibiotic resistance genes in the sediments of Lake Tai (China) deciphered by metagenomic assembly: A comparison with other global lakes. *Environ Int* 127:267–275. https://doi.org/10.1016/j.envint.2019.03.048

Collineau L, Bourely C, Rousset L, et al (2023) Towards one health surveillance of antibiotic resistance: Characterisation and mapping of existing programmes in humans, animals, food and the environment in France, 2021. *Eurosurveillance* 28:1–11. https://doi.org/10.2807/1560-7917.ES.2023.28.22.2200804

Dafale NA, Semwal UP, Rajput RK, Singh GN (2016) Selection of appropriate analytical tools to determine the potency and bioactivity of antibiotics and antibiotic resistance. *J Pharm Anal* 6:207–213. https://doi.org/10.1016/j.jpha.2016.05.006

Dafale NA, Srivastava S, Purohit HJ (2020a) Zoonosis: An emerging link to antibiotic resistance under "one health approach." *Indian J Microbiol* 60:139–152. https://doi.org/10.1007/s12088-020-00860-z

Dam S, Pagès JM, Masi M (2017) Dual regulation of the small RNA MicC and the quiescent porin OmpN in response to antibiotic stress in Escherichia coli. *Antibiotics* 6:. https://doi.org/10.3390/antibiotics 6040033

Dandachi I, Sokhn ES, Dahdouh EA, et al (2018) Prevalence and characterization of multi-drug-resistant gram-negative bacilli isolated from lebanese poultry: A nationwide study. *Front Microbiol* 9:1–11. https://doi.org/10.3389/fmicb.2018.00550

Darby EM, Trampari E, Siasat P, et al (2023) Molecular mechanisms of antibiotic resistance revisited. *Nat Rev Microbiol* 21:280–295. https://doi.org/10.1038/s41579-022-00820-y

Das S, Bombaywala S, Srivastava S, et al (2022) Genome plasticity as a paradigm of antibiotic resistance spread in ESKAPE pathogens. *Environ Sci Pollut Res* 29:40507–40519

Diallo OO, Baron SA, Abat C, et al (2020) Antibiotic resistance surveillance systems: A review. *J Glob Antimicrob Resist* 23:430–438. https://doi.org/10.1016/j.jgar.2020.10.009

Drew D, North RA, Nagarathinam K, Tanabe M (2021) Structures and general transport mechanisms by the major facilitator superfamily (MFS). *Chem Rev* 121:5289–5335. https://doi.org/10.1021/acs.chemrev.0c00983

Engin AB, Engin ED, Engin A (2023) Effects of co-selection of antibiotic-resistance and metal-resistance genes on antibiotic-resistance potency of environmental bacteria and related ecological risk factors. *Environ Toxicol Pharmacol* 98:104081. https://doi.org/10.1016/j.etap.2023.104081

Foxman B, Salzman E, Gesierich C, et al (2025) Wastewater surveillance of antibiotic resistant bacteria for public health action: Potential and challenges. *Am J Epidemiol* 194(5):1192–1199. https://doi.org/10.1093/aje/kwae419

Frieri M, Kumar K, Boutin A (2017) Antibiotic resistance. *J Infect Public Health* 10:369–378. https://doi.org/10.1016/j.jiph.2016.08.007

Fu Y, Zhu Y, Dong H, et al (2023) Effects of heavy metals and antibiotics on antibiotic resistance genes and microbial communities in soil. *Process Saf Environ Prot* 169:418–427. https://doi.org/10.1016/j.psep.2022.11.020

Fulaz S, Vitale S, Quinn L, Casey E (2019) Nanoparticle–Biofilm interactions: The role of the EPS matrix. *Trends Microbiol* 27:915–926. https://doi.org/10.1016/j.tim.2019.07.004

Gillieatt BF, Coleman NV (2024) Unravelling the mechanisms of antibiotic and heavy metal resistance co-selection in environmental bacteria. *FEMS Microbiol Rev* 48. https://doi.org/10.1093/femsre/fuae017

Gold HS, Moellering RC (1997) Drug therapy: Antimicrobial drug resistance. *Pneumologie* 51:276.

Grundmann H, Klugman KP, Walsh T, et al (2011) A framework for global surveillance of antibiotic resistance. *Drug Resist Updat* 14:79–87. https://doi.org/10.1016/j.drup.2011.02.007

Gupta S, Sreekrishnan TR, Ahammad SZ (2022) Effects of heavy metals on the development and proliferation of antibiotic resistance in urban sewage treatment plants. *Environ Pollut* 308:119649. https://doi.org/10.1016/j.envpol.2022.119649

Gwenzi W, Shamsizadeh Z, Gholipour S, Nikaeen M (2022) The air-borne antibiotic resistome: Occurrence, health risks, and future directions. *Sci Total Environ* 804:150154. https://doi.org/10.1016/j.scitotenv.2021.150154

Hall MC, Mware NA, Gilley JE, et al (2023) Oxidative stress induced by sub-lethal exposure to copper as a mediator in development of bacterial resistance to antibiotics. *J Hazard Mater* 8:1–14. https://doi.org/10.1021/acs.est.9b04834

Han B, Ma L, Yu Q, et al (2022) The source, fate and prospect of antibiotic resistance genes in soil: A review. *Front Microbiol* 13:1–14. https://doi.org/10.3389/fmicb.2022.976657

Hernando-Amado S, Blanco P, Alcalde-Rico M, et al (2016) Multidrug efflux pumps as main players in intrinsic and acquired resistance to antimicrobials. *Drug Resist Updat* 28:13–27. https://doi.org/10.1016/j.drup.2016.06.007

Holvoet K, Sampers I, Callens B, et al (2013) Moderate prevalence of antimicrobial resistance in escherichia coli isolates from lettuce, irrigation water, and soil. *Appl Environ Microbiol* 79:6677–6683. https://doi.org/10.1128/AEM.01995-13

Hubeny J, Harnisz M, Korzeniewska E, et al (2021) Industrialization as a source of heavy metals and antibiotics which can enhance the antibiotic resistance in wastewater, sewage sludge and river water. *PLoS One* 16:1–24. https://doi.org/10.1371/journal.pone.0252691

Huijbers PMC, Flach CF, Larsson DGJ (2019) A conceptual framework for the environmental surveillance of antibiotics and antibiotic resistance. *Environ Int* 130:104880. https://doi.org/10.1016/j.envint.2019.05.074

Iskandar K, Molinier L, Hallit S, et al (2021) Surveillance of antimicrobial resistance in low- and middle-income countries: A scattered picture. *Antimicrob Resist Infect Control* 10.

Jang S (2023) AcrAB–TolC, a major efflux pump in Gram negative bacteria: Toward understanding its operation mechanism. *BMB Rep* 56:326–334. https://doi.org/10.5483/BMBRep.2023-0070

Kavanaugh LG, Dey D, Shafer WM, Conn GL (2024) Structural and functional diversity of resistance-nodulation-division (RND) efflux pump transporters with implications for antimicrobial resistance. *Microbiol Mol Biol Rev* 88. https://doi.org/10.1128/mmbr.00089-23

Komijani M, Shamabadi NS, Shahin K, et al (2021) Heavy metal pollution promotes antibiotic resistance potential in the aquatic environment. *Environ Pollut* 274:116569. https://doi.org/10.1016/j.envpol.2021.116569

Kormos D, Lin K, Pruden A, Marr LC (2022) Critical review of antibiotic resistance genes in the atmosphere. *Environ Sci Process Impacts* 24:870–883. https://doi.org/10.1039/d2em00091a

Kümmerer K (2003) Promoting resistance by the emission of antibiotics from hospitals and households into effluent. *Clin Microbiol Infect* 9:1203–1214. https://doi.org/10.1111/j.1469-0691.2003.00739.x

Larsson DGJ (2008) Drug production facilities – An overlooked discharge source for pharmaceuticals to the environment. In: Kümmerer, K. (ed.), *Pharmaceuticals in the Environment*. Springer, Berlin, Heidelberg, pp. 37–42. https://doi.org/10.1007/978-3-540-74664-5_3

Larsson DGJ (2014) Pollution from drug manufacturing: Review and perspectives. *Philos Trans R Soc B Biol Sci* 369. https://doi.org/10.1098/rstb.2013.0571

Larsson DGJ, Flach CF (2022) Antibiotic resistance in the environment. *Nat Rev Microbiol* 20:257–269

Li J, Cao J, Zhu YG, et al (2018) Global survey of antibiotic resistance genes in air. *Environ Sci Technol* 52:10975–10984. https://doi.org/10.1021/acs.est.8b02204

Li Q, Chang W, Zhang H, et al (2019) The role of plasmids in the multiple antibiotic resistance transfer in ESBLs-producing Escherichia coli isolated from wastewater treatment plants. *Front Microbiol* 10. https://doi.org/10.3389/fmicb.2019.00633

Liguori R, Rommel SH, Bengtsson-Palme J, et al (2021) Microbial retention and resistances in stormwater quality improvement devices treating road runoff. *FEMS Microbes* 2:1–13. https://doi.org/10.1093/femsmc/xtab008

Lin Y, Wang L, Xu K, et al (2021) Revealing taxon-specific heavy metal-resistance mechanisms in denitrifying phosphorus removal sludge using genome-centric metaproteomics. *Microbiome* 9:1–17. https://doi.org/10.1186/s40168-021-01016-x

Madsen JS, Burmølle M, Hansen LH, Sørensen SJ (2012) The interconnection between biofilm formation and horizontal gene transfer. *FEMS Immunol Med Microbiol* 65:183–195. https://doi.org/10.1111/j.1574-695X.2012.00960.x

Maharana B, Mahalle S, Bhende R, Dafale NA (2024) Repercussions of prolonged pesticide use on natural soil microbiome dynamics using metagenomics approach. *Appl Biochem Biotechnol* 197(1):73–93. https://doi.org/10.1007/s12010-024-05033-y

Manaia CM, Macedo G, Fatta-Kassinos D, Nunes OC (2016) Antibiotic resistance in urban aquatic environments: Can it be controlled? *Appl Microbiol Biotechnol* 100:1543–1557. https://doi.org/10.1007/s00253-015-7202-0

Masi M, Réfregiers M, Pos KM, Pagès JM (2017) Mechanisms of envelope permeability and antibiotic influx and efflux in Gram-negative bacteria. *Nat Microbiol* 2. https://doi.org/10.1038/nmicrobiol.2017.1

Mazaheri Nezhad Fard R, Barton MD, Heuzenroeder MW (2011) Bacteriophage-mediated transduction of antibiotic resistance in enterococci. *Lett Appl Microbiol* 52:559–564. https://doi.org/10.1111/j.1472-765X.2011.03043.x

McEachran AD, Blackwell BR, Hanson JD, et al (2015) Antibiotics, bacteria, and antibiotic resistance genes: Aerial transport from cattle feed yards via particulate matter. *Environ Health Perspect* 123:337–343. https://doi.org/10.1289/ehp.1408555

McMurry LM, Aronson DA, Levy SB (1983) Susceptible Escherichia coli cells can actively excrete tetracyclines. *Antimicrob Agents Chemother* 24:544–551.

Michaelis C, Grohmann E (2023) Horizontal gene transfer of antibiotic resistance genes in biofilms. *Antibiotics* 12. https://doi.org/10.3390/antibiotics12020328

Milaković M, Vestergaard G, González-Plaza JJ, et al (2020) Effects of industrial effluents containing moderate levels of antibiotic mixtures on the abundance of antibiotic resistance genes and bacterial community composition in exposed creek sediments. *Sci Total Environ* 706:1–9. https://doi.org/10.1016/j.scitotenv.2019.136001

Munir M, Wong K, Xagoraraki I (2011) Release of antibiotic resistant bacteria and genes in the effluent and biosolids of five wastewater utilities in Michigan. *Water Res* 45:681–693. https://doi.org/10.1016/j.watres.2010.08.033

Murray LM, Hayes A, Snape J, et al (2024) Co-selection for antibiotic resistance by environmental contaminants. *NPJ Antimicrob Resist* 2:1–13. https://doi.org/10.1038/s44259-024-00026-7

Nappier SP, Liguori K, Ichida AM, et al (2020) Antibiotic resistance in recreational waters: State of the science. *Int J Environ Res Public Health* 17:1–28

Nava AR, Daneshian L, Sarma H (2022) Antibiotic resistant genes in the environment-exploring surveillance methods and sustainable remediation strategies of antibiotics and ARGs. *Environ Res* 215:114212. https://doi.org/10.1016/j.envres.2022.114212

Pal C, Bengtsson-Palme J, Rensing C, et al (2014) BacMet: Antibacterial biocide and metal resistance genes database. *Nucleic Acids Res* 42. https://doi.org/10.1093/nar/gkt1252

Pan M, Chu LM (2018) Occurrence of antibiotics and antibiotic resistance genes in soils from wastewater irrigation areas in the Pearl River Delta region, southern China. *Sci Total Environ* 624:145–152. https://doi.org/10.1016/j.scitotenv.2017.12.008

Parmar K, Dafale N, Pal R, et al (2018) An insight into phage diversity at environmental habitats using comparative metagenomics approach. *Curr Microbiol* 75:132–141. https://doi.org/10.1007/s00284-017-1357-0

Periasamy H, Joshi P, Palwe S, et al (2021) High prevalence of Escherichia coli clinical isolates in India harbouring four amino acid inserts in PBP3 adversely impacting activity of aztreonam/avibactam. *J Antimicrob Chemother* 75:1650–1651. https://doi.org/10.1093/JAC/DKAA021

Pratt LA, Silhavy TJ. (1994) OmpR mutants specifically defective for transcriptional activation. *J Mol Biol.* 243:579–594.

Qian X, Gu J, Sun W, et al (2019) Effects of passivators on antibiotic resistance genes and related mechanisms during composting of copper-enriched pig manure. *Sci Total Environ* 674:383–391. https://doi.org/10.1016/j.scitotenv.2019.04.197

Qiu D, Ke M, Zhang Q, et al (2022) Response of microbial antibiotic resistance to pesticides: An emerging health threat. *Sci Total Environ* 850:158057. https://doi.org/10.1016/j.scitotenv.2022.158057

Rehman MT, Faheem M, Khan AU (2015) An insight into the biophysical characterization of different states of cefotaxime hydrolyzing β-lactamase 15 (CTX-M-15). *J Biomol Struct Dyn* 33:625–638. https://doi.org/10.1080/07391102.2014.899925

Riquelme MV, Garner E, Gupta S, et al (2022) Wastewater based epidemiology enabled surveillance of antibiotic resistance. *Environ Sci Technol* 56(21):14982–14993. https://doi.org/10.1021/acs.est.1c08673

Rodriguez-Mozaz S, Vaz-Moreira I, Varela Della Giustina S, et al (2020) Antibiotic residues in final effluents of European wastewater treatment plants and their impact on the aquatic environment. *Environ Int* 140:105733. https://doi.org/10.1016/j.envint.2020.105733

Saleem Z, Sheikh S, Godman B, et al (2025) Increasing the use of the WHO AWaRe system in antibiotic surveillance and stewardship programmes in low-and middle-income countries. *JAC-Antimicrobial Resist* 7:1–20. https://doi.org/10.1093/jacamr/dlaf031

Sauve S, Williamson J, Polasa A, Moradi M (2023) Ins and outs of rocker switch mechanism in major facilitator superfamily of transporters. *Membranes (Basel)* 13. https://doi.org/10.3390/membranes13050462

Saxena P, Hiwrale I, Das S, et al (2021) Profiling of emerging contaminants and antibiotic resistance in sewage treatment plants: An Indian perspective. *J Hazard Mater* 408:124877. https://doi.org/10.1016/j.jhazmat.2020.124877

Schaenzer AJ, Wright GD (2020) Antibiotic resistance by enzymatic modification of antibiotic targets. *Trends Mol Med* 26:768–782. https://doi.org/10.1016/j.molmed.2020.05.001

Song W, Wemheuer B, Steinberg PD, et al (2021) Contribution of horizontal gene transfer to the functionality of microbial biofilm on a macroalgae. *ISME J* 15:807–817. https://doi.org/10.1038/s41396-020-00815-8

Su S, Li C, Yang J, et al (2020) Distribution of antibiotic resistance genes in three different natural water bodies-a lake, river and sea. *Int J Environ Res Public Health* 17:1–12. https://doi.org/10.3390/ijerph17020552

Tan L, Zhang C, Liu F, et al (2021) Three-compartment septic tanks as sustainable on-site treatment facilities? Watch out for the potential dissemination of human-associated pathogens and antibiotic resistance. *J Environ Manage* 300:113709. https://doi.org/10.1016/j.jenvman.2021.113709

Thomas KV., Dye C, Schlabach M, Langford KH (2007) Source to sink tracking of selected human pharmaceuticals from two Oslo city hospitals and a wastewater treatment works. *J Environ Monit* 9:1410–1418. https://doi.org/10.1039/b709745j

Tiedje JM, Fu Y, Mei Z, et al (2023) Antibiotic resistance genes in food production systems support one health opinions. *Curr Opin Environ Sci Heal* 34:100492. https://doi.org/10.1016/j.coesh.2023.100492

Toyofuku M, Inaba T, Kiyokawa T, et al (2016) Environmental factors that shape biofilm formation. *Biosci Biotechnol Biochem* 80:7–12. https://doi.org/10.1080/09168451.2015.1058701

Ugoeze K, Alalor C, Ibezim C, et al (2024) Environmental and human health impact of antibiotics waste mismanagement: A review. *Adv Environ Eng Res* 05:1–21. https://doi.org/10.21926/aeer.2401005

Velez R, Sloand E (2016) Combating antibiotic resistance, mitigating future threats and ongoing initiatives. *Mol Ecol* 25:1886–1889. https://doi.org/10.1111/jocn.13246

Wang HT, Gan QY, Li G, Zhu D (2024) Effects of Zinc Thiazole and Oxytetracycline on the microbial metabolism, antibiotic resistance, and virulence factor genes of soil, earthworm gut, and phyllosphere. *Environ Sci Technol* 58:160–170. https://doi.org/10.1021/acs.est.3c06513

Wang K, Zhuang T, Su Z, et al (2021a) Antibiotic residues in wastewaters from sewage treatment plants and pharmaceutical industries: Occurrence, removal and environmental impacts. *Sci Total Environ* 788:147811. https://doi.org/10.1016/j.scitotenv.2021.147811

Wang X, Lan B, Fei H, et al (2021b) Heavy metal could drive co-selection of antibiotic resistance in terrestrial subsurface soils. *J Hazard Mater* 411:124848. https://doi.org/10.1016/j.jhazmat.2020.124848

Wang Y, Tang W, Qiao J, Song L (2015) Occurrence and prevalence of antibiotic resistance in landfill leachate. *Environ Sci Pollut Res* 22:12525–12533. https://doi.org/10.1007/s11356-015-4514-7

Whittle EE, McNeil HE, Trampari E, et al (2021) Efflux impacts intracellular accumulation only in actively growing bacterial cells. *MBio* 12. https://doi.org/10.1128/mBio.02608-21

Yang Y, Liu G, Song W, et al (2019a) Plastics in the marine environment are reservoirs for antibiotic and metal resistance genes. *Environ Int* 123:79–86. https://doi.org/10.1016/j.envint.2018.11.061

Yasemi M, Jalali A, Asadzadeh M, Komijani M (2025) Organophosphate pesticides and their potential in the change of microbial population and frequency of antibiotic resistance genes in aquatic environments. *Chemosphere* 376:144296. https://doi.org/10.1016/j.chemosphere.2025.144296

Zhou Z, Shuai X, Lin Z, et al (2023) Association between particulate matter (PM)2·5 air pollution and clinical antibiotic resistance: A global analysis. *Lancet Planet Health* 7:e649–e659. https://doi.org/10.1016/S2542-5196(23)00135-3

Zhu G, Wang X, Yang T, et al (2021) Air pollution could drive global dissemination of antibiotic resistance genes. *ISME J* 15:270–281. https://doi.org/10.1038/s41396-020-00780-2

Zhu YG, Johnson TA, Su JQ, et al (2013) Diverse and abundant antibiotic resistance genes in Chinese swine farms. *Proc Natl Acad Sci USA* 110:3435–3440. https://doi.org/10.1073/pnas.1222743110

CHAPTER 3

# Use of a Biosimulator in the Identification of Microbiome of Hospital Intensive Care Units

*Sunil Thomas and Lawrence L. Livornese Jr.*

## 3.1 INTRODUCTION

Nosocomial infections, also referred to as hospital-acquired infections or healthcare-associated infections (HAIs), are infections that patients acquire while receiving care in healthcare settings such as hospitals, clinics, or long-term care facilities. These infections are neither present nor incubating at the time of a patient's admission but typically develop during the hospital stay, usually becoming clinically apparent 48 hours or more after admission (Haque et al., 2018; Thomas et al., 2025). HAIs can range from relatively mild infections to severe, life-threatening conditions, affecting patients across all age groups and clinical settings.

HAIs arise from multiple sources and transmission pathways. Khan et al. (2017) categorized the major reservoirs and routes of infection into three groups: (i) the patient's own endogenous microflora, which can cause cross-infection between different body sites; (ii) direct transmission of pathogens from patient to patient or from healthcare staff to patients, commonly via hands, saliva, or fecal contamination; and (iii) environmental sources, including contaminated water, food, medical equipment, surfaces, and waste (Livornese et al., 1992).

In addition to these routes, HAIs frequently occur in association with medical devices and implants. Central-line-associated bloodstream infections (BSIs), catheter-associated urinary tract infections (UTIs), and ventilator-associated pneumonia (VAP) are among the most common device-related infections. These infections underscore the importance of both strict hygiene practices and careful management of invasive devices to prevent pathogen transmission in healthcare settings.

Beyond their clinical impact, nosocomial infections have significant economic consequences for healthcare systems. They increase the length of hospital stays, require additional treatments, and consume extra healthcare resources, all of which contribute to substantially higher costs for hospitals and patients. Preventing HAIs through stringent infection control measures, surveillance, and hygiene protocols not only improves patient outcomes but also reduces financial burdens on the healthcare system. By minimizing the incidence of HAIs, hospitals can decrease expenses related to prolonged care, avoid regulatory penalties, and optimize the efficient use of beds and resources, highlighting the dual clinical and economic importance of effective infection prevention strategies.

HAIs pose significant concerns for patients, hospital administrators, and policymakers. In the United States, around 2 million patients experience HAIs annually, with an estimated 90,000 of them succumbing to these infections. The direct financial impact on hospitals due to HAIs is estimated to range from US$28 billion to 45 billion (Stone, 2009).

DOI: 10.1201/9781003637950-3

This chapter focuses on HAIs, exploring their prevalence, underlying causes, and the microorganisms commonly encountered in hospital intensive care units (ICUs). HAIs represent a major challenge in modern healthcare, as critically ill patients are particularly vulnerable to infections due to compromised immunity, invasive procedures, and prolonged hospital stays. The chapter highlights the most frequently implicated pathogens in ICU settings, including multidrug-resistant bacteria, fungi, and viruses, and examines the mechanisms by which these organisms spread through patient-to-patient contact, contaminated equipment, environmental surfaces, and healthcare personnel. In addition, the chapter introduces the use of a "biosimulator" to detect and analyze the ICU microbiome, providing a powerful tool to monitor microbial populations, evaluate the effectiveness of infection control strategies, and predict potential outbreak scenarios. By integrating clinical insights with technological innovations, this discussion emphasizes both the importance of understanding ICU microbial ecology and the practical applications of biosimulators in reducing HAIs, improving patient outcomes, and optimizing infection prevention measures in high-risk hospital environments.

## 3.2 SIGNIFICANCE OF HOSPITAL INTENSIVE CARE UNITS (ICUS)

The ICU plays a critical role in modern hospitals, serving as the primary setting for monitoring and treating patients with life-threatening conditions. Equipped with advanced technology and staffed by highly skilled professionals, the ICU is designed to deliver constant, specialized care that can respond rapidly to changes in a patient's condition. Over the years, the implementation of complex and precise medical interventions within this environment has been shown to significantly reduce mortality rates and limit the severity of complications. Beyond immediate treatment, evaluating the performance of ICU teams and systems is vital, as such assessments can highlight strengths, identify gaps, and ultimately contribute to improving both patient outcomes and overall satisfaction with care (Heidari et al., 2024).

Despite these critical efforts, patients in the ICU remain exceptionally vulnerable to HAIs. These infections arise not only because of the critical conditions that necessitate ICU admission but also due to the extensive use of invasive medical procedures and devices such as ventilators, catheters, and central lines. Moreover, many ICU patients experience immunosuppression—whether induced by their illness or by treatments administered to manage their condition—which further compromises their ability to resist infection. The combination of advanced interventions and weakened host defenses creates an environment where infectious agents can thrive and spread, placing ICU patients at a disproportionately high risk for HAIs (Blot et al., 2022).

Additional risk factors for HAIs among ICU patients include pre-existing comorbidities, malnutrition, frailty, and advanced age, all of which reduce physiological resilience and heighten susceptibility to complications. These vulnerabilities emphasize the importance of rigorous infection control practices, surveillance, and targeted preventive measures within the ICU environment. Addressing these risks not only protects individual patients but also strengthens the overall safety and effectiveness of the unit, ensuring that the ICU continues to fulfill its mission of preserving life and improving health outcomes for the most critically ill individuals.

## 3.3 NOSOCOMIAL INFECTIONS AND THEIR IMPACT IN CRITICAL CARE SETTINGS

Nosocomial infections—also known as hospital-acquired infections or HAIs—are infections that patients acquire while receiving care in a healthcare facility such as a hospital, clinic, or long-term care center. These infections are not present nor are they incubating, at the time of a patient's admission. Instead, they typically develop during the hospital stay, usually becoming clinically apparent 48 hours or more after admission (Haque et al., 2018; Thomas et al., 2025).

HAIs can affect nearly any part of the body and arise from a variety of sources. These include invasive procedures, surgical interventions, and the use of medical devices such as catheters, drains, and ventilators. Additionally, transmission may occur through contact with contaminated surfaces or healthcare personnel. Common types of nosocomial infections include UTIs, surgical site infections (SSIs), BSIs, and pneumonia—especially VAP (Haque et al., 2018).

The likelihood of developing an HAI increases with several risk factors: extended hospital stays, underlying chronic conditions, immunosuppressive therapies, and the overall severity of the patient's illness. These infections significantly compromise patient safety, often leading to prolonged hospitalization, increased healthcare costs, delayed recovery, higher morbidity, and in severe cases, mortality (Isigi et al., 2023).

Preventing HAIs requires comprehensive and consistent infection control practices. Key measures include rigorous hand hygiene, proper sterilization of medical instruments, use of personal protective equipment (PPE), and adherence to system-wide evidence-based protocols for the care and maintenance of invasive devices (Haque et al., 2018).

Among all healthcare settings, ICUs represent some of the highest-risk environments for nosocomial infections. ICUs serve as unique, high-stakes ecosystems where critically ill patients are treated using a range of invasive life-supporting devices and procedures. These units harbor diverse and complex microbial communities—many of which are still poorly understood. The ICU environment is shaped by frequent disinfection protocols, widespread use of broad-spectrum antibiotics, and prolonged patient stays. These factors not only disrupt native microbial communities but also foster the emergence of microorganisms with concerning traits such as multidrug resistance, hypervirulence, and enhanced biofilm formation (Calvo et al., 2024).

In this context, HAIs become one of the most serious complications of critical care. Many ICU-acquired infections are caused by antibiotic-resistant pathogens, making them particularly difficult to treat. For patients already in a vulnerable condition, such infections pose substantial risks—leading to extended recovery times, additional complications, and in many cases, life-threatening outcomes. The high frequency of invasive procedures, immunosuppression, and continuous exposure to high-risk microbes in the ICU further intensifies the severity and prevalence of HAIs in this population (Calvo et al., 2024).

## 3.4 MICROORGANISMS INVOLVED IN HEALTHCARE-ASSOCIATED INFECTIONS

HAIs, particularly those occurring in ICUs, are caused by a wide range of pathogenic microorganisms. Epidemiological data consistently highlight the predominance of Gram-negative bacteria as the leading agents responsible for ICU-acquired infections. These organisms have gained notoriety not only for their persistence in clinical settings but also for their increasing resistance to antibiotics, complicating both treatment and containment efforts.

Among Gram-negative pathogens, *Klebsiella* species, particularly *Klebsiella pneumoniae*, are frequently isolated from ICU patients. This is followed by *Escherichia coli*, *Pseudomonas aeruginosa*, and *Acinetobacter baumannii*—all of which are capable of causing severe, difficult-to-treat infections such as VAP, BSIs, and UTIs. Their ability to form biofilms, survive on surfaces, and acquire multidrug resistance mechanisms makes them particularly dangerous in high-risk environments like ICUs (Calvo et al., 2024).

In addition to Gram-negative organisms, Gram-positive bacteria also contribute significantly to the burden of HAIs. Notably, *Staphylococcus aureus*, including methicillin-resistant *S. aureus* (MRSA), and *Enterococcus* species, including vancomycin-resistant enterococci (VRE), remain persistent threats. These pathogens often colonize the skin or mucous membranes and can invade sterile sites through breaches in natural barriers caused by invasive devices or surgical procedures.

Interestingly, while the specific type of microorganism involved in an HAI is important for treatment decisions, current evidence suggests that pathogen identity alone does not consistently predict patient mortality. Instead, poor outcomes are more strongly associated with host-related factors, such as advanced age, multiple comorbidities, immunosuppression, and general frailty (Calvo et al., 2024). This highlights the dual challenge in ICU infection management: preventing the spread of drug-resistant organisms while simultaneously supporting vulnerable patients who are at greatest risk of adverse outcomes.

### 3.4.1 *Clostridioides difficile* and Antibiotic-Associated Diarrhea

*Clostridioides difficile* (formerly *Clostridium difficile*) is one of the most significant pathogens associated with HAIs, particularly in the context of antibiotic use and gastrointestinal disease. First identified in 1935, it was not recognized as a human pathogen until 1978, when it was established as the causative agent of pseudomembranous colitis. Today, it is widely acknowledged as the leading cause of antibiotic-associated diarrhea and is responsible for the majority of healthcare-associated diarrheal illnesses worldwide (Riddle and Dubberke, 2009).

The clinical presentation of *C. difficile* infection (CDI) spans a wide spectrum. In mild cases, patients may experience abdominal discomfort and self-limiting diarrhea, while severe cases can progress to fulminant colitis, toxic megacolon, septic shock, and even death. Trends in recent years indicate not only a rising incidence of CDI among hospitalized and ICU patients but also an increase in disease severity (Riddle and Dubberke, 2009).

The burden of CDI in hospitalized patients varies across institutions, with an estimated incidence of approximately 2% among all admissions. Critically ill patients face an even greater risk: the incidence in ICUs is nearly double at 4%. Alarmingly, up to 20% of ICU patients with symptomatic CDI develop fulminant colitis, a life-threatening condition associated with mortality rates as high as 60%. These statistics underscore the aggressiveness of CDI in vulnerable populations and the urgent need for improved strategies in prevention, early detection, and treatment.

In the ICU setting, CDI poses a formidable challenge. Its ability to form spores makes it particularly difficult to eradicate the pathogen in this environment. Patients are already immunocompromised and frequently exposed to multiple risk factors, including prolonged antibiotic use, invasive procedures, and complex comorbidities. Effective management requires not only prompt therapeutic intervention but also comprehensive approaches to infection control, including antibiotic stewardship, limiting pathogen transmission, and enforcing prevention protocols. Early recognition and rapid treatment are critical, as delays in therapy are strongly associated with poorer outcomes. Consequently, heightened clinical vigilance and robust diagnostic protocols remain central to reducing the devastating impact of CDI in critically ill populations.

#### 3.4.1.1 *Pathogenesis of* Clostridioides difficile *Infection (CDI)*

The virulence of *C. difficile* is primarily mediated through the production of exotoxins, rather than direct invasion of the colonic mucosa. The two major toxins—toxin A (TcdA) and toxin B (TcdB)—are central to disease progression. TcdA is traditionally regarded as an enterotoxin that induces neutrophil infiltration, fluid secretion, and mucosal inflammation, while TcdB is considered the primary cytotoxin, exhibiting roughly ten times greater potency in disrupting host cells.

Both toxins exert their effects by targeting the actin cytoskeleton of intestinal epithelial cells. They inactivate Rho GTPases, leading to the breakdown of actin filaments, disruption of tight junctions, cell rounding, apoptosis, and mucosal barrier damage. This damage causes the sloughing of epithelial cells into the intestinal lumen and triggers a robust inflammatory response, with the release of cytokines and recruitment of immune cells. The accumulation of fibrin, inflammatory cells, and necrotic debris on the damaged mucosa results in the formation of pseudomembranes, which are characteristic of pseudomembranous colitis.

These pathological features explain the often severe clinical manifestations of CDI and underscore why timely diagnosis and treatment are critical in preventing complications, particularly in the vulnerable ICU population (Riddle and Dubberke, 2009).

### 3.4.2 Viruses Involved in Healthcare-Associated Infections

Respiratory viruses pose a significant risk in the hospital environment, where vulnerable patients are often in close proximity to one another. Transmission can occur in multiple ways: through direct contact with infected visitors or family members, exposure to contagious healthcare workers, or interactions with other infected patients. Indirect transmission is also a major concern, as viruses can survive on contaminated surfaces and medical equipment, leading to spread via fomites. Furthermore, inadequate hand hygiene among healthcare providers can facilitate patient-to-patient transmission, amplifying the risk of outbreaks within healthcare facilities. The prevalence of specific respiratory viruses causing nosocomial infections often mirrors patterns of viral circulation within the surrounding community, highlighting the interconnectedness between hospital and community health (Chow and Mermel, 2017).

The spectrum of nosocomial respiratory viral infections is broad, encompassing both seasonal and opportunistic pathogens. Common culprits include influenza viruses, which are responsible for significant seasonal morbidity and mortality, and respiratory syncytial virus (RSV), which is especially dangerous for infants, elderly patients, and immunocompromised patients. Coronaviruses, including those responsible for past and recent global outbreaks, have also increased the potential for rapid hospital spread. In addition, adenoviruses and enteroviruses can cause severe respiratory illness in hospitalized patients, sometimes leading to prolonged outbreaks within wards or ICUs. Because of their ability to spread rapidly and cause significant complications in vulnerable populations, these viruses underscore the critical importance of strict infection prevention measures, such as hand hygiene, use of PPE, patient isolation, and vigilant surveillance for early detection and containment (Wan et al., 2023; Chow and Mermel, 2017).

### 3.4.3 Healthcare-Associated Fungal Infections

Fungi possess remarkable adaptability that enables them to survive and proliferate in human hosts. They can alter their vegetative morphology, existing either as filaments or unicellular forms depending on the environment. Some fungi develop a high degree of melaninization, which hinders the penetration of antifungal drugs, while others thrive under extreme conditions that contribute to their pathogenic potential. Human fungal infections arise when pathogenic fungi invade and colonize host tissues, giving rise to a wide range of clinical manifestations (Shah et al., 2024).

The pathogenesis of fungal infections involves several interconnected processes, including adhesion, invasion, immune evasion, host tissue damage, and fungal persistence. Adhesion is often mediated by fungal surface proteins binding to host cell receptors. Evidence for this comes from deletion mutants of pathogens that demonstrate reduced adherence and virulence. Among the best-known fungal adhesins are glycosylphosphatidylinositol (GPI)-modified cell wall proteins, with RodA from *Aspergillus fumigatus* being extensively studied for its interactions with host cells. Invasion provides fungi with a means of crossing barriers such as the vascular endothelium and pulmonary epithelium. Hydrolytic enzymes—including proteinases, lipases, and phospholipases—facilitate this process, while intracellular invasion offers protection from phagocytes and access to host nutrients. Fungal pathogens have also evolved sophisticated mechanisms to evade host immunity, such as secreting proteases, toxins, and superantigens. *Cryptococcus neoformans*, once not considered cytotoxic, has been shown to damage host cells through lytic exocytosis, organelle dysfunction, phagolysosomal damage, and cytoskeletal alterations. Other pathogens, such as *C. neoformans*, *Candida glabrata*, and *Histoplasma capsulatum*, can survive phagocytic uptake by blocking phagolysosomal maturation or acidification, ensuring persistence within host tissues (Shah et al., 2024).

Clinically, a relatively small number of fungal species are responsible for most healthcare-associated outbreaks. These include *Aspergillus*, *Candida*, *Pneumocystis jirovecii*, and Mucorales. In recent years, however, *Candida auris* has emerged as a major concern due to its multidrug resistance and its role in severe outbreaks of invasive infections across four continents. Outbreaks caused by less common fungal species, such as *Exserohilum rostratum* and *Sarocladium kiliense*, have been linked to the contamination of medical products, while the source of an outbreak caused by *Saprochaete clavata* remains unclear (Bougnoux et al., 2018). Transmission pathways in healthcare settings include contaminated hands of healthcare workers and persistence on environmental surfaces. Although long-term carriage among healthcare workers appears rare, *C. auris* has been isolated from medical equipment contaminated through the skin shedding of colonized patients. As a result, considerable attention has been given to screening high-risk patients at the time of admission, and evaluating disinfectant efficacy for both skin decolonization and environmental decontamination. The CDC recommends strict adherence to standard and contact precautions, including thorough daily and terminal disinfection of patient rooms and shared medical equipment to reduce the spread of *C. auris* (Bougnoux et al., 2018).

## 3.5 TERMINAL ROOM DECONTAMINATION IN HEALTHCARE SETTINGS

Multiple studies have provided strong evidence that advanced no-touch disinfection technologies, such as ultraviolet (UV) light devices and hydrogen peroxide vapor systems, are effective tools in reducing microbial contamination in healthcare environments. Laboratory-based experiments have demonstrated that these systems can successfully inactivate a wide range of microbes, including multidrug-resistant organisms (MDROs), when plated on carrier materials and placed within hospital rooms. Beyond controlled experiments, their efficacy has also been observed in real-world conditions, where they have been shown to decontaminate high-touch surfaces in patient rooms naturally contaminated with MDROs, including *C. difficile*, MRSA, and vancomycin-resistant *Enterococcus* (VRE) (Weber et al., 2016).

More importantly, an increasing number of clinical studies have extended these findings to patient outcomes. When used as part of terminal disinfection—after a patient is discharged and before a new patient is admitted—UV light and hydrogen peroxide systems have been shown to significantly reduce the risk of colonization and HAIs in subsequent occupants of those rooms. By providing a more thorough level of environmental cleaning than manual disinfection alone, these technologies address the problem of human error and variability in routine cleaning practices. As hospitals continue to face rising threats from multidrug-resistant pathogens, incorporating UV and hydrogen peroxide systems into infection prevention protocols represents an important strategy to enhance patient safety, reduce transmission, and strengthen overall infection control programs (Weber et al., 2016).

## 3.6 USE OF A BIOSIMULATOR FOR CELL AND MICROBIAL STUDIES

A "biosimulator" is a specially designed petri dish with engravings that was developed to promote the adherence of cells that typically do not attach to conventional culture surfaces. Within 48 hours of placement, these non-adherent cells begin to anchor themselves along the engravings, displaying an organized polarity. Interestingly, most cells position themselves either above the engravings in the upper region of the biosimulator or below the engravings in the lower region, suggesting that the unique structure of the biosimulator influences spatial orientation and growth patterns (Thomas, 2020).

Further experiments have subsequently demonstrated that when environmental microorganisms were cultured in a biosimulator, they exhibited weak adhesion but remarkably high proliferation rates. This finding suggested that the biosimulator not only provides an alternative surface for cellular adherence but also creates an environment conducive to microbial expansion. Advanced analysis using 16S rRNA profiling confirmed this observation, revealing that the biosimulator was significantly more effective at inducing the proliferation of a wide range of microbial phyla compared with traditional culture methods. Such efficiency highlights its potential as a valuable tool for expanding and characterizing complex microbial communities (Thomas, 2020).

In addition to its ability to promote the adherence and proliferation of microorganisms, the biosimulator can also serve as a powerful tool for studying the metabolome, providing critical insights into the functional output of microbial communities. When the microbiome is cultured in the biosimulator for 72 hours, the microorganisms release a variety of metabolites, most notably short-chain fatty acids (SCFAs), which play an essential role in microbial ecology as well as host health. These SCFAs—such as acetate, propionate, and butyrate—are important signaling molecules that influence energy metabolism, gut health, immune function, and even neurological processes. The biosimulator creates a controlled microenvironment that not only supports the proliferation of diverse microbial phyla but also facilitates the natural release and accumulation of these metabolic products in measurable concentrations. Through analytical techniques such as mass spectrometry, the SCFAs and other metabolites released during culture can be identified and quantified, providing a detailed metabolomic profile of the microbial community. This capacity to link microbial identity with functional metabolic output makes the biosimulator especially valuable, as it bridges the gap between microbiome composition and microbiome activity (Thomas et al., 2023). Beyond environmental studies, this approach could be extended to clinical research, where characterizing the metabolome of ICU or hospital-associated microbiomes could reveal critical biomarkers of infection, antibiotic resistance, or dysbiosis, paving the way for new diagnostic strategies and therapeutic interventions.

Building on these findings, researchers proposed that the biosimulator could extend beyond environmental applications and be used in clinical settings. Specifically, its capacity to promote microbial proliferation and enable more comprehensive identification suggests it could be applied to study the microbiome of hospital ICUs. Since ICUs are high-risk environments where patients are vulnerable to infections, understanding the microbial landscape with greater precision could have critical implications. By using the biosimulator in this context, researchers may gain deeper insights into pathogenic risks, antibiotic resistance, and microbial dynamics, ultimately improving infection control and patient safety strategies.

### 3.6.1 Use of a Biosimulator to Understand the Microbiome of Hospital ICU

In a recent study we explored the use of the biosimulator to elucidate the microbiome composition within ICUs in a hospital setting. The biosimulator, with a nutrient-rich bacterial growth medium, was placed in the corner of the rooms of ICUs for 24h and then incubated for three days under aerobic and anaerobic conditions. Using 16S rRNA sequencing, we profiled the ICU microbiome from multiple samples. Our findings showed that ICU microbiomes closely mirrored those of patients, with microorganisms in the ICU exhibiting stronger interrelationships than in control conditions. The combined use of the biosimulator and profiling offers an effective approach for analyzing and understanding microbiome changes in healthcare settings, particularly in high-risk areas, such as ICUs.

In our study, we obtained microbial samples from biosimulators housed in both hospital ICU and office environments. For each environment, biosimulator data were replicated under both aerobic and anaerobic conditions, and bacterial communities were profiled using 16S rRNA marker gene sequencing.

At the phylum level, Firmicutes were the most abundant group across hospital samples. Specifically, 9 of 12 hospital samples cultured under aerobic conditions were dominated by Firmicutes, while the remaining three showed higher representation of Bacteroidota, Actinobacteriota, or Proteobacteria. In contrast, office samples revealed a greater presence of Proteobacteria, with three of five containing more than 10% relative abundance under both aerobic and anaerobic conditions. Bacteroidota were consistently less abundant in hospital samples, although their levels were not significantly different between aerobic and anaerobic conditions. Collectively, all other phyla contributed less than 1% mean relative abundance.

At the family level, Bacillaceae and Staphylococcaceae were most abundant in hospital samples. By contrast, Burkholderiaceae and Comamonadaceae were less prevalent in hospital samples than in office samples. No bacterial family showed significant differences between aerobic and anaerobic conditions overall. Within the ICU, however, distinct trends emerged: Micrococcaceae, Moraxellaceae, and Wecksellaceae predominated under aerobic conditions, whereas Clostridiaceae dominated under anaerobic conditions, particularly in rooms housing patients with CDIs. At the genus level, *Staphylococcus* dominated office samples across both oxygen conditions. In the ICU, aerobic samples commonly contained *Ralstonia*, *Staphylococcus*, *Enterococcus*, *Bacillus*, *Chryseobacterium*, and *Micrococcus*, while anaerobic samples featured *Staphylococcus*, *Enterococcus*, *Bacillus*, *Clostridium*, and *Actinomyces*.

Microbial diversity metrics further highlighted differences between ICU and office environments. Alpha diversity analyses revealed that ICU samples had lower richness and Shannon diversity than office samples, indicating less diverse and less evenly distributed microbial communities. Faith's phylogenetic diversity (Faith's PD) was also reduced in ICU samples, suggesting a narrower evolutionary breadth. Beta diversity comparisons demonstrated pronounced dissimilarities between ICU and office microbiota, with UniFrac analyses confirming that environmental setting strongly shapes microbial composition. Differential Abundance Analysis (DAA) revealed that *Streptococcus*, *Blautia*, *Ralstonia*, and *Faecalibacterium* were significantly enriched in office environments compared to ICU samples.

Among pathogens of particular concern, *C. difficile* emerged as one of the most significant findings. This Gram-positive, spore-forming bacterium is a leading cause of HAIs, thriving in ICUs where antibiotic use disrupts gut microbiota. Its spores enable long-term persistence in hospital environments, with infections ranging from mild diarrhea to life-threatening colitis, disproportionately affecting elderly and immunocompromised patients. In our study, *C. difficile* was detected in ICU biosimulators, with spore transfer via healthcare personnel identified as a likely transmission pathway. These findings emphasize the urgent need for continuous air and surface decontamination in ICU environments.

Our broader investigation confirmed that ICU microbiomes are uniquely enriched in pathogenic taxa, despite being less diverse than office microbiomes. Organisms such as *Actinomyces*—which can cause nosocomial wound infections—were detected in ICU samples. Families including Microbacteriaceae, Micrococcaceae, Bacillaceae, Enterococcaceae, and Staphylococcaceae were also abundant, many of which are associated with device-related infections such as central-line BSIs, catheter-associated UTIs, and VAP. Rigorous contamination controls ensured the accuracy of these findings, ruling out artifacts from laboratory reagents.

The biosimulator demonstrated strong potential as a tool for profiling microbial dynamics in hospital ICUs. By replicating environmental conditions and enabling detailed sequencing-based analysis, it provides insights into pathogen behavior and infection risks that can inform more effective infection control strategies. Looking forward, integration of the biosimulator with genomic sequencing, smart sensor technologies, and AI-driven predictive modeling could enable real-time monitoring of microbial shifts, early detection of outbreaks, and the precise targeting of MDROs. Such advances would not only improve patient outcomes and safety but also significantly reduce the economic burden of HAIs on healthcare systems (Thomas et al., 2025).

## 3.7 CONCLUSION

Hospital-acquired infections continue to pose a serious burden to healthcare systems, especially in critical care settings such as ICUs where patients are highly susceptible. Timely detection, effective prevention, and rapid management remain essential to minimizing their impact on vulnerable populations. The present study highlights the value of the biosimulator as an innovative approach for characterizing and monitoring microbiome composition in healthcare environments. Through 16S rRNA sequencing, we demonstrated strong correlations between microbial communities present in ICU environments and those colonizing patients, underscoring the interconnected nature of hospital ecology and patient health. These findings establish the biosimulator not only as an effective research tool for analyzing microbiome dynamics but also as a promising technology for guiding practical infection control strategies. Looking forward, broader application of this method could strengthen our capacity to track emerging pathogens, reduce the spread of MDROs, and ultimately enhance both patient outcomes and the overall safety and efficiency of healthcare delivery.

## ACKNOWLEDGMENTS

The author (ST) acknowledges Wawa Foundation, Sharpe-Strumia Research Foundation, and the Abraham Thomas Foundation for providing the resources for this work.

## REFERENCES

Blot S, Ruppé E, Harbarth S, Asehnoune K, Poulakou G, Luyt CE, Rello J, Klompas M, Depuydt P, Eckmann C, Martin-Loeches I, Povoa P, Bouadma L, Timsit JF, Zahar JR. Healthcare-associated infections in adult intensive care unit patients: Changes in epidemiology, diagnosis, prevention and contributions of new technologies. *Intensive Crit Care Nurs.* 2022; 70:103227.

Bougnoux, ME., Brun, S. Zahar, JR. Healthcare-associated fungal outbreaks: New and uncommon species, new molecular tools for investigation and prevention. *Antimicrob Resist Infect Control.* 2018; 7: 45.

Calvo M, Stefani S, Migliorisi G. Bacterial infections in intensive care units: Epidemiological and microbiological aspects. *Antibiotics (Basel).* 2024; 13(3):238.

Chow EJ, Mermel LA. Hospital-acquired respiratory viral infections: Incidence, morbidity, and mortality in pediatric and adult patients. *Open Forum Infect Dis.* 2017; 4(1):ofx006.

Haque M, Sartelli M, McKimm J, Abu Bakar M. Health care-associated infections – An overview. *Infect Drug Resist.* 2018; 11:2321–2333.

Heidari S, Tavakkoli-Moghaddam R, Salimi B, Mehdizadeh-Somarin Z, Hamid M. An integrated approach for evaluating and improving the performance of hospital ICUs based on ergonomic and work-motivational factors. *Comput Biol Med.* 2024; 168:107773.

Isigi SS, Parsa AD, Alasqah I, Mahmud I, Kabir R. Predisposing factors of nosocomial infections in hospitalized patients in the United Kingdom: Systematic review. *JMIR Public Health Surveill.* 2023; 9:e43743.

Khan HA, Baig FK, Mehboob R. Nosocomial infections: Epidemiology, prevention, control and surveillance. *Asian Pac J Trop Biomed.* 2017; 7: 478–482.

Livornese LL Jr., Dias S, Samuel G, Romanowski B, Taylor S, May P, Pitsakis P, Woods G, Kaye D, Levinson ME, Johnson CC. Hospital-acquired infection with vancomycin resistant enterococcus faecium transmitted by electronic thermometers. *Ann Intern Med.* 1992; 117(2): 112–116.

Riddle DJ, Dubberke ER. Clostridium difficile infection in the intensive care unit. *Infect Dis Clin North Am.* 2009; 23(3):727–43.

Shah K, Deshpande M, Shah P. Healthcare-associated fungal infections and emerging pathogens during the COVID-19 pandemic. *Front Fungal Biol.* 2024; 5:1339911.

Stone PW. Economic burden of healthcare-associated infections: An American perspective. *Expert Rev Pharmacoecon Outcomes Res.* 2009; 9(5):417–422.

Thomas S. An engraved surface induces weak adherence and high proliferation of nonadherent cells and microorganisms during culture. *Biotechniques*. 2020; 69(2):113–125.

Thomas S, Bittinger K, Livornese LL Jr. Utilizing the biosimulator to analyze the environmental microbiome within the intensive care units of a hospital. *Biotechniques*. 2025; 77(2):66–75.

Thomas S, Dilbarov N, Kelly J, Mercogliano G, Prendergast GC. Diet effects on colonic health influence the efficacy of Bin1 mAb immunotherapy for ulcerative colitis. *Sci Rep*. 2023; 13(1):11802.

Wan T, Lauring AS, Valesano AL, Fitzsimmons WJ, Bendall EE, Kaye KS, Petrie JG. Investigating epidemiologic and molecular links between patients with community- and hospital-acquired influenza A: 2017–2018 and 2019–2020, Michigan. *Open Forum Infect Dis*. 2023; 10(2):ofad061.

Weber DJ, Rutala WA, Anderson DJ, Chen LF, Sickbert-Bennett EE, Boyce JM. Effectiveness of ultraviolet devices and hydrogen peroxide systems for terminal room decontamination: Focus on clinical trials. *Am J Infect Control*. 2016; 44(5 Suppl):e77–e84.

CHAPTER 4

# Public Transit Microbiome

## A CENTRAL POINTER FOR HUMAN AND ENVIRONMENTAL HEALTH

*Shashi Prabha Kumari, Sunila Hooda, Prerna Diwan, and Rakesh Kumar Gupta*

### HIGHLIGHTS

1. The importance of studying the air microbiome, especially the public transit microbiome, has been highlighted.
2. All the public transit microbiome studies conducted globally till now emphasize its significance in public health.
3. Air serves as a medium of mass dissemination for potentially pathogenic taxa and antibiotic resistance genes (ARGs).
4. The understanding of dynamic interactions between the human microbiome and the aerobiome is crucial for developing future strategies to address public health concerns.

### 4.1 INTRODUCTION

The assemblage of microorganisms, including bacteria, archaea, viruses, fungi, algae, and protozoa, together with their genomes and the surrounding environmental conditions (habitat), i.e., both the biotic and abiotic components, comprises the microbiome (Marchesi and Ravel, 2015). In terms of the environmental microbiome, soil (Fierer, 2017) and water sources (Fresia et al., 2019) act as reservoirs of microbial life and are important contributors to the air microbiome, alongside urban and rural areas, vegetation, animals, anthropogenic sources, and other sources. Aerosolization by sea spray production, aeolian soil erosion, and mechanical disturbances introduces microbe-laden bioaerosols into the atmosphere (Michaud et al., 2018; Núñez et al., 2021). The airborne microorganisms (bioaerosols) are transported globally with air currents and inside airborne water droplets (Després et al., 2012; Griffin et al., 2017). Their dimensions and aerodynamic properties dictate how long they can reside in the atmosphere, ranging from days to weeks (Després et al., 2012). The aerosolized microbes trapped in the dust particles can be transported over large distances and even to different continents during dust storm events, with a higher proportion of microbial cells maintaining their viability (Mazar et al., 2016; Meola et al., 2015). Bioaerosols are effective carriers of pathogens from soil and water sources to humans, animals, and plants, far and wide, facilitating rapid disease dispersion (Lee, 2011).

Nonetheless, the aerobiome remains underexplored, possibly due to the low air biomass (around $10^5$ cells per cubic m of air), requiring stringent sampling procedures and consideration of location and meteorological variables (Dommergue et al., 2019; Prussin et al., 2015). Though several studies have documented the varying air microbiome of built environments, including homes, schools (Jin et al., 2025),



DOI: 10.1201/9781003637950-4

hospitals (Bobrovska et al., 2025), and mass transit networks (Danko et al., 2021; Grydaki et al., 2021; Leung et al., 2021; Sharma et al., 2024), as well as outdoor air (Chen et al., 2024), our understanding of the aerobiome is still in its infancy. Employing shotgun metagenomics will enable greater taxonomic resolution as well as the functional screening of all the genes present in the air sample, including antibiotic resistance determinants, transposons, and other biosynthetically relevant genes. Furthermore, there is significant evidence regarding the direct or indirect interactions between the air microbiome and the human microbiome, including the skin, oral, respiratory, and gastrointestinal microbiome, with both positive and negative effects. The airborne microorganisms are potentially involved in priming the immune system by triggering immunological responses at the respiratory surfaces, conferring protection against asthma and allergy (Ege et al., 2011; O'Connor et al., 2018; Tischer et al., 2016). Occupational work exposure alters the skin, oral, and respiratory microbiomes of the workers, either transiently (short-duration exposure) or possibly permanently in cases of long-term exposure (Kraemer et al., 2018; Lai et al., 2017; Shukla et al., 2017).

In the wake of the coronavirus pandemic, recent years have seen a rise in investigations of the airborne viruses, particularly SARS coronaviruses, their detection, persistence in the air, dissemination while maintaining infectivity, and their mitigation (Mortazavi et al., 2023; Z. Wang et al., 2025). Studying the air microbiome and its associated factors will enhance our understanding of airborne microbial dynamics, enabling us to devise strategies and prevent future viral infection outbreaks. Public transit systems around the world transport millions of people daily, with their most crowded stations handling hundreds of thousands of commuters every day. Any diminutive dissemination or release of antibiotic-resistant bacteria in the air would not take much time to spread among the larger community, posing uncontrollable severe epidemics. Therefore, determination of the microbiological quality of air at various indoor and outdoor overcrowded subway stations and the prevalence of ARGs in the microbial community suspended in the subway station environment are necessary to formulate strategies to contain any potential threat arising from their spread in the community.

The current review focuses on the constituents of the air microbiome, and its variation in urban and rural areas, especially emphasizing the public transit microbiome and resistome as well as how they modulate the human microbiome. We aim to outline the recent progress of research on the air microbiome in the context of mass transit networks, as well as the existing gaps and pitfalls associated with such studies.

## 4.2 COMPOSITION OF AIR MICROBIOME

The consortia of microbes (bacteria, fungi, and viruses) present in the medium of air constitute airborne microorganisms. These microorganisms, including their fragments as well as other biological materials such as nucleic acids, toxins, allergens, pollen, and plant/animal debris, in the size range of 50 nm to 10 μm constitute bioaerosols, with potential health and environmental effects (Kim et al., 2018; Lee, 2011; Moelling, 2021). Air microbiome profiling studies have been conducted for a long time, even before the advent of metagenomics, to determine the composition of suspended microorganisms and associated effects in a particular region (Bowers et al., 2009; DeLeon-Rodriguez et al., 2013; Leung et al., 2014; Pastuszka et al., 2000; Zhu et al., 2003). Culture-dependent studies have determined that the air contains around $10^2$–$10^6$ bacterial colony-forming units (CFU) and $10^2$–$10^3$ fungal spores per cubic meter (Bowers et al., 2009; Jo & Seo, 2005; Lighthart, 2000; Pastuszka et al., 2000; Zhu et al., 2003). The small sizes of bacterial cells aid in their suspension in the atmosphere for comparatively longer durations, thus enhancing their collection and detection efficiency (Burrows et al., 2009). Additionally, more than 50% of the culturable bacterial bioaerosols were less than 4.7 μm in size (respirable range) (Brągoszewska et al., 2017). Viable bacterial cells make up to 20% of 0.25–1.00 μm-sized air particles, with lower fungal cell concentrations (less by an order of magnitude), as revealed by microscopy and quantitative PCR (based on average small subunit ribosomal ribonucleic acid (SSU rRNA) copy number) (DeLeon-Rodriguez et al., 2013). Another quantification study involving fluorescent dyeing (SYBR Gold) of a 0.2 μm membrane filter detected $10^5$ bacterium-like particles per cubic meter of air within a size range of 0.5–5 μm (Prussin et al., 2015). Airborne viral particles remain under-studied compared to their bacterial and fungal counterparts, owing to their small sizes and other technical challenges

(Prussin et al., 2015). It has been estimated that a cubic meter of indoor and outdoor air can contain up to $10^6$ and $10^7$ virus-like particles (0.02 and 0.5 μm in size), respectively (Bowers et al., 2009; Griffin et al., 2001; Prussin et al., 2015; Whon et al., 2012). Furthermore, the virus–bacteria ratio (VBR) ranged from 0.7 to 1.4 depending on indoor and outdoor environments (Prussin et al., 2015), and 1.6–3.6 over different land-use types (residential, forest, and industrial), in the near-surface atmosphere (Whon et al., 2012).

The air microbiome is subject to seasonal and geographical variability, influenced by temperature, relative humidity, wind patterns, rainfall, $CO_2$ levels, and other anthropogenic factors (Figure 4.1) (Lee et al., 2017; Núñez et al., 2021). Certain studies reported higher microbial concentration and diversity in the summer and spring seasons compared to the winters (Bowers et al., 2013; Brągoszewska et al., 2017; Genitsaris et al., 2017; Ma et al., 2023; Núñez et al., 2021; Pollegioni et al., 2023), while one study reported the contrary (Lee et al., 2017). The contrasting observations were attributed to the different potential sources of bioaerosol generation as well as meteorological parameters unique to a particular region, such as wind patterns that play a substantial role in the transport of such aerosols (Bowers et al., 2013; Genitsaris et al., 2017; Lee et al., 2017; Núñez et al., 2021). The seasonal variation could reflect changes in the relative significance of contributions from various environmental sources to the air microbiome, depending on the season, such as leaf sources (less in winter), agricultural activities, etc. (Bowers et al., 2013). However, bacterial concentrations in the near-surface atmosphere show minimal short-term variation with respect to different meteorological variables (Bowers et al., 2011a). This is also supported by the work of Matthias-Maser et al. (2000), which demonstrated that the abundances of smaller bioaerosol particles (>2 μm, bacteria) do not change significantly with atmospheric conditions compared to the larger bioaerosol

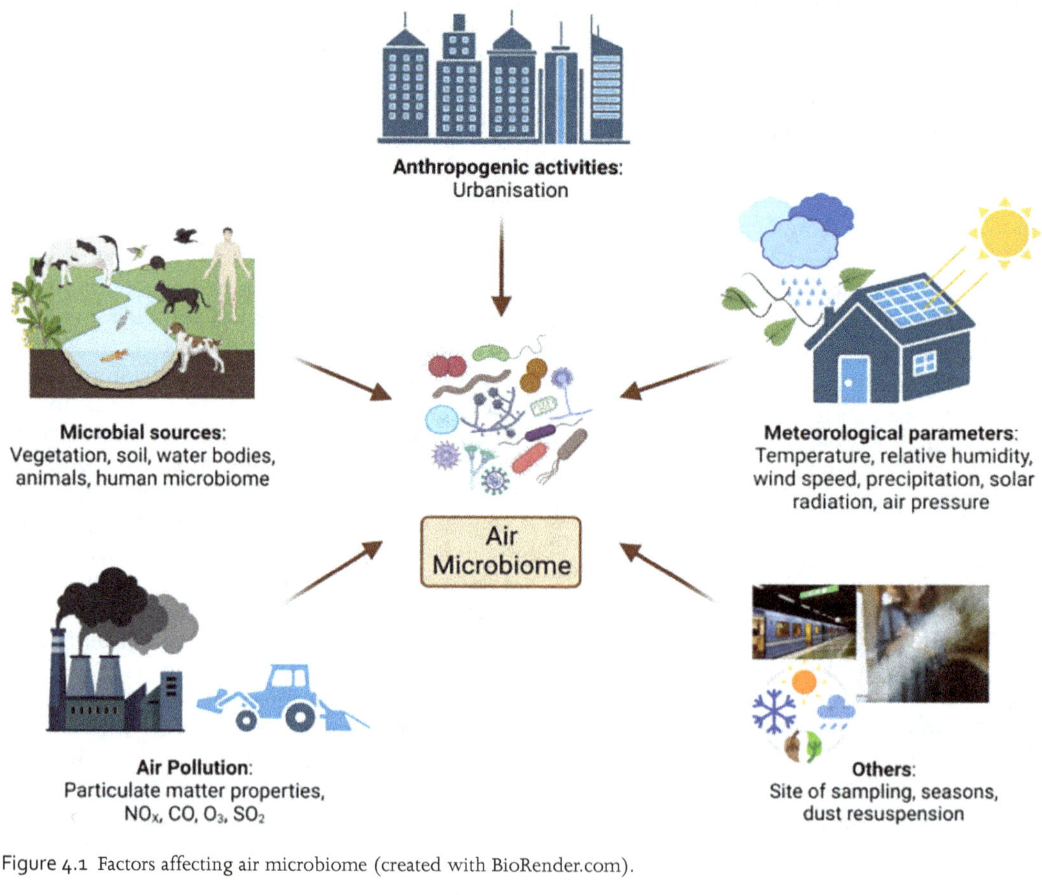


Figure 4.1 Factors affecting air microbiome (created with BioRender.com).

particles, including pollen, fungal spores, and plant debris (Matthias-Maser et al., 2000).

The aerobiome also varies spatially and temporally substantially, following a diel cycle with more day-and-night variations than inter-day or inter-month variations, a possible result of an interplay of various environmental factors (temperature, moisture content, dry and wet deposition, gaseous composition) (Gusareva et al., 2020, 2019). However, the said factors also contribute to the long-term stability of the air microbiome, as evident by the characterization of 795 core microbial taxa by Gusareva et al. (2019) in their 13-month study of the tropical air ecosystem of Singapore (Gusareva et al., 2019).

Earlier culture-based studies using impactors for sampling identified numerous viable airborne bacterial species, including *Microbacterium saperdae*, *Bacillus subtilis*, *Paenibacillus macerans*, *Staphylococcus epidermidis*, *Escherichia coli*, *Arthrobacter aurescens*, *Acinetobacter radioresistens*, *Xanthomonas maltophilia*, *Nocardia* spp., and *Ochrobactrum anthropi* (Gram-positive bacteria predominant), and fungal species such as *Candida*, *Cladosporium herbarum*, *Cladosporium cladosporioides*, *Rhodotorula*, *Myrothecium*, *Alternaria alternata*, *Absidia glauca*, and *Penicillium aurantiogriseum* in the indoor and outdoor air (up to $10^3$ CFU/$m^3$ for both bacteria and fungi) (Pastuszka et al., 2000; Zhu et al., 2003). In an urban area, bioaerosols collected using impingers followed by amplicon sequencing (16S rRNA gene) revealed higher abundances of Gram-negative bacteria than Gram-positive ones, including *Flavobacterium*, *Flexibacter*, *Pseudomonas*, *Rhizobium*, *Methylobacterium*, *Sphingomonas*, *Hymenobacter*, *Arthrobacter*, and Enterobacteriaceae ; plant cells: *Pinus*, *Alnus*, *Abies*, *Equisetum*, *and Poa*; fungi: 97% Ascomycota (Hypocreales order) and 3% Basidiomycota (Hymenomycetes), with considerable temporal variability (Fierer et al., 2008). Another study collected air particles on cellulose nitrate filters and subjected them to pyrosequencing. Bacterial (β- and γ-Proteobacteria) and fungal (Dothideales and Eurotiomycetes) species were in equivalent abundances (Bowers et al., 2009).

The metagenomic analysis of the particles captured by electret filters of high-volume air samplers exemplified the seasonal variation and elucidated the summer (6–24°C) West Siberian air microbiome composition as (i) Ascomycota and Basidiomycota fungi: wood-rotting saprophytes (*Dichomitus squalens*, *Trametes pubescens*, *Phlebiopsis gigantea*, *Phlebia centrifuga*, *Stereum hirsutum*, etc.), phytopathogens (*Botrytis cinerea*, *A. alternata*, *Rhizoctonia solani*, *Puccinia striiformis*, etc.), and common mushroom fruiting body-forming fungi (*Pleurotus ostreatus*, *Agaricus bisporus*, etc.); (ii) air-, soil-, and water-associated Actinobacteria and Proteobacteria: *Knoellia aerolata*, *Skermanella aerolata*, *Altererythrobacter troitsensis*, *Sphingomonas astaxanthinifaciens*, *Rubellimicrobium mesophilum*, etc.; (iii) typical local plant taxa: *Helianthus annuus*, *Aegilops tauschii*, etc. The winter (−21 to −6°C) air microbiome was revealed to be primarily comprising cold- or stress-tolerant bacterial and fungal species: *Wallemia mellicola*, *Rachicladosporium antarcticum*, *Arthrobacter* sp. L77, *Kocuria polaris*, *Hymenobacter roseosalivarius*, *Aspergillus*, *Saccharopolyspora rectivirgula*, etc., as well as biomass of cereal plants (Poaceae) (Gusareva et al., 2020).

## 4.3 MICROBIOME IN RURAL AND URBAN AREAS

The urban and rural air microbiomes arise from distinct contributing sources, with urban air largely shaped by dense human occupancy, traffic emissions, built environments, and industrial activities, whereas the rural air microbiome is predominantly influenced by agricultural practices, livestock farming, composting sites, soil dust resuspension, and the close human–animal interface, leading to differences in microbial composition, particle size distribution, and associated health risks (Yang et al., 2025; Yu et al., 2025). Urbanization tends to promote disequilibrium in the local biotic environment, favoring the richness of certain species, ultimately leading to homogenization of plant and animal taxa as well as land-use types across urban areas (McKinney, 2006). However, studies have not yet found direct effects of urbanization on airborne microbial communities, but they were revealed to be homogeneous and not spatially variable across urbanized areas compared to rural areas (rural air comprises comparatively more different types of microbes) (Barberán et al., 2015). Furthermore, there are subtle differences in the structure of the microbial community based on the land-use type and vegetation (a significant contributor to aerosolization apart from skin). The air of forest and agricultural areas was found to be more soil-influenced, whereas that of semi-urban areas resembled more the taxa associated with the leaf-surface habitat, with potential influences on dispersal patterns of the microbes in the atmosphere (Bowers et al., 2011a, 2011b; Spring et al., 2021). The local vegetation serves as an important driver of outdoor air and, consequently, the indoor air microbiome, with the taxonomic composition showing a partial dependence on the vegetation type (phyllosphere microorganisms are plant species-specific) (Bowers et al., 2011a; Lymperopoulou et al., 2016; Ruiz-Gil et al., 2020).

Air pollution-led elevation of suspended particulate matter concentrations, conjoint with low-stress factor levels (UV radiation, desiccation), sustains diverse airborne microorganisms, including dead cells and extracellular nucleic acids, and supports transient microbial interactions (Moelling and Broecker, 2020; Qin et al., 2020). Mobile genetic elements harboring the ARGs encourage horizontal gene transfer within bacterial communities, increasing the overall airborne ARG burden (Qin et al., 2020). The $PM_{10}$ microbiome in the city of Rome exhibited seasonally diverse taxa including enrichment of *Nocardioides*, *Blastococcus*, *Kocuria*, *Cladosporium*, *Alternaria*, and *Mycosphaerella* (resistant to summer environmental conditions) during summers, as well as, *Acetomicrobium*, *Petrotoga*, *Penicillium*, *Trametes*, and *Aspergillus* during winters (Pollegioni et al., 2023). $PM_{2.5}$ and $PM_{10}$ hosted similar taxonomic and functional profiles, with pathogenic taxa such as *Pseudomonas aeruginosa* and *Stenotrophomonas maltophilia* exhibiting increased relative abundances at higher AQI levels in the megacity of Beijing (Qin et al., 2020). However, size-fractionated particulate matter sampling and subsequent air microbiome elucidation may only partially represent the aerobiome, as different size fractions are associated with different microbial consortia (Bertolini et al., 2013; Franzetti et al., 2011). The microbiome of Milan's (Italy) atmosphere (urban area) was revealed to be largely temporally variable, with the community profile dependent on the seasonally predominant sources, following the analysis of total suspended PM. Rhizobiales, Sphingomonadales, and chloroplast sequences prevailed during the summer seasons, reflecting plant sources, while Pseudomonadales, Burkholderiales, and Actinobacteriales of soil origin predominated the colder seasons, similar to the findings by Bowers et al. (2013) in their Greeley urban air study (Bertolini et al., 2013; Bowers et al., 2013).

The urban air near the surface is majorly constituted by a core set of microorganisms globally, mainly originating from the soil, such as bacteria: *Sphingomonas*, *Clostridium*, *Paracoccus*, *Nocardioides*, *Kocuria*, and *Corynebacterium*, with relative abundance alterations depending upon the sampling protocol and geographical parameters (Núñez et al., 2021). Characterization of the air microbiome of various large cities revealed Proteobacteria to be the most abundant bacterial phylum, followed by Actinobacteria, Firmicutes, Bacteroidetes, and Cyanobacteria (Bertolini et al., 2013; Cao et al., 2014; Gou et al., 2016; Kumari et al., 2024; Lee et al., 2017; Pollegioni et al., 2023; Whon et al., 2012). The microorganisms present in the air in the respirable range are primarily non-pathogenic; nonetheless, bacterium *Streptococcus pneumoniae*, fungus *Aspergillus fumigatus*, and virus human adenovirus C, responsible for allergies as well as respiratory diseases, could also be detected. However, the authors concluded that these pollutant-associated microorganisms pose a negligible risk to public health and that increased vegetation would prevent the aerosolization of harmful terrestrial and facal-associated microbes (Cao et al., 2014). On the contrary, recent findings highlight the potential risks posed by increasing concentrations of opportunistic and pathogenic airborne microorganisms such as *Acinetobacter lwoffii*, *Alternaria infectoria*, *Ceriporia lacerata*, *Clostridium perfringens*, *Staphylococcus* spp., *Pseudomonas* spp., and *A. fumigatus*, consistent with the worsening air quality, impacting human health (Jiang et al., 2022; Mkilima, 2025; Pollegioni et al., 2023).

The air mycobiome of urban areas in China, Greece, and Bulgaria consisted of the phyla Ascomycota and Basidiomycota, including common predominant genera such as *Aspergillus*, *Alternaria*, *Malassezia*, *Cladosporium*, *Penicillium*, and *Epicoccum*. Other detected genera included *Nigrospora*, *Meyerozyma*, *Cryptococcus*, *Pleospora*, and *Penidiella* in China, *Aureobasidium*, *Candida*, *Fusarium*, and *Filobasidium* in Greece, and *Talaromyces*, *Bjerkandera*, *Trametes*, *Stereum*, *Coniothyrium*, and *Herpotrichia* in Bulgaria. These fungal genera have potential implications for both human and plant health (Gou et al., 2016; Ivanova et al., 2025; Metaxatos et al., 2025). The urban air virome characterized metagenomically in Seoul, South Korea, was found to follow the seasonal fluctuation trend (temperature- and humidity-influenced), similar to its bacterial and fungal counterparts. More than 90% of the identified sequences corresponded to ssDNA viruses comprising the following families: Geminiviridae, Circoviridae, Microviridae, Nanoviridae, and Inoviridae, followed by dsDNA viruses and an ssDNA satellite. Among the geminivirus-related viruses (>90% abundance), 71% of the viral sequences were designated as *Sclerotinia sclerotiorum* hypovirulence-associated DNA virus 1 (SsHADV-1)-related viruses, highlighting the contribution of plant sources to the aerial viral assemblages. More than 50% of the sequenced metagenome remained uncharacterized, suggesting the occurrence of novel species in the air which demands extensive studies in this context (Whon et al., 2012).

The facal bacteria in the air, which could originate from rural animal farms or sewage water, have the potential to be transported over long distances while maintaining viability (Cao et al., 2014; Moelling and Broecker, 2020). Investigations of air microbiota of agricultural areas in Vaud, Switzerland (wheat harvesting region) revealed the occurrence of mycotoxigenic and allergenic taxa such as *Cladosporium*, *Alternaria*, *Epicoccum*, and *Fusarium* originating from grain dust, posing a health hazard for grain workers (Pellissier et al., 2016). Air samples from a rural area (Greeley) contained a higher bacterial richness (72%), compared to an urban area (Denver: 67%), with notable differences in the relative abundances of members of Bacteroidetes, Actinobacteria, Proteobacteria, and Firmicutes, explainable by the varying land-use type (Bowers et al., 2013). Another recent study from a rural area in China revealed that total bioaerosol concentrations were significantly lower than in urban areas, but the rural air hosted a higher relative abundance of potentially pathogenic bacteria, with distinct indoor (*Vibrio*) and outdoor (*Prevotella*, *Faecalibacterium*, *Bacteroides*) bacterial communities (Yang et al., 2025). Furthermore, the surfaces of walled-in built environments in urban areas were determined to be derived more from the human microbiome than the open or rural homes equipped with better airflow (Ruiz-Calderon et al., 2016).

## 4.4 PUBLIC TRANSIT MICROBIOME

The subway or metro systems are used by millions of people daily as a popular form of public transportation in urban and suburban areas. The large-scale commuter traffic in an enclosed space underlines the importance of assessing the microbiological air quality of the mass transit systems. They represent oligotrophic yet highly dynamic built environments that support complex airborne microbial communities, including bacteria, fungi, and viruses, shaped by human occupancy, environmental conditions, and infrastructure design (Figure 4.2). Knowledge of microbial load, diversity, and the temporal and spatial dynamics of bioaerosols is essential for tracking the dispersal of pathogens in crowded and heavily populated public places. The Metagenomics and Metadesign of Subways and Urban Biomes (MetaSUB) Consortium is an international initiative to carry out large-scale, multi-faceted urban air microbiome studies concerning the identification and characterization of novel species, antibiotic resistance, and virulence genes along with novel biosynthetic gene clusters. It aims to standardize air sampling and subsequent data analysis pipelines to facilitate comparison studies, which are otherwise impossible owing to methodological variations. The associated members are working toward refining community health, highlighting the antimicrobial resistance (AMR) burden by building a global air microbiome dataset, easily accessible to public health officials (MetaSUB International Consortium, 2016).

Danko et al. (2021) performed an international mass transit microbiome (urban metagenome) study covering 60 cities over three years (4728 surface samples from benches, kiosks, and railings) employing a shotgun metagenomic sequencing approach. They defined the 4424 taxa discovered in the overall sample set under the term 'urban pan-microbiome,' consisting of 31 core (positive occurrence in >97% samples) and 1145 sub-core (>70% samples) microbial species, which mainly were bacterial (Proteobacteria, Actinobacteria, Firmicutes, Bacteroidetes – varied relative abundances) except *Saccharomyces cerevisiae*. A resident of human skin, *Cutibacterium acnes*, showed the highest relative abundance across all samples. The microbial profile of surfaces such as doorknobs, railings, etc., resembled the human skin more by virtue of more contact than windows and floor. The taxonomic diversity decreased significantly with respect to latitude from the equator, consistent with the effects of temperature (Gohli et al., 2019). A considerable proportion of the obtained metagenomic DNA corresponded to unclassified microbial taxa, with a probability of their uniqueness to the urban environment, presenting a gap that warrants future exploratory studies. Analysis of metagenome-assembled genomes revealed additional unclassified 1302 bacterial, 2 archaeal, and 10,928 viral genomes along with 838,532 CRISPR arrays (3245 annotated). Microbial diversity was determined to be contingent on climate and geography, evidenced by variation with city and year of sampling (Danko et al., 2021).

Besides the consortium, several culture-dependent and independent (employing NGS) studies have attempted to comprehensively characterize the microbial community dominating the environments of public transit systems, especially subways (principal mode of transport) around the world (Tables 4.1 and 4.2, Figure 4.3). Integrating insights from global microbiome profiling reveals that despite low nutrient availability and continuous air movement, subway systems harbor unexpectedly high microbial diversity with pronounced spatial and seasonal structuring.

The taxonomic diversity of the air microbiome of the Hong Kong Mass Transit Railway network was characterized using an impactor for air sampling from indoor (underground) as well as outdoor (above-ground) subway stations, followed by 16S rRNA gene sequencing (V4 region) and statistical analyses. The microbial community comprises bacterial phyla: Proteobacteria (23.4%), Actinobacteria (15.6%), Firmicutes (6.3%), and Deinococcus–Thermus (1.6%), and archaeal phyla: Euryarchaeota and Crenarchaeota (<0.1%). The detected bacterial genera primarily constituted skin/oral commensals: *Micrococcus*, *Enhydrobacter*, *Propionibacterium*, *Staphylococcus*, and *Corynebacterium* genera, in addition to soil-borne: *Sphingobium* and *Acinetobacter* and water-borne: *Blastomonas* genera. The microbial profile of outdoor stations differed from that of indoor ones owing to architecture (mechanical ventilation inhibiting much inflow of outer air). The microbial community variation was also found to be subject to the day and time of sampling (peak/ non-peak hours) as well as environmental factors of temperature and relative humidity with a more pronounced effect on outdoor lines (Leung et al., 2014).

A large-scale study using dry air samplers (electret filters) spanned public transit systems of six cities, including New York City, Stockholm, London, Oslo, Hong Kong, and Denver, covering multiple continents and encompassed profiling of microbiome as well as resistome. They identified 468 taxa up to species level, constituting 96.5% bacteria, 3.21% virus, 0.19% fungi, and 0.043% archaea, in terms of relative abundance. Overall, 17 species were detected in more than 75% of the samples (core microbiome), notably comprising host and environmentally originated bacteria: *C. acnes*, *Micrococcus luteus*, *Propionibacterium granulosum*, *Staphylococcus hominis*, *Enhydrobacter aerosaccus*, and *Kocuria rhizophila*. The potential reservoirs of the public transit air microbiome and resistome comprise human skin, water, soil, and various surfaces in the transit systems, thus evidencing substantial geographical variability (Leung et al., 2021). Biosynthetic gene clusters along with genes involved in antimicrobial activity, such as polyketide synthases, non-ribosomal peptide synthetases, and bacteriocins, were also detected in the air samples (Leung et al., 2021). The subway air virome at these locations exhibited high diversity and strong biogeographical structuring, with thousands of vOTUs identified, many of which were taxonomically unclassified, indicating substantial viral novelty. The virome was dominated by virulent viruses, with frequent detection of human-associated papillomaviruses and widespread Caulimoviridae, alongside viral genes involved in replication, integration, and antibiotic resistance, underscoring the adaptive strategies of airborne viruses in oligotrophic transit environments (Lei et al., 2025). The Shanghai subway air virome constituted a very small fraction of the total microbiome (~0.3%) and was dominated primarily by bacteriophages, with marked seasonal shifts from Lenarviricota and Artverviricota in spring to a strong predominance of Uroviricota in summer, while only a limited number of viral species were shared between seasons, and no highly pathogenic human viruses were detected (Liu et al., 2024).

Characterization of the surface microbiome of underground metro stations (turnstiles) and trains (handrails) in Mexico City (16S rRNA V3–V4 gene sequencing) revealed 1058 bacterial genera: predominantly 15% *Cutibacterium*, 13% *Corynebacterium*, 9% *Streptococcus*, and 5% *Staphylococcus* likely to originate from dust, skin, and saliva. The turnstile samples hosted comparatively higher bacterial diversity by reason of more outside exposure and less direct human skin contact (more of clothes), thus favoring dust- and soil-borne taxa. Interestingly, the metagenomically derived OTUs also corresponded to unicellular algae: Chlorophyta and Stramenopiles, as well as plants including *Zea*, *Phaseolus vulgaris*, *Cucumis sativus*, *Carica papaya*, *Lactuca sativa*, *H. annuus*, *Pisum sativum*, *Coffea arabica*, *Nicotiana tabacum*, *Eucalyptus globulus*, *Abies*, *Populus alba*, *Pinus*, *Juniperus virginiana*, and *Cycas taitungensis*. The possible sources of such cells could be pollen, plant debris, and plant-derived products (Hernández et al., 2020).

Gohli et al. (2019) performed subway microbiome studies in Oslo, Norway (16 stations), comparing air and surface microbial composition with respect to seasonal variation. They followed the methodology of air sampling using electret filters and surface swabbing of benches, kiosks, and railways, followed by 16S amplicon sequencing and analysis. Actinobacteria and Proteobacteria were predominant in the air of Norway subway stations, similar to other cities (Danko et al., 2021; Gohli et al., 2019; Leung et al., 2014). The abundances of the top 20 phyla were similar in the microbial communities of the air and surface, potentially evidencing deposition and aerosolization/resuspension of air particles (Gohli et al., 2019). They reported the taxonomic diversity to be negatively proportional to humidity to a small extent while positively proportional to temperature, contrary to the observations by Leung et al. (2014). The attributable reasons are differential survival rates of bacteria as well as consideration of a broader range of temperature and humidity in their study (Gohli et al., 2019; Leung et al., 2014). The alpha diversity of air samples also showed a more positive correlation to temperature than surface samples (also higher during summer and spring), exemplifying the effects of temperature-induced air circulation patterns and microbial viability (Cáliz et al., 2018; Gohli et al., 2019).

Certain unique bacterial genera, such as *Allorhizobium*, *Amaricoccus*, *Barrientosiimonas*, *Acidibacter*, *Epilithonimonas*, and *Accumulibacter*, were detected in the air of the Sukhumvit subway station in Bangkok, Thailand, apart from the common genera found globally in subways that are influenced by human presence and the environment. The authors offered insights into the potential usage of metabarcoding to infer the functional profiles of the airborne microorganisms (reportedly varying geographically) that may enhance the understanding of the biological effects of air quality on the microbial cells suspended in the air (Siriarchawatana et al., 2023). At the same subway station, the core mycobiome of bioaerosols was found to be composed of *Cladosporium*, *Malassezia*, *Yarrowia*, *Cutaneotrichosporon*, *Aspergillus*, and *Leptobacillium* genera, with Ascomycota and Basidiomycota being identified as the dominant phyla. Significant temporal variations were observed in the predominant fungal genera depending upon the environmental variables of humidity, temperature, and particulate matter. The airborne fungal community, including the allergenic and opportunistic genera, was revealed to exist in co-occurrence networks. They are involved in complex and dynamic positive and negative interactions that are affected by relative humidity and $PM_{<10}$, with implications for biotechnological applications and human health (Harnpicharnchai et al., 2023). Furthermore, the Beijing subway air harbored a highly diverse and seasonally dynamic mycobiome dominated by Ascomycota and Basidiomycota, with persistent detection of Mucoromycota across all months, including peaks in summer and winter. Core fungal taxa such as *Alternaria*, *Aspergillus*, *Cladosporium*, *Candida*, *Meyerozyma*, and *Rhinocladiella* were abundant, with higher diversity and pathogenic potential in high-traffic stations and during periods of elevated $PM_{2.5}$, highlighting their potential public health relevance (Zhou et al., 2025).

The interplay of the surface type (materials promoting adherence and growth), indoor/outdoor exposure, and human body interactions shape the microbial community structure and diversity of the subway air, irrespective of different train lines and stations. This is evident by grips/horizontal poles and seat surfaces hosting higher abundances of oral (*Rothia*, *Veillonella*), skin (*Propionibacterium acnes*), and vaginal commensals (clothing mediated – *Corynebacterium*, *Gardnerella vaginalis*, *Staphylococcus*, *Lactobacillus*), respectively. Moreover, the vinyl seats corresponded to lower levels of microbial species, compared to the polyester ones. *Propionibacterium* bacteriophages, Merkel cell polyomavirus, *Malassezia globosa*, and environmental origin, such as Alphaproteobacteria (*Sphingomonas* and *Methylobacterium*), Rhodobacteraceae, and Methylocystaceae, were also prevalent in the Boston subway air. The dataset was not found to contain significant antibiotic resistance and virulence genes, declaring the transit system safe from the public health point of view (Hsu et al., 2016).

The bioaerosols of New York City subway platforms hosted a similar bacterial profile, with Staphylococcaceae being the most abundant family, in addition to environmentally sourced ones (Micrococcaceae and Moraxellaceae). Eukaryotic DNA sequences pertaining to fungi, plants, insects, rodents, and protists were detected. The fungal sequences were primarily identified as Agaricomycetes (mushrooms),

Figure 4.2 Overview of public transit microbiome: sources, influencing factors, and health effects (created with BioRender.com).

Eurotiales (mildews), Capnodiales (saprophytes), Saccharomycetales (yeasts), and wood-rotting fungi (wooden track platforms) (Robertson et al., 2013). The internal air microbiome profile of the subway showed many similarities to that of outside air, contrary to the study of the Hong Kong metro network (Leung et al., 2014), as a result of the absence of local air conditioning and the train-pumping process facilitating proper air mixing at regular intervals (Robertson et al., 2013).

A culture-based study of air particles collected using electret filters followed by 16S rRNA identification of the isolates in an underground subway station in Oslo, Norway, revealed higher bacterial concentration in the air during the day than at night, dominated by *Micrococcus*, *Bacillus*, and *Staphylococcus* species, evidencing anthropogenic origin (Dybwad et al., 2012). In addition, *Candida*, *Penicillium*, *Cladosporium*, *Aspergillus*, and *Alternaria* dominated the airborne fungal population in tunnel and surface metro stations of Cairo, Egypt, and underground subway stations of Mexico City, aside from Actinomycetes, with potential health implications (Awad, 2002; Hernández-Castillo et al., 2014). An earlier cultivation-dependent study of air sampled through impactors in a crowded underground concourse in Tokyo also noted a similar higher prevalence of Gram-positive bacteria, especially *S. epidermidis* and *Micrococcus* spp., reflecting influences of numerous pedestrians and relative humidity (Seino et al., 2005). Bacterial numbers increased with the depth of the underground stations, consequence of higher temperatures supporting metabolic activities and improper ventilation promoting microbial confinement (Hernández-Castillo et al., 2014).

Collectively, across public transit systems, genera such as *Staphylococcus*, *Acinetobacter*, *Corynebacterium*, *Cutibacterium*, *Pseudomonas*, *Micrococcus*, *Kocuria*, *Dietzia*, *Enhydrobacter*, *Brevundimonas*, *Massilia*, *Propionibacterium*, and *Streptococcus* are consistently reported in air and surface microbiome studies (Table 4.1, Figure 4.3), reflecting their broad ecological adaptability and frequent introduction through human occupancy.

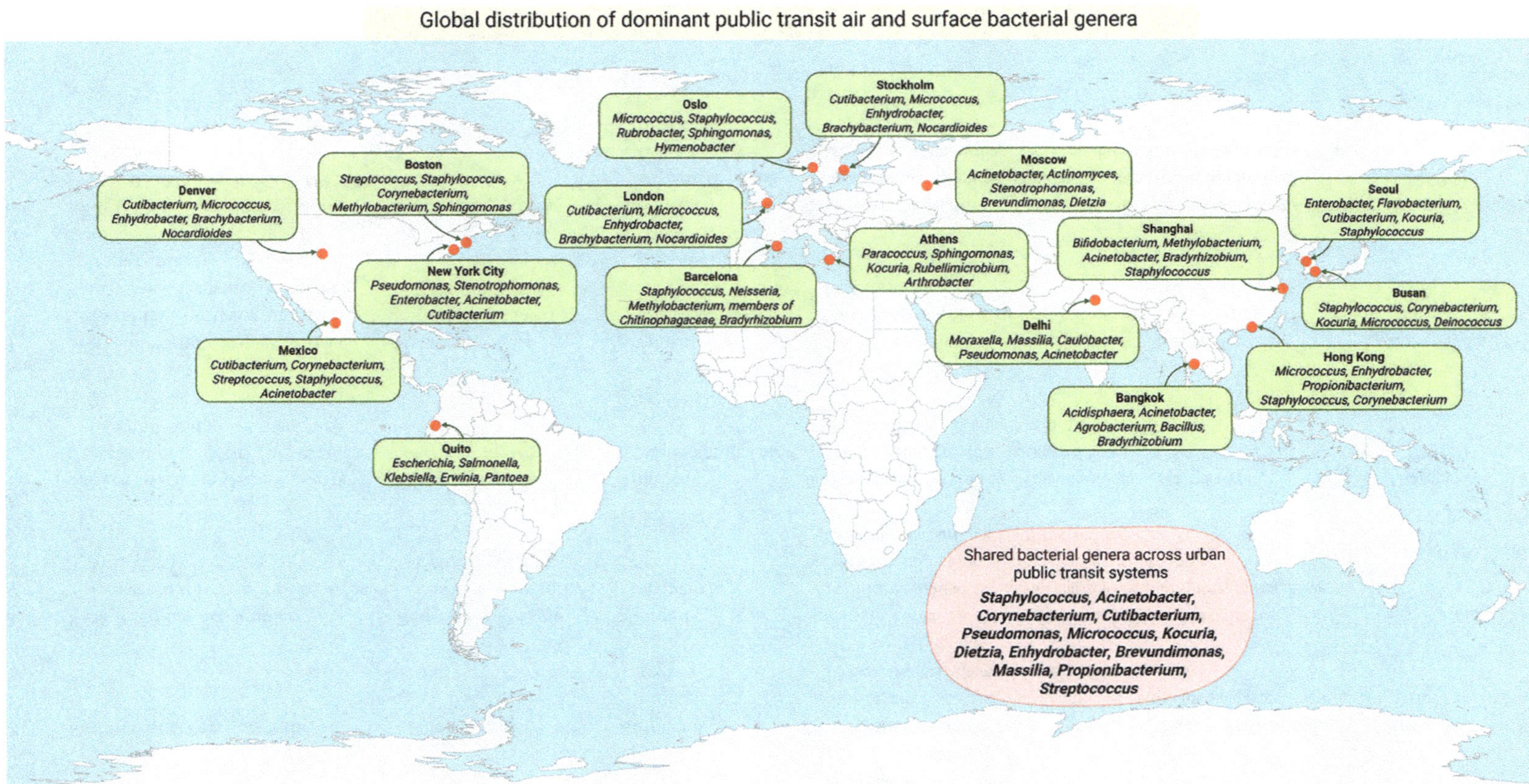


Figure 4.3 Global map showing major cities where public transit microbiome has been investigated and their reported dominant bacterial genera, along with the common bacterial genera consistently reported across studies. City locations are marked geographically, while adjacent labels summarize the predominantly detected genera, highlighting the widespread recurrence of human-associated and environmental taxa in public transit systems worldwide.

TABLE 4.1

*Studies characterizing public transit microbiome based on metagenomic sequencing approach*

| S. No. | Method of air sampling/device used | Geographical location of the public transit system (subway stations) | Method of study (diversity elucidation) | Microbial diversity | References |
|---|---|---|---|---|---|
| 1 | BioSampler (SKC Inc., PA, USA) | Delhi, India (railway stations) | Shotgun sequencing | *Moraxella*, *Massilia*, *Caulobacter*, *Pseudomonas*, *Acinetobacter*, *Sphingomonas*, *Stutzerimonas*, *Methylobacterium*, *Novosphingobium*, *Barrientosiimonas*, *Stenotrophomonas* | Kumari et al. (2025) |
| 2 | Surface swabbing (fan baffles) | Beijing, China | Amplicon sequencing: ITS2 region | Mycobiome: *Fusarium*, *Alternaria*, *Aspergillus*, *Chaetomium*, *Cladosporium*, *Meyerozyma*, *Nigrospora*, *Amphinema*, *Botryotrichum*, *Epicoccum*, *Parengyodontium*, *Phoma*, *Rhinocladiella* | Zhou et al. (2025) |
| 3 | SASS 3100 dry air samplers (Research International, Monroe, WA, USA) | Denver, Hong Kong, London, New York, Oslo, Stockholm | Shotgun sequencing | Virome: Caudoviricetes, Revtraviricetes, Papovaviricetes; Papillomaviridae (Genera *Alphapapillomavirus*, *Betapapillomavirus*, *Dyodeltapapillomavirus*, *Dyothetapapillomavirus*, and *Gammapapillomavirus*), Caulimoviridae | Lei et al. (2025) |
| 4 | Surface swabbing (hanging grip, vertical pole, doorknob, escalator handrail, and ticket machine) | Shanghai, China | Amplicon sequencing: 16S rRNA (V3–V4 regions) | *Bifidobacterium*, *Acinetobacter*, *Propionibacterium*, *Staphylococcus*, *Enhydrobacter*, *Prevotella*, *Pelomonas*, *Bacteroides*, *Lactobacillus*, *Faecalibacterium*, *Blautia*, *Escherichia* | Zhang et al. (2025) |
| 5 | Wet-wall vortex sampler; surface swabbing (escalator and seat handrails, turnstiles, restroom door handles) | Shanghai, China (railway station) | Amplicon sequencing: 16S rRNA (V3–V4 regions) | *Methylobacterium*, *Bradyrhizobium*, *Pseudomonas*, *Blautia*, *Prevotella*, *Corynebacterium*, *Streptococcus*, *Cutibacterium*, *Acinetobacter*, *Faecalibacterium* | Yan et al. (2025) |
| 6 | Mini volume air sampler (Model PAS201, Air Metrics Co.) (PM10 samples) | Seoul, South Korea | Amplicon sequencing: 16S rRNA (V3–V4 regions) | *Enterobacter*, *Flavobacterium*, *Tissierella*, *Flaviflexus*, *Clostridium*, *Cytobacillus*, *Agrobacterium*, *Urmitella*, *Pseudomonas*, *Sporosarcina*, *Paenibacillus*, *Leucobacter*, *Lysinibacillus*, *Schnuerera*, *Brevundimonas*, *Bacillus*, *Acinetobacter* | Sharma et al. (2024) |

| | | | | | |
|---|---|---|---|---|---|
| 7 | Wet-wall cyclone bioaerosol sampler (BaST-Pump200) | Shanghai, China | Shotgun sequencing | Bacteria: *Staphylococcus epidermidis*, *Shewanella chilikensis*, *Staphylococcus hominis*, *Agrobacterium* sp. DE0009, *Staphylococcus pettenkoferi*<br>Fungi: *Malassezia restricta*, *Komagataell aphaffii*, *Alternaria alternata*, *Penicillium rubens*, *Phyophthora infestans*<br>Viruses: *Escherichia virus MS2*, *Gibbon ape leukemia virus*, *Woolly monkey sarcoma virus*, *Koala retrovirus*, *Propionibacterium virus ATCC29399BT*<br>*Archaea*: *Natronorubrum aibiense*, *Halorussus litoreus*, *Haloterrigena daqingensis*, *Natronorubrum sediminis*, *Natronococcus ssp LS1_42* | Liu et al. (2024) |
| 8 | AirDNA samplers (customized) and polyethersulfone membrane filters (47 mm diameter, 0.45 μm pore size) | Bangkok, Thailand | Amplicon sequencing: ITS2 region | Mycobiome: *Yarrowia*, *Cladosporium*, *Hortaea*, *Cutaneotrichosporon*, *Leptobacillium*, *Aspergillus*, *Malassezia*, *Candida*, *Kurtzmaniella*, *Penicillium*, *Nigrospora*, *Fusarium*, *Pseudopithomyces*, *Curvularia*, *Vishniacozyma*, *Phaeosphaeria*, *Cyphellophora*, *Pleurotus*, *Epicoccum*, *Clavispora* | Harnpicharnchai et al. (2023) |
| 9 | AirDNA samplers (customized) and polyethersulfone membrane filters (47 mm diameter, 0.45 μm pore size) | Bangkok, Thailand | Amplicon sequencing: 16S rRNA (V3–V4 regions) | *Acidisphaera*, *Acinetobacter*, *Agrobacterium*, *Bacillus*, *Bradyrhizobium*, *Brevibacterium*, *Brevundimonas*, *Corynebacterium*, *Cutibacterium*, *Kocuria*, *Kytococcus*, *Lawsonella*, *Micrococcus*, *Moraxella*, *Paracoccus*, *Pseudomonas*, *Rhizobium*, *Sphingomonas*, *Staphylococcus*, *Stenotrophomonas* | Siriarchawatana et al. (2023) |
| 10 | Surface swabbing (entry and exit turnstiles) | Quito, Ecuador | Sanger sequencing: 16S rRNA amplicons (V1–V3 regions) | *Escherichia coli*, *Salmonella enterica*, *Klebsiella*, *Erwinia*, *Pantoea*, *Pseudomonas*, *Acinetobacter*, *Staphylococcus*, *Streptococcus*, *Mesorhizobium*, *Cutibacterium*, *Micrococcus*, *Corynebacterium* | Hernández-Alomía et al. (2023) |
| 11 | Surface swabbing (frequently human-exposed) | Busan, Boston, Mexico, Moscow | Amplicon sequencing: 16S rRNA (V3–V4 regions) | *Corynebacterium*, *Streptococcus*, *Staphylococcus*, *Kocuria*, *Cutibacterium*, *Brevundimonas*, *Dietzia*, *Janibacter*, *Micrococcus*, *Rothia* | Kim et al. (2023) |
| 12 | Air purifiers (Medium and HEPA filters) | Seoul, South Korea | Amplicon sequencing: 16S rRNA (V3–V4 regions) and ITS (Internal transcribed spacer region) | Bacteria: *Cutibacterium*, *Kocuria*, *Blastococcus*, *Sphingomonas*, *Staphylococcus*, *Deinococcus*, *Micrococcus*, *Enhydrobacter*, *Corynebacterium*<br>Fungi: *Aspergillus*, *Candida*, *Nigrospora*, *Penicillium*, *Schizopora*, | Hassan et al. (2022) |

(Continued)

Table 4.1 (Continued)

| S. No. | Method of air sampling/device used | Geographical location of the public transit system (subway stations) | Method of study (diversity elucidation) | Microbial diversity | References |
|---|---|---|---|---|---|
| 13 | Surface swabbing (frequently human-exposed) | Busan, South Korea | Amplicon sequencing: 16S rRNA (V3–V4 regions) and ITS (internal transcribed spacer region) | Bacteria: *Staphylococcus*, *Corynebacterium*, *Kocuria*, *Micrococcus*, *Paracoccus*, *Enhydrobacter*, *Deinococcus*, *Skermanella*, *Roseomonas*, *Pseudomonas*, *Bacillus*, *Streptococcus*<br>Fungi: *Ascomycota*, *Penicillium*, *Aspergillus*, *Malassezia*, *Densospora*, *Alternaria*, *Cladosporium*, *Agaricomycetes*, *Curvularia*, *Naganishia*, *Didymella* | Kim et al. (2022) |
| 14 | SASS 2300 sampler (Research International Inc., Monroe, WA, USA); virtual SASS 4000 impactor (aerosol); Surface swabbing (turnstiles, checkout areas) | Moscow | 16S rRNA gene sequencing (V2, V3, V4, V6, V7, V8, and V9 regions); cultivation-based | *Acinetobacter*, *Actinomyces*, *Aeromicrobium*, *Brevundimonas*, *Chryseobacterium*, *Corynebacterium*, *Cupriavidus*, *Dietzia*, *Lactobacillus*, *Methylobacterium*, *Peptoniphilus*, *Pseudomonas*, *Staphylococcus*, *Stenotrophomonas*, *Streptococcus* | Pochtovyi et al. (2022) |
| 15 | Surface swabbing (high-contact surfaces) | 60 cities from North America, East Asia, Europe, Sub-Saharan Africa, South America, the Middle East, and Oceania | Shotgun sequencing | Core taxa: *Staphylococcus* (*S. epidermidis*, *S. hominis*), *Lactococcus lactis*, *Streptococcus* (*S. mitis*, *S. oralis*), *Stenotrophomonas maltophilia*, *Acinetobacter johnsonii*, *Moraxella osloensis*, *Pseudomonas* (*P. koreensis*, *P. putida*, *P. stutzeri*), *Paracoccus yeei*, *Brevundimonas naejangsanensis*, *Sphingobium yanoikuyae*, *Sphingomonas melonis*, *Actinomyces oris*, *Blastococcus saxobsidens*, *Cutibacterium acnes*, *Micrococcus luteus*, *Kocuria* (*K. flava*, *K. palustris*, *K. rhizophila*) | Danko et al. (2021) |
| 16 | SASS 3100 dry air samplers (Research International, Monroe, WA, USA) | Denver, Hong Kong, London, New York, Oslo, Stockholm | Shotgun sequencing | *C. acnes*, *M. luteus*, *Enhydrobacter aerosaccus*, *Brachybacterium*, *Nocardioides*, *Dietzia*, *Corynebacterium pseudogenitalium*, *Pseudomonas*, *Propionibacterium granulosum*, *Massilia*, *Gardenerella vaginalis*, *K. rhizophila*, *Kocuria* spp., *Acinetobacter*, *S. hominis* | Leung et al. (2021) |

| | | | | | |
|---|---|---|---|---|---|
| 17 | Low-volume gravimetric air sampler (ENCO PM, TCR TECORA, Italy): 0.45 μm polytetrafluoroethylene (PTFE) filters | Athens | 16S rRNA gene amplicon sequencing | Bacteria: *Paracoccus*, *Sphingomonas*, *Kocuria*, *Rubellimicrobium*, *Arthrobacter*, *Acinetobacter*, *Corynebacterium*, *Modestobacter*, *Staphylococcus*, *Massilia*, *Blastococcus*, *Nocardioides*, *Hymenobacter*, *Roseomonas*, *Deinococcus*, *Geodermatophilus*, *Methylobacterium*, *Diplorickettsia*, *Skermanella*<br>Fungi: *Cladosporium*, *Mycosphaerella*, *Antrodia*, Helotiales, Ascomycota, *Sistotrema*, *Bjerkandera*, *Penicillium*, Pleosporales, *Stereum*, Polyporales, Agaricomycetes, *Sporobolomyces*, *Alternaria*, Filobasidiales, *Aspergillus*, *Coriolopsis* | Grydaki et al. (2021) |
| 18 | Surface swabbing (station turnstiles, vertical handrails inside train) | Mexico | 16S rRNA gene sequencing (V3–V4 region) | *Cutibacterium*, *Corynebacterium*, *Streptococcus*, *Staphylococcus*, *Acinetobacter*, *Halomonas*, *Rothia*, *Actinomyces*, *Kocuria*, *Leuconostoc*, *Methylobacterium*, *Prevotella*, *Brachybacterium*, *Pseudomonas*, *Shewanella*, *Veillonella*, *Paracoccus*, *Micrococcus*, *Arthrobacter*, *Anaerococcus*, *Haemophilus*, *Lactococcus* | Hernández et al. (2020) |
| 19 | Surface swabbing (bench, wall, floor, information stand, railings) | Moscow | 16S rRNA gene sequencing (V4 region) | *Stenotrophomonas* (*S. acidaminiphila*), *Pseudomonas* (*P. stutzeri*, *P. veronii*), *Dietzia*, *Brevundimonas* (*B. diminuta*), *Staphylococcus*, *Rhodococcus*, *Erwinia*, *Arsenicicoccus*; Intrasporangiaceae family | Klimenko et al. (2020) |
| 20 | SASS 3100 dry air samplers (Research International, Monroe, WA, USA); surface swabbing (railings, kiosks, benches) | Oslo, Norway | 16S rRNA gene amplicon sequencing | *Micrococcus*, *Staphylococcus*, *Rubrobacter*, *Sphingomonas*, *Hymenobacter*, *Arthrobacter*, *Corynebacterium*, *Nocardioides*, *Psychrobacter*, *Blastococcus*, *Kocuria*, *Streptococcus*, *Pseudonocardia*, *Flavobacterium*, *Cutibacterium*, *Lactobacillus*, *Deinococcus*, *Massilia* (relative abundance >1%) | Gohli et al. (2019) |
| 21 | Surface swabbing (station and train car surfaces) | Boston | 16S amplicon sequencing; shotgun metagenomic sequencing | Corynebacteriaceae, Staphylococcaceae, Streptococcaceae, Moraxellaceae, Micrococcaceae, Pseudomonadaceae, Tissierellaceae, Actinomycetaceae, Enterobacteriaceae, Lactobacillaceae, Sphingomonadaceae, Methylobacteriaceae, Cytophagaceae, Oxalobacteraceae; *Malassezia globosa*, *Propionibacterium* phages, Merkel cell polyomavirus, *Staphylococcus* phages | Hsu et al. (2016) |
| 22 | Coriolis μ impinger (Bertin Technologies, Montigny-le-Bretonneux, France) | Barcelona | Quantitative PCR and 454 pyrosequencing | *Staphylococcus*, *Neisseria*, *Methylobacterium*, members of Chitinophagaceae, *Bradyrhizobium*, *Paracoccus*, *Sphingomonas*, *S. epidermidis*, Influenza A and B, Rhinovirus, *Aspergillus fumigatus* | Triadó-Margarit et al. (2017) |

(Continued)

Table 4.1 (Continued)

| S. No. | Method of air sampling/device used | Geographical location of the public transit system (subway stations) | Method of study (diversity elucidation) | Microbial diversity | References |
|---|---|---|---|---|---|
| 23 | Surface swabbing (high-contact surfaces at stations and on trains) | New York City | Shotgun sequencing | *Pseudomonas stutzeri*, *Stenotrophomonas maltophilia*, *Enterobacter cloacae*, *Acinetobacter radioresistans*, *Acinetobacter nosocomialis*, *Lysinibacillus sphaericus*, *Enterococcus casseliflavus*, *Brevundimonas diminuta*, *Acinetobacter lwoffii*, *Bacillus cereus*, *Enterobacteria* phage phiX174, Epsilon 15 like virus, *Erwinia* phage, *Stenotrophomonas* phage phiSMA7, *Staphylococcus* phage PVL, *Lactococcus* phage ul36 | Afshinnekoo et al. (2015) |
| 24 | Leland Legacy portable sample pumps (SKC Inc., PA, USA); Sioutas Cascade Impactor (SKC Inc., PA, USA) | Hong Kong | 16S rRNA sequencing (V4 region) | *Micrococcus* (4.9%), *Enhydrobacter* (3.1%), *Propionibacterium* (2.9%), *Staphylococcus* (1.8%), and *Corynebacterium* (1.5%), *Sphingobium*, *Blastomonas*, *Clostridium cellulovorans*, *Synechococcus*, *Leuconostoc*, *Acinetobacter*, *Xanthomonas* | Leung et al. (2014) |
| 25 | Omni 3000 fluid impinger (InnovaPrep LLC, Drexel, MO) | New York City | SSU rRNA gene amplification and Sanger sequencing, 454 pyrosequencing | Staphylococcaceae, Moraxellaceae, Micrococcaceae, Enterobacteriaceae, Corynebacteriaceae, Microbacteriaceae, Flavobacteriaceae, Pseudomonadaceae, Rhodobacteraceae, Sphingomonadaceae, Xanthomonadaceae, Comamonadaceae, Nocardioidaceae, Propionibacteriaceae, Prevotellaceae, Streptococcaceae, Bacillaceae, Planococcaceae, Cytophagaceae, Oxalobacteraceae, Carnobacteriaceae, Cyanobacteria; Fungi: Agaricomycetes, Eurotiales, Capnodiales, Saccharomycetales, Trichocomaceae, *Wallemia*, Davidiellaceae, Dothideomycetes | Robertson et al. (2013) |

TABLE 4.2

*Studies characterizing public transit microbiome using cultivation-based approach*

| S. No. | Method of air sampling/device used | Geographical location of the public transit system (subway stations) | Method of study (diversity elucidation) | Microbial diversity | References |
|---|---|---|---|---|---|
| 1 | Andersen cascade two-stage impactors | Mexico City | Cultivation-based and identification by staining and morphology | Bacteria: *Bacillus*, *Streptococcus*; Fungi: *Aspergillus* spp., *Penicillium* spp., *Alternaria* spp. | Hernández-Castillo et al. (2014) |
| 2 | SASS 3100 (Research International, Monroe, WA) | Oslo, Norway | 16S rRNA sequencing-based identification of cultivated isolates | Total cultivation: *Micrococcus* (32%), *Staphylococcus* (20%), *Bacillus* (18%), *Pseudomonas* (4%), *Microbacterium* (3%), and *Streptomyces* (3%); bacterial spores: *Bacillus* (68%), *Staphylococcus* (15%), *Paenibacillus* (9%), *Viridibacillus*, *Tumebacillus*, *Brevibacillus*, *Lysinibacillus* | Dybwad et al. (2012) |
| 3 | Passive sedimentation on petri plates containing different media | St. Petersburgh, Russia | Cultivation-based and identification after sporulation | Fungal genera prevailing: *Acremonium*, *Aspergillus*, *Cladosporium*, and *Penicillium* | Bogomolova and Kirtsideli (2009) |
| 4 | All Glass Impinger (AGI-30) samplers | Cairo, Egypt | Cultivation-based and microscopic identification | Fungal spores: *Aspergillus* spp. (*A. fumigatus*, *A. niger*, etc.), *Cladosporium*, *Penicillium*, *Alternaria*, *Monilia*, *Fusarium*, *Sporotrichum*, *Botrytis*, *Mucor*, *Fusidium*, *Spicaria*, *Cunninghamella*, *Helminthosporium*; Yeasts | Awad (2002) |

Many of these taxa are typical skin-, respiratory-, or environment-associated bacteria, highlighting the strong influence of passenger density, human shedding, resuspension from surfaces, and ventilation dynamics on microbial assemblages in transit environments. Their recurrent detection underscores public transit systems as dynamic interfaces where human-associated and environmental microbes converge, with implications for exposure assessment, microbial persistence, and potential health relevance in densely populated urban settings.

## 4.5 FUNCTIONAL MAPPING TO PUBLIC TRANSIT RESISTOME

The term 'resistome' stands for all the ARGs (acquired and intrinsic), their precursors, and mechanisms that may require specific alterations or evolution for expression, conferring resistance (protoresistance), present in both pathogenic and non-pathogenic bacteria (Perry et al., 2014). The concept of environmental resistome originated from the soil, leading to the identification of numerous ARGs from soil and aquatic sources via a metagenomic approach (Berglund et al.,

2019). The environmental ARGs can be transmitted to humans by the food chain, direct contact, or bioaerosols after aerosolization into the air, threatening public health (Larsson et al., 2018). The resistome hosted by the human commensal microflora is potentially disseminated by activities such as sneezing, coughing, talking, exhalation, and skin cell shedding (Gwenzi et al., 2022). Several studies have investigated the airborne resistome in indoor (Li et al., 2020) and outdoor environments (Zhou et al., 2023), including public transit systems (Danko et al., 2021; Hernández-Alomía et al., 2023; Kumari et al., 2025). The mass transit systems are important reservoirs of human-associated and ambient ARGs, acting both as sources and sinks due to the considerable interplay between anthropogenic factors and environmental parameters.

A global metagenomic study of transit systems by Danko et al. (2021) utilized MegaRES ontology and alignment software to map sequenced reads to known ARGs but was unable to identify a consistent core set of ARGs. The AMR genes were low in abundance, widely heterogeneous, and distributed unevenly throughout the cities depicting varied antibiotic usage.The genes against beta-lactam and MLS (macrolides, lincosamides, streptogramins) (56% of ARG-positive samples) classes of antibiotics included the most identified ones. Multi-drug efflux pumps, 23S rRNA methyltransferases, and EF-Tu inhibition were the most prevalent AMR mechanisms detected in 25% of the ARG-positive samples. Three large clusters and various smaller clusters of these genes were detected in individual samples, possibly of different antibiotic classes, representing plasmids carrying multiple ARGs, or ecological competition dictating the acquisition of resistance genes by microbes (Danko et al., 2021). Another recent seasonal airborne resistome profiling of the Indian public transit system using the same ARG database revealed 111 distinct ARGs spanning 17 antibiotic classes, with consistently high abundances of beta-lactam, aminoglycoside, macrolide-lincosamide-streptogramin, and multi-drug resistance genes, and a pronounced enrichment of clinically important ARGs during autumn. Notably, high-risk resistance determinants including carbapenemases (*bla*, *NDM*, *kpc*, *oxa*), methicillin resistance (*mecA*), colistin resistance (icr, mcr), and tigecycline-associated genes (*ramA*, *tetX*) were detected, highlighting the public transit air as a potential reservoir and dissemination pathway for clinically relevant AMR (Kumari et al., 2025). Moreover, genome-resolved analyses of urban aerobiome revealed extensive plasmid- and phage-mediated horizontal gene transfer of antibiotic resistance across diverse bacterial taxa. Railway stations emerged as key urban hotspots where intense human activity and enriched mobile genetic elements synergistically accelerate the spread of ARGs through metabolically active airborne bacteria (Bai et al., 2025).

Resistome profiling using ShortBRED revealed 527 antibiotic resistance protein families, immensely heterogeneous among the air samples from transit networks of six different cities, with 1.3% AR families identified in ≥75% samples (core resistome), conferring resistance against tetracycline, aminoglycoside, fluoroquinolone, macrolide, and elfamycin antibiotics. The resistome also varied geographically, similar to the microbiome. Source tracking of the aerial resistome showed that it overlapped with that of human skin (35.5 ± 15.6%), soil (31.9 ± 19.1%), wastewater (15.6 ± 12.8%), oral cavity (1.1 ± 3.0%), gastrointestinal tract (0.8 ± 2.5%), and the adjacent public transit surfaces, highlighting the influences of the outdoor environment and commuters in shaping the airborne ARGs. Moreover, the genes were located both in the chromosomal and plasmid regions and contained integron components (integrases, att sites) (Leung et al., 2021). Another study detected a low abundance of ARGs in the Boston transit system, using ShortBRED and the common resistance mechanisms comprising efflux pumps, modifying or replacing antibiotic targets, inactivating the antibiotic, and modifying nucleic acid machinery (Hsu et al., 2016). Tetracycline resistance genes (*tetRRPP*, *tetA-B*, *tetM*, and *tetMWOS*) were highly abundant in a Hong Kong subway line connecting to mainland China, possibly sourced from swine farms (Kang et al., 2018). Extended-spectrum beta-lactamase genes, including *blaTEM-1* and *blaCTX-M-1*, and metallo-beta-lactamase genes, including *blaOXA-181* and *mecA*, along with their sub-variants present in different Gram-negative and Gram-positive bacterial species were identified in the subway air of Quito, Ecuador (Hernández-Alomía et al., 2023). Fluoroquinolone, glycopeptide, and tetracycline ARGs were abundantly present in the bioaerosols from the Sukhumvit subway station of Bangkok (reflecting prevalent local antibiotic usage), majorly contributed by the commensal and transient human-associated bacteria (Harnpicharnchai et al., 2024).

Besides the metagenomic approach, a few mass transit microbiome studies performed cultivation-based determination of antibiotic resistance. The cultivated bacterial species were resistant to chloramphenicol, streptomycin, tetracycline, kanamycin, ampicillin, and nalidixic acid (Afshinnekoo et al., 2015; Dybwad et al., 2012; Pochtovyi et al., 2022). The mobility of AR determinants enables their transport to pathogenic and opportunistically pathogenic bacteria, increasing the overall AMR burden (Dybwad et al., 2012). Future research is warranted to ascertain the viability of airborne microorganisms, whether carrying AR determinants or not, in order to fully acknowledge the transmission potential of pathogenic species through public transit air (Leung et al., 2021).

## 4.6 LINKING ENVIRONMENTAL MICROBIOME TO HUMAN MICROBIOME

A healthy human microbiome is determined by the healthy environmental microbiome, and both interact dynamically with a continuous exchange between surfaces (Figure 4.4) (Lloyd-Price et al., 2016). The immediate environment of an infant plays a major role in the initial establishment of the microbiome specific to body sites (Gilbert et al., 2018; Lloyd-Price et al., 2016). Likewise, the microbial community living on and in the human body undergoes continuous minor changes with considerable environmental influence, along with other factors (genotype, hygiene, etc.) (Dekaboruah et al., 2020). To exemplify, a study by Lai et al. (2017) displayed the contribution of the work environment of an animal research facility to the skin, oral, and nasal microbiome of the workers (Lai et al., 2017). Earlier reports have demonstrated the transient presence or colonization of livestock-associated methicillin-resistant *Staphylococcus aureus* strain (LA-MRSA) in the nares of pig farm workers (Fang et al., 2014; Frana et al., 2013). However, further studies need to explore the effects and the underlying mechanisms of occupational microbial exposure on workers' health, as well as the long-term implications of such changes (Lai et al., 2017). A concept of transmission of pathogenic and non-pathogenic microbes between humans, animals, and the environment exists, and 'One Health approaches' aim to discern their influences on one another through microbiome sequencing and predictive models, with the ultimate goal of enhancing public health (Trinh et al., 2018).

Such environmental transfer of microbiota can occur through various sources, including diet, drinking water, fomites, pets, or direct exposure to air. Limited evidences suggest the association of drinking water microbiome with gut microflora variations, depending on the origin and intake (for example, increased diversity in well water drinkers) (Vanhaecke et al., 2022). The airborne microorganisms are potentially associated with particulate matter particles suspended in the air, with sizes up to 10 μm, and have been widely studied concerning pathogens and human health (Cao et al., 2014; Moelling and Broecker, 2020; Yu et al., 2025). The inhalable particulate matter in polluted air ($PM_{2.5}$) is known to cause respiratory, cardiovascular, gastrointestinal, and neurodegenerative diseases as well as type 2 diabetes, autism, and lung cancer, among others, facilitated by its ability to penetrate deeper into the lung tissues and deposition (Gavito-Covarrubias et al., 2024; M. Wang et al., 2025). Ultrafine particles ($PM_1$ and $PM_{0.1}$) can penetrate through cells and readily translocate throughout the body, resulting in adverse systemic health outcomes (Li et al., 2016). The airborne microbiota transported via these air particles has the potential to interact with the normal flora of the human body, modifying it for better or for worse, with scales weighing more toward the worse. For instance, bacterial pathogens such as pneumococci and fungal allergens carried by the suspended fine particles can enter the human respiratory system and even the lungs and blood, leading to respiratory and cardiovascular diseases (Rao et al., 2020).

Concerning the subway environment, commuters enable community-wide dissemination and exchange of microorganisms by transferring their skin commensal microflora to the subway surfaces and contracting new diversity from surfaces that host both human- and environmental origin microbes via touch. The aerosolization of these deposited biological particles, as well as direct aerial shedding of microbe-laden particles by the human body, compounded with settling, facilitates exchange via air. The hands of the passengers have been demonstrated to resemble each other and the subway surface microbiome (variations in transient microbiota) more while preserving their microbial fingerprints (highly abundant taxa). Moreover, cleaning high-contact surfaces, such as poles and handrails, lowers the microbial burden but is effective only for brief

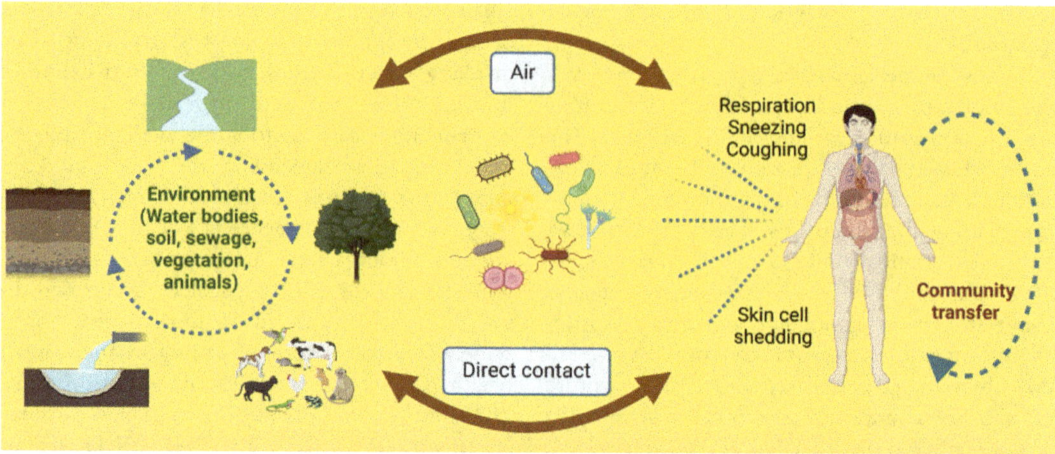


Figure 4.4 Human–environment–microbiome nexus (created with BioRender.com).

intervals before microorganisms resettling to similar concentrations occur. Though the skin acts as a barrier, the transfer of potentially pathogenic species to the mucosa (oral/nasal/conjunctival) through hands results in microbial self-inoculation and disease transmission (Kang et al., 2018; Peimbert and Alcaraz, 2022; Vargas-Robles et al., 2020).

Furthermore, from the overall ecosystem health perspective, the impacts of the spatial and temporal distribution of the microorganisms on the microbiome of humans and natural communities, in general, need to be considered as well. Incorporating the socio-ecological aspects such as biodiversity patterns aids in a better representation of the co-evolving microbiomes and a comprehensive understanding of the human–environment–microbiome nexus. The pattern-oriented models (data/theory-based) developed on the metacommunity approach (multiple interacting communities) would enable the identification of key diversity drivers, dispersal mechanisms, stable and transient states, and variability factors with respect to static natural conditions or particular population outcomes. In simpler terms, the functional diversity of the microbiome needs to be emphasized in conjunction with the taxonomic diversity (Cullen et al., 2020).

### 4.6.1 Association with Respiratory Microbiome

Microorganisms colonizing the nasal, oropharyngeal, nasopharyngeal, and lung cavities constitute the respiratory microbiome, which plays a significant role in the development of the local immune system and diseases such as allergy, asthma, etc. (Zhao et al., 2023). Inhalation introduces dust and associated microbiota into the airways; the influences of the latter on the normal flora remain to be discerned. Although mucociliary clearance facilitates the removal of most inhaled particles and microbes from the airways, elevated mucus production and impaired clearance processes may increase the risk of infection for patients with chronic respiratory inflammation (Soumana and Carlsten, 2021). A mattress dust microbiome study revealed influences from outdoor farm environments and human factors, with bacterial richness similar to that of the nasal microbiome in school-age children, highlighting the potential link between air and the respiratory microbiome (Birzele et al., 2017).

Mouth-breathing introduces microorganisms into the oral mucosa and saliva, triggering potent immunological responses through cell interactions that may play a significant role in the development of non-communicable diseases (Ooi et al., 2020). Studies have explored the influences of the air microbiome on childhood asthma and allergic diseases. Exposure to house pest allergens (such as house dust) in the early years (0–3 years) has been linked to allergic sensitization and recurrent wheezing, potentially contributing to the development of childhood allergic diseases. On the other hand, allergen exposure in combination with specific bacterial families and fungal diversity (microbe-rich indoor/outdoor environment) yielded a protective effect against atopy and wheezing in children (Lynch et al.,

2014; Tischer et al., 2016). The bacterial families present as normal flora in the human body, including Lachnospiraceae, Prevotellaceae, and Ruminococcaceae, were found to elicit protective effects via immunomodulatory metabolite production, with the possible occurrence of this phenomenon at airway mucosal surfaces, in addition to the gastrointestinal tract (Lynch et al., 2014). Additionally, *Kocuria* sp. (produces macrolide kocurin, activity against potential pathogens: *Staphylococcus* (Palomo et al., 2013)), *Alloiococcus*, *Bifidobacterium*, *Acinetobacter*, *Listeria monocytogenes*, *Corynebacterium* sp., *Bacillus* sp., and fungal taxon *Eurotium* are some other microbial species present in the immediate environment of children (air and home dust of poor urban areas or farming regions) with potentially protective effects against asthma (Ege et al., 2011; O'Connor et al., 2018). Stein et al. elucidated the protective role of sustained exposure to house dust-associated microbial endotoxins against allergic asthma by regulating the innate immune response in experimental murine models (Stein et al., 2016). Furthermore, a case–control study conducted in Nigeria revealed that the microbiome of airborne dust in homes is associated with a lower incidence of lower respiratory tract infections in children under five years of age (Fakunle et al., 2023). The genera under phyla Proteobacteria, Actinobacteria, Firmicutes, Chloroflexi, and Ascomycota were positively associated, while phyla Bacteriodota and Deinococcota were negatively associated with childhood lower respiratory tract infections (Fakunle et al., 2023; Fu et al., 2020).

The indoor airborne particulate matter harbors a distinct microbiome shaped by building characteristics and human occupancy, which differs from that of outdoor air and includes bacteria associated with the skin and respiratory system, carrying AMR genes (Gilbert and Hartmann, 2024; Solazzo et al., 2025). Importantly, the composition of the indoor air microbiome correlated with that of the upper respiratory microbiome of occupants, supporting the role of airborne particles as potential vectors linking environmental microbes with the human respiratory tract. Specifically, indoor airborne particles were enriched in human-associated and opportunistic taxa, such as *Paracoccus*, *S. aureus*, and *Klebsiella pneumoniae*, while outdoor particles were dominated by plant- and soil-associated bacteria. Common bacterial taxa, including *M. luteus*, *Kocuria rosea*, and *Moraxella osloensis*, showed consistent presence across both environments, with positive correlations observed between indoor particle concentration and respiratory abundances of *S. aureus*, *K. pneumoniae*, and *Cutibacterium* species (Solazzo et al., 2025).

The bacteriome hosted by the nasal cavities of pig farm workers was reported to be richer and more diverse compared to cow farm workers and a non-exposed control group. A probable explanation put forward for the observed phenomenon is the higher pig farm airborne dust concentration hosting a rich aerobiome (diverse aerosolized microbes) that modifies and enriches the normal microbiota of the nasal cavity (Kraemer et al., 2018). Similarly, in an urban setting, the nasal and oral microbiome of dairy farmers was revealed to constitute many unique genera, as well as the dominance of phylum Bacteroidetes and families, including Ruminococcaceae, Carnobacteriaceae, and Lachnospiraceae, representative of the dairy environment (higher bioaerosol content), in comparison to non-farmers (Shukla et al., 2017). Increased biodiversity in the nares of dairy farmers confers protection by maintaining immune homeostasis and preventing colonization by opportunistic species, as evidenced by the lower burden of *Staphylococcus* spp. (Shukla et al., 2017). However, the stability and importance of such changes, with underlying mechanisms, are a subject of further study.

### 4.6.2 Association with Gut Microbiome

The smaller-sized air particles entering the airways are retained in the lungs, while the larger-sized fraction (>6 μm) undergoes mucociliary clearance making its way into the gastrointestinal tract. The ingestion of the particles sequestered in alveolar macrophages (long-term retained particles) and the mucus layer through the oropharynx, also accounts for intestinal exposure to pollutants, besides food and water sources (Kreyling et al., 1999; Möller et al., 2004). The airborne microbial cells attached to the particulate matter can potentially be transported to the gut, where they may interact with the gut microbiome, either promoting stability or contributing to disease. Though the effects of air pollutants on the gut microbiome have been discretely studied, how airborne microorganisms interact with the gastrointestinal microbiome remains to be discerned (Ananthakrishnan et al., 2011; Kish et al., 2013; Mutlu et al., 2018). *Clostridium* spp. and *Bacillus* spp. in the indoor dust microbiome were revealed to elicit potential protective effects on the gut microbiome of children in a study conducted in Shanghai, China. The authors also demonstrated that the indoor air microbiome and associated metabolites are significantly linked to gut microbiota composition and diversity, with indoor indole derivatives showing a positive association with protective gut microbial profiles and overall gut health. Together, the results suggested that indoor airborne microorganisms and metabolites may influence gut microbiota via systemic pathways, supporting a functional connection between the air microbiome and the gut ecosystem (Zhang et al., 2023). Nonetheless, the current understanding of the modulating effects of the air microbiome is limited to a few studies on the respiratory microbiome and needs to be extended to the gastrointestinal microbiome to truly comprehend the direct and indirect effects of the environmental microbiome on the human microbiome.

## 4.7 IDENTIFYING CHALLENGES IN STUDYING PUBLIC TRANSIT MICROBIOME

The density of microorganisms in the air is lower compared to the soil and water environments, as reported by different isolated studies (Bowers et al., 2009; Jo & Seo, 2005; Lighthart, 2000; Prussin et al., 2015; Zhu et al., 2003), making it challenging to obtain a sufficient amount of DNA for conducting metagenomic studies.The sampling of thousands of liters of air enables the collection of a substantial number of microbes suspended in the air, which amounts to a considerable quantity of the initial genetic material, facilitating further DNA-based studies. Amplification methodologies can be used to multiply the genetic content of a sample but may introduce biases in the estimation of the population census (Marine et al., 2014). Therefore, high-efficiency air sampling devices are cardinal to draw in more air and subsequently entrap the suspended microorganisms optimally in shorter durations to accumulate sufficient biomass for further downstream processes. There are various types of air sampling devices available commercially following different mechanisms of particle entrapment in different media (Table 4.3). The sampling equipment should be carefully chosen according to the target of the study and downstream applications.

The composition of airborne microorganisms is subject to constant variations with respect to meteorological, temporal, and spatial parameters, i.e., it varies depending upon the season, day or night, location of sampling (land-use type), etc. (Bertolini et al., 2013; Bowers et al., 2013, 2011a). The continual changes make it difficult to establish reproducible background composition levels for subsequent comparative studies, examining the influences of environmental and anthropogenic factors, as well as the monitoring of pathogens concerning public health (Behzad et al., 2015). In the built environment of mass transit systems, collecting additional metadata, such as the type of ventilation (natural or mechanical), the number of commuters, and whether the stations are underground or above-ground, is also essential to explore their correlation with the diversity of airborne microbes. The measurements of airflow would enable the identification of potential transmission routes and source tracking of the existing aerobiome.

After appropriate sampling of air, DNA extraction is another challenging task as different approaches and methodologies generate varying DNA yields representing differing microbial diversity (Yuan et al., 2012). The microbial community composition elucidated varies depending on the air sampler used (impactor, impinger, or filter-based), as well as the DNA extraction protocol, especially for the lower-abundant species (Fahlgren et al., 2011; Hoisington et al., 2014; Li, 2011). The isolated nucleic acid is subjected to either amplicon sequencing (amplifying and sequencing marker genes: 16S and 18S rRNA) or shotgun metagenomic sequencing, where the DNA is fragmented, sequenced, reassembled de novo (into contigs), and annotated. The sequencing process can introduce bias by over-representing some abundant species and overlooking the less abundant ones, which can be partially remedied by deep sequencing and increasing read depth (reads per nucleotide) to ensure a nearly precise depiction of the community composition. The contig assembly, gene prediction, and annotation pose a significant bioinformatics-related challenge even after the wide availability of required software and analysis tools. One major shortcoming is the limited availability of reference sequences in databases concerning the aerobiome, which hinders novel gene identification. The gap can only be filled by conducting more metagenomic research to contribute to more

TABLE 4.3
*Types of available air samplers*

| S. No. | Type of sampler | Principle of particle entrapment | Examples of commercially available air samplers | Advantages | Disadvantages | References |
|---|---|---|---|---|---|---|
| 1. | Solid impactor | Air flows through small orifices and bends sharply allowing the particles to get entrained on a solid substrate such as agar plates, filters, gelatin-coated slides, etc. by inertial impaction | Cascade impactors: Andersen single-stage and multi-stage, University of Washington Mark III, Mercer, and Micro orifice uniform deposit impactors (MOUDI) | Size fractionation possible.<br>Reduced desiccation. | Bounce-off and particle re-entrainment into the air.<br>Decreased viability with increased impaction velocity. | Cao et al. (2011), Finlayson-Pitts and Pitts (2000), and Lago et al. (2018) |
| | | | Suction slit impactor: Burkard air sampler | Longer duration sampling for pollen and spores. | | |
| | | | Slit-to-agar impactors: Mattson-Garvin sampler | Determination of viable cells per unit of air. | | |
| | | | Cyclone separator (NIOSH biosampler) | Approximately 265L/min collection rate, operable for longer durations. | Possibly damage the collected viruses. | |
| 2. | Virtual impactor | Inertial impaction of airstream against a still airmass, withdrawn through a filter for particle collection | Dichotomous sampler | Avoids bounce-off and particle re-entrainment into the air.<br>High flow rate and lower cut-off size. | – | Finlayson-Pitts and Pitts (2000) and Lim et al. (2020) |

(*Continued*)

Table 4.3 (Continued)

| S. No. | Type of sampler | Principle of particle entrapment | Examples of commercially available air samplers | Advantages | Disadvantages | References |
|---|---|---|---|---|---|---|
| 3. | Liquid impinger | Airstream passes through single or multiple nozzles – inertial impaction into liquid | All glass impinger (AGI), SKC Biosampler | Direct collection of air particles in a suitable liquid medium.<br>Viability retained (depending upon the collection medium).<br>Biosampler – better collection efficiency, minimal bubbling, and re-entrainment (swirling motion). | Reaerosolization may occur.<br>Liquid evaporation during prolonged sampling.<br>Flow rates up to 12.5 L/min.<br>AGI-30 impinger – maximum run time of 30 min. | Kesavan et al. (2010) and Lin et al. (2000) |
| 4. | Wetted wall cyclones (WWC) | Deposition of bioaerosol particles on the cyclone wall coated with a thin film of sampling liquid (centrifugal force) | SASS 2300 air sampler | Prolonged sampling time.<br>Particle retention and constant sample fuid volume.<br>Flow rate – >300 L/min for 0.5–10 μm particles. | Reaerosolization may occur.<br>Viability loss. | Kesavan et al. (2010), Kesavan and Schepers (2006), Lin et al. (2000), Mcfarland et al. (2010), and Sung et al. (2017) |
| | | | Coriolis micro air sampler<br>Omni 3000 air sampler | Allows rapid microbiological analysis (fluid samples).<br>Can detect aerial viruses (Coriolis).<br>Flow rate up to 300 L/min, operable for minutes to hours. | Material loss during prolonged sampling (>30 min). | |

| | | | | | | |
|---|---|---|---|---|---|---|
| 5. | Filter-based | Inertial impaction on filter | Gelatin filter, polycarbonate, nitrocellulose | Higher physical sampling efficiency.<br>Retains viability for shorter durations.<br>Gelatin filters – efficient virus collection.<br>Polycarbonate – thermally stable, biologically and chemically inert. | Efficiency may be influenced by the elution process of collected particles.<br>Desiccation may occur.<br>Possibly lower biological sampling efficiency as microbes may be drawn past the filter along with air. | Dybwad et al. (2014), Haddrell and Thomas (2017), and Verreault et al. (2008) |
| | | Electret filters: comprises polymer disc with electric field frozen into each fiber, which induces a charge in aerosols passing through the filter, capturing it | SASS 3100 dry air sampler | Longer collection periods up to days.<br>Highly efficient for sub-micron size bioaerosols.<br>High air flow rate (50–300 L/min).<br>Compact and lightweight. | Desiccation may occur.<br>May affect cell viability.<br>May injure microbial cells. | |
| 6. | Electrostatic precipitation | Incoming air particles are charged electrically and subjected to precipitating electric field, followed by collection onto solid surfaces | Electrostatic precipitators | High collection efficiency.<br>Compatible with various analysis processes. | Desiccation may occur. | Haddrell and Thomas (2017), Mainelis et al. (2010), and Roux et al. (2016) |

extensive genomic databases, which would facilitate the corresponding metagenomic data analysis. Although diverse software tools and analysis pipelines have been developed (Liu et al., 2025; Mukherjee et al., 2023), further advancements in bioinformatics are needed to articulate the complex metagenomic data, which would ease the study of the environmental microbiome, particularly in the air (Behzad et al., 2015).

Furthermore, the research so far has focused heavily on the bacterial community, with less attention given to their fungal, viral, and archaeal counterparts, i.e., measuring DNA exclusively while neglecting RNA viruses, as well as the transcriptional activity of bacteria and archaea. The studies should incorporate the identification of all the categories of microorganisms to accurately reflect the microbial community and deduce their interaction with each other, as well as the interplay of extrinsic variables. Lastly, the effects of the community behavioral changes on the microbiome brought on by the pandemic, such as wearing face masks, hand sanitization, installation of antimicrobial devices like UV lamps, antibacterial coatings, and probiotic-based sanitation, are emerging areas of exploration (D'Accolti et al., 2023; Peimbert and Alcaraz, 2022).

The standardization of air sampling and processing methodologies, in concert with the constant progress in gene sequencing and computational technologies, and curating more comprehensive genomic databases, would address these challenges to enable thorough characterization of the public transit microbiome, as well as their impacts on human health and the ecosystem.

## 4.8 CONCLUSION

There are millions of microorganisms in the air we breathe, but little is known about them, their functions, and effects. The advent of next-generation sequencing technology and metagenomics has facilitated enormous growth in studies characterizing the environmental microbiome, particularly in the air. Microorganisms suspended in the air of mass transit systems have been identified in numerous cities worldwide. Various studies explore the spatial and temporal variability of the aerobiome along with the associated factors, contributing sources and sinks. Air also acts as a medium of mass dispersion for antibiotic resistance genes, as documented by certain studies reviewed here. However, the majority of existing literature on the microbiome of public transit networks is limited to elucidating the taxonomic composition. Very few research studies employing shotgun metagenomics investigate the functional profiles and the resistome of the public transit air. Different studies utilize different investigative methodologies to achieve their objectives. Each methodology provides significant knowledge about the transit microbiome, but the variations hinder comparative studies across different metropolitan cities.

'One Health microbiome relationships' investigates the interactions of the human, animal, and environmental microbiomes, and their effects on one another. In accordance, future research exploring the direct or indirect effects of airborne microbes on the human skin, respiratory, and gut microbiome is paramount to understanding how the environmental microbiome modulates the human commensal microflora and its implications. It is also essential to explore the relationship between the aerial microbiome and associated ecological parameters, including geographical and meteorological variables. For the mitigation of infectious disease outbreaks, such as the recent SARS-CoV-2 pandemic or other multidrug-resistant pathogens, characterizing the ecological and pollution parameters associated with higher exposure risks is essential. Overcoming the challenges associated with air sampling, DNA sequencing, and data analysis will enable a comprehensive understanding of the multiple facets of the air microbiome, serving as a reference for formulating public health policies that minimize commuter exposure to allergens and pathogens.

**Author contributions**: Shashi Prabha Kumari: Investigation, Writing – original draft, Writing – review and editing, Sunila Hooda: Conceptualization, Writing – review and editing, Supervision, Prerna Diwan: Conceptualization, Writing – review and editing, Supervision, Rakesh Kumar Gupta: Conceptualization, Supervision, Writing – review and editing, Funding acquisition, Resources, Project administration.

**Availability of data and materials**: No new data were created or analyzed in this study. Data sharing is not applicable to this chapter.

**Funding**: This work was supported by the Indian Council of Medical Research (Grant number: 2020-9584).

**Declaration of competing interest**: The authors declare that they have no known competing financial interests or personal relationships that could have appeared to influence the work reported in this chapter.

## REFERENCES


Afshinnekoo, E. et al., 2015. Geospatial resolution of human and bacterial diversity with city-scale metagenomics. *Cell Syst.* 1(1), 72–87. https://doi.org/10.1016/j.cels.2015.01.001

Ananthakrishnan, A.N. et al., 2011. Ambient air pollution correlates with hospitalizations for inflammatory bowel disease. *Inflamm. Bowel Dis.* 17(5), 1138–1145. https://doi.org/10.1002/ibd.21455

Awad, A.H.A., 2002. Environmental study in subway metro stations in Cairo, Egypt. *J. Occup. Health* 44(2), 112–118. https://doi.org/10.1539/joh.44.112

Bai, H. et al., 2025. Phages and plasmids mediate antibiotic resistance gene transfer in urban airborne bacteria. *J. Hazard. Mater.* 501, 140779. https://doi.org/10.1016/j.jhazmat.2025.140779

Barberán, A. et al., 2015. Continental-scale distributions of dust-associated bacteria and fungi. *Proc. Natl. Acad. Sci. U. S. A.* 112(18), 5756–61. https://doi.org/10.1073/pnas.1420815112

Behzad, H. et al., 2015. Challenges and opportunities of airborne metagenomics. *Genome Biol. Evol.* 7(5), 1216. https://doi.org/10.1093/GBE/EVV064

Berglund, F. et al., 2019. Identification and reconstruction of novel antibiotic resistance genes from metagenomes. *Microbiome* 7(1). https://doi.org/10.1186/S40168-019-0670-1

Bertolini, V. et al., 2013. Temporal variability and effect of environmental variables on airborne bacterial communities in an urban area of Northern Italy. *Appl. Microbiol. Biotechnol.* 97(14), 6561–70. https://doi.org/10.1007/s00253-012-4450-0

Birzele, L.T. et al., 2017. Environmental and mucosal microbiota and their role in childhood asthma. *Allergy Eur. J. Allergy Clin. Immunol.* 72(1), 109–119. https://doi.org/10.1111/ALL.13002

Bobrovska, S. et al., 2025. Hospital air sampling enables surveillance of respiratory virus infections and genomes. *Sci. Total Environ.* 977, 179346. https://doi.org/10.1016/j.scitotenv.2025.179346

Bogomolova, E., Kirtsideli, I., 2009. Airborne fungi in four stations of the St. Petersburg underground railway system. *Int. Biodeterior. Biodegradation* 63(2), 156–160. https://doi.org/10.1016/j.ibiod.2008.05.008

Bowers, R.M. et al., 2009. Characterization of airborne microbial communities at a high-elevation site and their potential to act as atmospheric ice nuclei. *Appl. Environ. Microbiol.* 75(15), 5121–5130. https://doi.org/10.1128/AEM.00447-09

Bowers, R.M. et al., 2011a. Spatial variability in airborne bacterial communities across land-use types and their relationship to the bacterial communities of potential source environments. *ISME J.* 5(4), 601–12. https://doi.org/10.1038/ismej.2010.167

Bowers, R.M. et al., 2011b. Sources of bacteria in outdoor air across cities in the midwestern United States. *Appl. Environ. Microbiol.* 77(18), 6350–6. https://doi.org/10.1128/AEM.05498-11

Bowers, R.M. et al., 2013. Seasonal variability in bacterial and fungal diversity of the near-surface atmosphere. *Environ. Sci. Technol.* 47(21), 12097–12106. https://doi.org/10.1021/es402970s

Brągoszewska, E. et al., 2017. Concentration and size distribution of culturable bacteria in ambient air during spring and winter in Gliwice: A typical urban area. *Atmosphere (Basel)*. 8(12), 239. https://doi.org/10.3390/atmos8120239

Burrows, S.M. et al., 2009. Bacteria in the global atmosphere – Part 1: Review and synthesis of literature data for different ecosystems. *Atmos. Chem. Phys.* 9(23), 9263–9280. https://doi.org/10.5194/acp-9-9263-2009

Cáliz, J. et al., 2018. A long-term survey unveils strong seasonal patterns in the airborne microbiome coupled to general and regional atmospheric circulations. *Proc. Natl. Acad. Sci. U. S. A.* 115(48), 12229–12234. https://doi.org/10.1073/pnas.1812826115

Cao, G. et al., 2011. Development of an improved methodology to detect infectious airborne influenza virus using the NIOSH bioaerosol sampler. *J. Environ. Monit.* 13(12), 3321. https://doi.org/10.1039/c1em10607d

Cao, C. et al., 2014. Inhalable microorganisms in Beijing's PM2.5 and PM10 pollutants during a severe smog event. *Environ. Sci. Technol.* 48(3), 1499–507. https://doi.org/10.1021/es4048472

Chen, Y. et al., 2024. Indoor/Outdoor airborne microbiome characteristics in residential areas across four seasons and its indoor purification. *Environ. Int.* 190, 108857. https://doi.org/10.1016/j.envint.2024.108857

Cullen, C.M. et al., 2020. Emerging priorities for microbiome research. *Front. Microbiol.* 11(February). https://doi.org/10.3389/fmicb.2020.00136

D'Accolti, M. et al., 2023. Shaping the subway microbiome through probiotic-based sanitation during the COVID-19 emergency: A pre–post case–control study. *Microbiome* 11(1), 64. https://doi.org/10.1186/s40168-023-01512-2

Danko, D. et al., 2021. A global metagenomic map of urban microbiomes and antimicrobial resistance. *Cell* 184(13), 3376–3393.e17. https://doi.org/10.1016/j.cell.2021.05.002

Dekaboruah, E. et al., 2020. Human microbiome: An academic update on human body site specific surveillance and its possible role. *Arch. Microbiol.* 202(8), 2147–2167. https://doi.org/10.1007/s00203-020-01931-x

DeLeon-Rodriguez, N. et al., 2013. Microbiome of the upper troposphere: Species composition and prevalence, effects of tropical storms, and atmospheric implications. *Proc. Natl. Acad. Sci. U. S. A.* 110(7), 2575–2580. https://doi.org/10.1073/PNAS.1212089110/

Després, V. et al., 2012. Primary biological aerosol particles in the atmosphere: A review. *Tellus B Chem. Phys. Meteorol.* 64(1), 15598. https://doi.org/10.3402/tellusb.v64i0.15598

Dommergue, A. et al., 2019. Methods to investigate the global atmospheric Microbiome. *Front. Microbiol.* 10(FEB), 243. https://doi.org/10.3389/FMICB.2019.00243

Dybwad, M. et al., 2012. Characterization of airborne bacteria at an underground subway station. *Appl. Environ. Microbiol.* 78(6), 1917. https://doi.org/10.1128/AEM.07212-11

Dybwad, M. et al., 2014. Comparative testing and evaluation of nine different air samplers: End-to-end sampling efficiencies as specific performance measurements for bioaerosol applications. *Aerosol Sci. Technol.* 48(3), 282–295. https://doi.org/10.1080/02786826.2013.871501/SUPPL_FILE/UAST_A_871501_SM5434.ZIP

Ege, M.J. et al., 2011. Exposure to environmental microorganisms and childhood asthma. *N. Engl. J. Med.* 364(8), 701–709. https://doi.org/10.1056/NEJMOA1007302/SUPPL_FILE/NEJMOA1007302_DISCLOSURES.PDF

Fahlgren, C. et al., 2011. Diversity of airborne bacteria in samples collected using different devices for aerosol collection. *Aerobiologia (Bologna)*. 27(2), 107–120. https://doi.org/10.1007/S10453-010-9181-Z/FIGURES/3

Fakunle, A.G. et al., 2023. Childhood lower respiratory tract infections linked to residential airborne bacterial and fungal microbiota. *Environ. Res.* 231, 116063. https://doi.org/10.1016/j.envres.2023.116063

Fang, H.W. et al., 2014. Livestock-associated methicillin-resistant staphylococcus aureus ST9 in pigs and related personnel in Taiwan. *PLoS One* 9(2), e88826. https://doi.org/10.1371/JOURNAL.PONE.0088826

Fierer, N. et al., 2008. Short-term temporal variability in airborne bacterial and fungal populations. *Appl. Environ. Microbiol.* 74(1), 200. https://doi.org/10.1128/AEM.01467-07

Fierer, N., 2017. Embracing the unknown: Disentangling the complexities of the soil microbiome. *Nat. Rev. Microbiol.* 15(10), 579–590. https://doi.org/10.1038/nrmicro.2017.87

Finlayson-Pitts, B.J., Pitts, J.N., 2000. Analytical methods and typical atmospheric concentrations for gases and particles. in: *Chemistry of the Upper and Lower Atmosphere*. Elsevier, pp. 547–656. https://doi.org/10.1016/B978-012257060-5/50013-7

Frana, T.S. et al., 2013. Isolation and characterization of methicillin-resistant staphylococcus aureus from pork farms and visiting veterinary students. *PLoS One* 8(1). https://doi.org/10.1371/JOURNAL.PONE.0053738

Franzetti, A. et al., 2011. Seasonal variability of bacteria in fine and coarse urban air particulate matter. *Appl. Microbiol. Biotechnol.* 90(2), 745–753. https://doi.org/10.1007/s00253-010-3048-7

Fresia, P. et al., 2019. Urban metagenomics uncover antibiotic resistance reservoirs in coastal beach and sewage waters. *Microbiome* 7(1), 1–9. https://doi.org/10.1186/S40168-019-0648-Z/FIGURES/4

Fu, X. et al., 2020. Associations between respiratory infections and bacterial microbiome in student dormitories in Northern China. *Indoor Air* 30(5), 816–826. https://doi.org/10.1111/ina.12677

Gavito-Covarrubias, D. et al., 2024. Epigenetic mechanisms of particulate matter exposure: Air pollution and hazards on human health. *Front. Genet.* 14, 1306600. https://doi.org/10.3389/fgene.2023.1306600

Genitsaris, S. et al., 2017. Variability of airborne bacteria in an urban Mediterranean area (Thessaloniki, Greece). *Atmos. Environ.* 157, 101–110. https://doi.org/10.1016/j.atmosenv.2017.03.018

Gilbert, J.A., Hartmann, E.M., 2024. The indoors microbiome and human health. *Nat. Rev. Microbiol.* 22(12), 742–755. https://doi.org/10.1038/s41579-024-01077-3

Gilbert, J.A. et al., 2018. Current understanding of the human microbiome. *Nat. Med.* 24(4), 392. https://doi.org/10.1038/NM.4517

Gohli, J. et al., 2019. The subway microbiome: Seasonal dynamics and direct comparison of air and surface bacterial communities. *Microbiome* 7(1), 1–16. https://doi.org/10.1186/S40168-019-0772-9/FIGURES/4

Gou, H. et al., 2016. Assessment of microbial communities in PM1 and PM10 of Urumqi during winter. *Environ. Pollut.* 214, 202–210. https://doi.org/10.1016/j.envpol.2016.03.073

Griffin, D.W. et al., 2001. African desert dust in the Caribbean atmosphere: Microbiology and public health. *Aerobiologia (Bologna)*. 17(3), 203–213. https://doi.org/10.1023/A:1011868218901

Griffin, D.W. et al., 2017. Global-scale atmospheric dispersion of microorganisms. In *Microbiology of aerosols* (eds A.-M. Delort and P. Amato). Chichester: Wiley, pp. 155–194. https://doi.org/10.1002/9781119132318.ch2c

Grydaki, N. et al., 2021. Bioaerosols in the Athens Metro: Metagenetic insights into the PM10 microbiome in a naturally ventilated subway station. *Environ. Int.* 146, 106186. https://doi.org/10.1016/j.envint.2020.106186

Gusareva, E.S. et al., 2019. Microbial communities in the tropical air ecosystem follow a precise diel cycle. *Proc. Natl. Acad. Sci. U. S. A.* 116(46), 23299–23308. https://doi.org/10.1073/PNAS.1908493116/-/DCSUPPLEMENTAL

Gusareva, E.S. et al., 2020. Taxonomic composition and seasonal dynamics of the air microbiome in West Siberia. *Sci. Reports* 10(1), 1–7. https://doi.org/10.1038/s41598-020-78604-8

Gwenzi, W. et al., 2022. The air-borne antibiotic resistome: Occurrence, health risks, and future directions. *Sci. Total Environ.* 804, 150154. https://doi.org/10.1016/J.SCITOTENV.2021.150154

Haddrell, A.E., Thomas, R.J., 2017. Aerobiology: Experimental considerations, observations, and future tools. *Appl. Environ. Microbiol.* 83(17), 809–826. https://doi.org/10.1128/AEM.00809-17

Hamidou Soumana, I., Carlsten, C., 2021. Air pollution and the respiratory microbiome. *J. Allergy Clin. Immunol.* 148(1), 67–69. https://doi.org/10.1016/j.jaci.2021.05.013

Harnpicharnchai, P. et al., 2023. Temporal dynamics in diversity and composition of the air mycobiome and dominant allergenic fungi in the subway environment. *Environ. DNA* 5(6), 1265–1288. https://doi.org/10.1002/edn3.449

Harnpicharnchai, P. et al., 2024. Interplay of xenobiotic-degrading and antibiotic-resistant microorganisms among the microbiome found in the air, handrail, and floor of the subway station. *Environ. Res.* 247, 118269. https://doi.org/10.1016/j.envres.2024.118269

Hassan, Z.U. et al., 2022. Seasonal variations of the airborne microbial assemblages of the Seoul subway, South Korea from 16S and ITS gene profiles with chemical analysis. *Sci. Rep.* 12(1), 18456. https://doi.org/10.1038/s41598-022-21120-8

Hernández, A.M. et al., 2020. Station and train surface microbiomes of Mexico City's metro (subway/underground). *Sci. Rep.* 10(1), 8798. https://doi.org/10.1038/s41598-020-65643-4

Hernández-Alomía, F. et al., 2023. Beta-lactam antibiotic resistance genes in the microbiome of the public transport system of Quito, Ecuador. *Int. J. Environ. Res. Public Health* 20(3), 1900. https://doi.org/10.3390/ijerph20031900

Hernández-Castillo, O. et al., 2014. Aerobiological study in the Mexico City subway system. *Aerobiologia (Bologna).* 4(30), 357–367. https://doi.org/10.1007/S10453-014-9334-6

Hoisington, A.J. et al., 2014. Impact of sampler selection on the characterization of the indoor microbiome via high-throughput sequencing. *Build. Environ.* 80, 274–282. https://doi.org/10.1016/j.buildenv.2014.04.021

Hsu, T. et al., 2016. Urban transit system microbial communities differ by surface type and interaction with humans and the environment. *mSystems* 1(3), 1–18. https://doi.org/10.1128/mSystems.00018-16

Ivanova, M. et al., 2025. Composition and occurrence of airborne fungi in two urbanized areas of the city of Sofia, Bulgaria. *Appl. Microbiol.* 5(3), 96. https://doi.org/10.3390/applmicrobiol5030096

Jiang, S. et al., 2022. Airborne microbial community structure and potential pathogen identification across the PM size fractions and seasons in the urban atmosphere. *Sci. Total Environ.* 831, 154665. https://doi.org/10.1016/j.scitotenv.2022.154665

Jin, X. et al., 2025. Potential effects of indoor microbiome characteristics on health: A systematic review. *Public Health* 240, 137–147. https://doi.org/10.1016/j.puhe.2025.01.030

Jo, W.-K., Seo, Y.-J., 2005. Indoor and outdoor bioaerosol levels at recreation facilities, elementary schools, and homes. *Chemosphere* 61(11), 1570–1579. https://doi.org/10.1016/j.chemosphere.2005.04.103

Kang, K. et al., 2018. The environmental exposures and inner- and intercity traffic flows of the metro system may contribute to the skin microbiome and resistome. *Cell Rep.* 24(5), 1190–1202.e5. https://doi.org/10.1016/J.CELREP.2018.06.109

Kesavan, J. et al., 2010. Sampling and retention efficiencies of batch-type liquid-based bioaerosol samplers. *Aerosol Sci. Technol.* 44(10), 817–829. https://doi.org/10.1080/02786826.2010.497513

Kesavan, J.S., Schepers, D.R., 2006. Characteristics and sampling efficiencies of OMNI 3000 aerosol samplers. Maryland: Edgewood Chemical Biological Center. https://doi.org/10.21236/ADA457464

Kim, J. et al., 2022. Dust-associated bacterial and fungal communities in indoor multiple-use and public transportation facilities. *Atmosphere (Basel).* 13(9), 1373. https://doi.org/10.3390/atmos13091373

Kim, J. et al., 2023. Investigation of surface bacterial diversities and compositions in the global subway facilities. *Atmosphere (Basel).* 14(1), 140. https://doi.org/10.3390/atmos14010140

Kim, K.-H. et al., 2018. Airborne bioaerosols and their impact on human health. *J. Environ. Sci.* 67, 23–35. https://doi.org/10.1016/j.jes.2017.08.027

Kish, L. et al., 2013. Environmental particulate matter induces murine intestinal inflammatory responses and alters the gut microbiome. *PLoS One* 8(4), e62220. https://doi.org/10.1371/journal.pone.0062220

Klimenko, N.S. et al., 2020. Co-occurrence patterns of bacteria within microbiome of Moscow subway. *Comput. Struct. Biotechnol. J.* 18, 314–322. https://doi.org/10.1016/j.csbj.2020.01.007

Kraemer, J.G. et al., 2018. Influence of Pig farming on the human nasal microbiota: Key role of airborne microbial communities. *Appl. Environ. Microbiol.* 84(6). https://doi.org/10.1128/AEM.02470-17

Kreyling, W.G. et al., 1999. Anatomic localization of 24- and 96-h particle retention in canine airways. *J. Appl. Physiol.* 87(1), 269–284. https://doi.org/10.1152/jappl.1999.87.1.269

Kumari, S.P. et al., 2024. An efficient method for isolating bacterial DNA from bioaerosols for metagenomics. *Aerosol Sci. Technol.* 59(5), 595–607. https://doi.org/10.1080/02786826.2024.2410037

Kumari, S.P. et al., 2025. Seasonal variations and functional insights into the urban air microbiome across public transit environments at railway stations in Delhi, India. *Sci. Total Environ.* 1009, 181062. https://doi.org/10.1016/j.scitotenv.2025.181062

Lai, P.S. et al., 2017. Impact of environmental microbiota on human microbiota of workers in academic mouse research facilities: An observational study. *PLoS One* 12(7). https://doi.org/10.1371/JOURNAL.PONE.0180969

Larsson, D.G.J. et al., 2018. Critical knowledge gaps and research needs related to the environmental dimensions of antibiotic resistance. *Environ. Int.* 117, 132–138. https://doi.org/10.1016/J.ENVINT.2018.04.041

Lee, B.U., 2011. Life comes from the air: A short review on bioaerosol control. *Aerosol Air Qual. Res.* 11(7), 921–927. https://doi.org/10.4209/aaqr.2011.06.0081

Lee, J.Y. et al., 2017. Airborne bacterial communities in three East Asian cities of China, South Korea, and Japan. *Sci. Rep.* 7(1), 5545. https://doi.org/10.1038/s41598-017-05862-4

Lei, H. et al., 2025. Global biogeography of airborne viruses in public transit systems and their host interactions. *Microbiome* 13(1), 193. https://doi.org/10.1186/s40168-025-02173-z

Leung, M.H.Y. et al., 2014. Indoor-air microbiome in an urban subway network: Diversity and dynamics. *Appl. Environ. Microbiol.* 80(21), 6760–6770. https://doi.org/10.1128/AEM.02244-14

Leung, M.H.Y. et al., 2021. Characterization of the public transit air microbiome and resistome reveals geographical specificity. *Microbiome* 9(1), 112. https://doi.org/10.1186/S40168-021-01044-7

Li, K., 2011. Molecular comparison of the sampling efficiency of four types of airborne bacterial samplers. *Sci. Total Environ.* 409(24), 5493–8. https://doi.org/10.1016/j.scitotenv.2011.09.010

Li, N. et al., 2016. A work group report on ultrafine particles (AAAAI) why ambient ultrafine and engineered nanoparticles should receive special attention for possible adverse health outcomes in humans. *J. Allergy Clin. Immunol.* 138(2), 386. https://doi.org/10.1016/J.JACI.2016.02.023

Li, N. et al., 2020. Airborne antibiotic resistance genes in Hong Kong kindergartens. *Environ. Pollut.* 260, 114009. https://doi.org/10.1016/J.ENVPOL.2020.114009

Lighthart, B., 2000. Mini-review of the concentration variations found in the alfresco atmospheric bacterial populations. *Aerobiologia (Bologna)*. 16(1), 7–16. https://doi.org/10.1023/A:1007694618888

Lim, J.H. et al., 2020. Development of a multi-slit virtual impactor as a high-volume bio-aerosol sampler. *Sep. Purif. Technol.* 250, 117275. https://doi.org/10.1016/J.SEPPUR.2020.117275

Lin, X. et al., 2000. Survival of airborne microorganisms during swirling aerosol collection. *Aerosol Sci. Technol.* 32(3), 184–196. https://doi.org/10.1080/027868200303722

Liu, S. et al., 2025. Analysis of metagenomic data. *Nat. Rev. Methods Prim.* 5(1), 5. https://doi.org/10.1038/s43586-024-00376-6

Liu, Y. et al., 2024. Characterization of microbial communities in urban subway: Connotation for indoor environment quality and public health. *Air Qual. Atmos. Heal.* 17(7), 1401–1413. https://doi.org/10.1007/s11869-024-01515-4

Lloyd-Price, J. et al., 2016. The healthy human microbiome. *Genome Med.* 8(1), 51. https://doi.org/10.1186/s13073-016-0307-y

Lymperopoulou, D.S. et al., 2016. Contribution of vegetation to the microbial composition of nearby outdoor air. *Appl. Environ. Microbiol.* 82(13), 3822–3833. https://doi.org/10.1128/AEM.00610-16

Lynch, S.V. et al., 2014. Effects of early-life exposure to allergens and bacteria on recurrent wheeze and atopy in urban children. *J. Allergy Clin. Immunol.* 134(3), 593–601.e12. https://doi.org/10.1016/j.jaci.2014.04.018

Ma, L. et al., 2023. Seasonal variation characteristics of inhalable bacteria in bioaerosols and antibiotic resistance genes in Harbin. *J. Hazard. Mater.* 446, 130597. https://doi.org/10.1016/j.jhazmat.2022.130597

Mainelis, G. et al., 2010. Design and collection efficiency of a new electrostatic precipitator for bioaerosol collection. *Aerosol Sci. Technol.* 36(11), 1073–1085. https://doi.org/10.1080/02786820290092212

Marchesi, J.R., Ravel, J., 2015. The vocabulary of microbiome research: A proposal. *Microbiome* 3(1). https://doi.org/10.1186/S40168-015-0094-5

Marine, R. et al., 2014. Caught in the middle with multiple displacement amplification: The myth of pooling for avoiding multiple displacement amplification bias in a metagenome. *Microbiome* 2(1), 1–8. https://doi.org/10.1186/2049-2618-2-3/TABLES/3

Matthias-Maser, S. et al., 2000. The size distribution of primary biological aerosol particles in cloud water on the mountain Kleiner Feldberg/Taunus (FRG). *Atmos. Res.* 54(1), 1–13. https://doi.org/10.1016/S0169-8095(00)00039-9

Mazar, Y. et al., 2016. Effect of dust storms on the atmospheric microbiome in the Eastern Mediterranean. *Environ. Sci. Technol.* 50(8), 4194–4202. https://doi.org/10.1021/ACS.EST.5B06348

Mcfarland, A.R. et al., 2010. Wetted wall cyclones for bioaerosol sampling. *Aerosol Sci. Technol.* 44(4), 241–252. https://doi.org/10.1080/02786820903555552

McKinney, M.L., 2006. Urbanization as a major cause of biotic homogenization. *Biol. Conserv.* 127(3), 247–260. https://doi.org/10.1016/J.BIOCON.2005.09.005

Meola, M. et al., 2015. Bacterial composition and survival on Sahara dust particles transported to the European Alps. *Front. Microbiol.* 6(DEC), 1454. https://doi.org/10.3389/FMICB.2015.01454/BIBTEX

MetaSUB International Consortium, 2016. The metagenomics and metadesign of the subways and urban biomes (MetaSUB) international consortium inaugural meeting report. *Microbiome* 4(1), 24. https://doi.org/10.1186/s40168-016-0168-z

Metaxatos, A. et al., 2025. Unveiling microbial diversity in Greek urban air and recreational seawater using DNA barcoding. *Atmosphere (Basel).* 16(9), 1082. https://doi.org/10.3390/atmos16091082

Michaud, J.M. et al., 2018. Taxon-specific aerosolization of bacteria and viruses in an experimental ocean-atmosphere mesocosm. *Nat. Commun.* 9(1), 1–10. https://doi.org/10.1038/s41467-018-04409-z

Mkilima, T., 2025. Urban aerosols, the microbiome, and the rise of "urban atopy." *Aerosol Sci. Eng.* 00331–2. https://doi.org/10.1007/s41810-025-00331-2

Moelling, K., 2021. Within-host and between-host evolution in SARS-CoV-2—New variant's source. *Viruses* 13(5), 751. https://doi.org/10.3390/v13050751

Moelling, K., Broecker, F., 2020. Air microbiome and pollution: Composition and potential effects on human health, including SARS Coronavirus infection. *J. Environ. Public Health* 2020, 1–14. https://doi.org/10.1155/2020/1646943

Möller, W. et al., 2004. Mucociliary and long-term particle clearance in the airways of healthy nonsmokers. *J Appl Physiol*, 97. https://doi.org/10.1152/japplphysiol.00970.2003

Mortazavi, H. et al., 2023. Detection of SARS-CoV-2 in the indoor air and surfaces of subway trains in Mashhad, Iran. *Brazilian J. Microbiol.* 54(3), 1865–1873. https://doi.org/10.1007/s42770-023-01089-w

Mukherjee, A. et al., 2023. Metagenomic approaches for understanding microbial communities in contaminated environments: Bioinformatic tools, case studies and future outlook, in: Vineet Kumar, Muhammad Bilal, Sushil Kumar Shahi, Vinod Kumar Garg (eds.), *Metagenomics to Bioremediation*. Academic Press, pp. 103–156. https://doi.org/10.1016/B978-0-323-96113-4.00003-2

Mutlu, E.A. et al., 2018. Inhalational exposure to particulate matter air pollution alters the composition of the gut microbiome. *Environ. Pollut.* 240, 817–830. https://doi.org/10.1016/j.envpol.2018.04.130

Núñez, A. et al., 2021. Seasonal changes dominate long-term variability of the urban air microbiome across space and time. *Environ. Int.* 150, 106423. https://doi.org/10.1016/j.envint.2021.106423

O'Connor, G.T. et al., 2018. Early-life home environment and risk of asthma among inner-city children. *J. Allergy Clin. Immunol.* 141(4), 1468–1475. https://doi.org/10.1016/j.jaci.2017.06.040

Ooi, D.S.-Q. et al., 2020. Developmental origins of health and disease: Impact of environmental dust exposure in modulating microbiome and its association with non-communicable diseases. *J. Dev. Orig. Health Dis.* 11(6), 545–556. https://doi.org/10.1017/S2040174420000549

Palomo, S. et al., 2013. Sponge-derived Kocuria and Micrococcus spp. as sources of the new thiazolyl peptide antibiotic kocurin. *Mar. Drugs* 11(4), 1071. https://doi.org/10.3390/MD11041071

Pastuszka, J.S. et al., 2000. Bacterial and fungal aerosol in indoor environment in Upper Silesia, Poland. *Atmos. Environ.* 34(22), 3833–3842. https://doi.org/10.1016/S1352-2310(99)00527-0

Peimbert, M., Alcaraz, L.D., 2022. Where environmental microbiome meets its host: Subway and passenger microbiome relationships. *Mol. Ecol.* 32(10), 2602–2618. https://doi.org/10.1111/mec.16440

Pellissier, L. et al., 2016. Airborne and grain dust fungal community compositions are shaped regionally by plant genotypes and farming practices. *Appl. Environ. Microbiol.* 82(7), 2121. https://doi.org/10.1128/AEM.03336-15

Perry, J.A. et al., 2014. The antibiotic resistome: What's new? Curr. *Opin. Microbiol.* 21, 45–50. https://doi.org/10.1016/J.MIB.2014.09.002

Pochtovyi, A.A. et al., 2022. Microbiological characteristics of some stations of Moscow subway. *Biology (Basel).* 11(2). https://doi.org/10.3390/BIOLOGY11020170/S1

Pollegioni, P. et al., 2023. Variability of airborne microbiome at different urban sites across seasons: A case study in Rome. *Front. Environ. Sci.* 11. https://doi.org/10.3389/fenvs.2023.1213833
Prieto Lago, A. et al., 2018. Viral and bacterial environmental detection in livestock farms. *Ref. Modul. Earth Syst. Environ. Sci.* https://doi.org/10.1016/B978-0-12-409548-9.11252-7
Prussin, A.J. et al., 2015. Total concentrations of virus and bacteria in indoor and outdoor air. *Environ. Sci. Technol. Lett.* 2(4), 84–88. https://doi.org/10.1021/acs.estlett.5b00050
Qin, N. et al., 2020. Longitudinal survey of microbiome associated with particulate matter in a megacity. *Genome Biol.* 21(1), 55. https://doi.org/10.1186/s13059-020-01964-x
Rao, Y. et al., 2020. Community structure and influencing factors of airborne microbial aerosols over three chinese cities with contrasting social-economic levels. *Atmosphere (Basel).* 11(4), 317. https://doi.org/10.3390/atmos11040317
Robertson, C.E. et al., 2013. Culture-independent analysis of aerosol microbiology in a metropolitan subway system. *Appl. Environ. Microbiol.* 79(11), 3485. https://doi.org/10.1128/AEM.00331-13
Roux, J.M. et al., 2016. Development of a new portable air sampler based on electrostatic precipitation. *Environ. Sci. Pollut. Res.* 23(9), 8175–8183. https://doi.org/10.1007/S11356-015-5522-3/FIGURES/7
Ruiz-Calderon, J.F. et al., 2016. Walls talk: Microbial biogeography of homes spanning urbanization. *Sci. Adv.* 2(2), e1501061. https://doi.org/10.1126/sciadv.1501061
Ruiz-Gil, T. et al., 2020. Airborne bacterial communities of outdoor environments and their associated influencing factors. *Environ. Int.* 145, 106156. https://doi.org/10.1016/J.ENVINT.2020.106156
Seino, K. et al., 2005. An evidential example of airborne bacteria in a crowded, underground public concourse in Tokyo. *Atmos. Environ.* 39(2), 337–341. https://doi.org/10.1016/J.ATMOSENV.2004.09.030
Sharma, S. et al., 2024. The composition of the bacterial communities collected from the PM10 samples inside the Seoul subway and railway station. *Sci. Rep.* 14(1), 6478. https://doi.org/10.1038/s41598-023-49848-x
Shukla, S.K. et al., 2017. The nasal microbiota of dairy farmers is more complex than oral microbiota, reflects occupational exposure, and provides competition for staphylococci. *PLoS One* 12(8). https://doi.org/10.1371/JOURNAL.PONE.0183898
Siriarchawatana, P. et al., 2023. Temporal, compositional, and functional differences in the microbiome of Bangkok subway air environment. *Environ. Res.* 219, 115065. https://doi.org/10.1016/j.envres.2022.115065
Solazzo, G. et al., 2025. The microbiome of total suspended particles and its influence on the respiratory microbiome of healthy office workers. *Ecotoxicol. Environ. Saf.* 291, 117874. https://doi.org/10.1016/j.ecoenv.2025.117874
Spring, A.M. et al., 2021. Land use effects on airborne bacterial communities are evident in both near-surface and higher-altitude air. *Diversity* 13(2), 85. https://doi.org/10.3390/d13020085
Stein, M.M. et al., 2016. Innate immunity and asthma risk in Amish and Hutterite farm children. *N. Engl. J. Med.* 375(5), 411–421. https://doi.org/10.1056/NEJMOA1508749/SUPPL_FILE/NEJMOA1508749_DISCLOSURES.PDF
Sung, G. et al., 2017. Highly efficient in-line wet cyclone air sampler for airborne virus detection †. *J. Mech. Sci. Technol.* 31(9), 4363–4369. https://doi.org/10.1007/s12206-017-0835-4
Tischer, C. et al., 2016. Urban dust microbiome: Impact on later atopy and wheezing. *Environ. Health Perspect.* 124(12), 1919–1923. https://doi.org/10.1289/EHP158
Triadó-Margarit, X. et al., 2017. Bioaerosols in the Barcelona subway system. *Indoor Air* 27(3), 564–575. https://doi.org/10.1111/ina.12343
Trinh, P. et al., 2018. One health relationships between human, animal, and environmental microbiomes: A mini-review. *Front. Public Heal.* 6, 235. https://doi.org/10.3389/fpubh.2018.00235
Vanhaecke, T. et al., 2022. Drinking water source and intake are associated with distinct gut microbiota signatures in US and UK populations. *J. Nutr.* 152(1), 171–182. https://doi.org/10.1093/jn/nxab312
Vargas-Robles, D. et al., 2020. Passenger-surface microbiome interactions in the subway of Mexico City. *PLoS One* 15(8), e0237272. https://doi.org/10.1371/JOURNAL.PONE.0237272
Verreault, D. et al., 2008. Methods for sampling of airborne viruses. *Microbiol. Mol. Biol. Rev.* 72(3), 413. https://doi.org/10.1128/MMBR.00002-08
Wang, M. et al., 2025. Particulate matter air pollution as a cause of lung cancer: Epidemiological and experimental evidence. *Br. J. Cancer* 132(11), 986–996. https://doi.org/10.1038/s41416-025-02999-2
Wang, Z. et al., 2025. SARS-CoV-2 detection in heating, ventilation, and air conditioning filters of university lecture rooms. *ACS ES&T Air* 2(7), 1280–1291. https://doi.org/10.1021/acsestair.5c00056
Whon, T.W. et al., 2012. Metagenomic characterization of airborne viral DNA diversity in the near-surface atmosphere. *J. Virol.* 86(15), 8221–8231. https://doi.org/10.1128/JVI.00293-12

Yan, R. et al., 2025. Microbial diversity and environmental determinants at Shanghai Hongqiao railway station: A comprehensive microbial assessment. *Environ. Pollut.* 366, 125534. https://doi.org/10.1016/j.envpol.2024.125534

Yang, L. et al., 2025. Insights into bacteria characteristics and potential pathogen in rural indoor households in Fenwei Plain, China. *Atmos. Environ.* 347, 121111. https://doi.org/10.1016/j.atmosenv.2025.121111

Yu, X. et al., 2025. Bioparticle sources, dispersion, and influencing factors in rural environmental air. *Aerobiology* 3(2), 4. https://doi.org/10.3390/aerobiology3020004

Yuan, S. et al., 2012. Evaluation of methods for the extraction and purification of DNA from the human microbiome. *PLoS One* 7(3). https://doi.org/10.1371/JOURNAL.PONE.0033865

Zhang, L. et al., 2025. Bacterial diversity and biomarkers screening of station and carriage surface in Shanghai metro system, China. *Curr. Res. Microb. Sci.* 8, 100374. https://doi.org/10.1016/j.crmicr.2025.100374

Zhang, M. et al., 2023. Impact of environmental characteristics on children's gut microbiota – A pilot study in assessing the role of indoor microbiome and metabolites. *Environ. Res.* 234, 116114. https://doi.org/10.1016/j.envres.2023.116114

Zhao, L. et al., 2023. The human respiratory microbiome: Current understandings and future directions. *Am. J. Respir. Cell Mol. Biol.* 68(3), 245–255. https://doi.org/10.1165/rcmb.2022-0208TR

Zhou, X. et al., 2025. Dynamic airborne mycobiome in the metropolitan city transit system is driven by seasonality and station type. *Microbiol. Spectr.* 13(11), e01626–25. https://doi.org/10.1128/spectrum.01626-25

Zhou, X.-Y. et al., 2023. City-scale distribution of airborne antibiotic resistance genes. *Sci. Total Environ.* 856, 159176. https://doi.org/10.1016/j.scitotenv.2022.159176

Zhu, H. et al., 2003. Experimental study of indoor and outdoor airborne bacterial concentrations in Tempe, Arizona, USA. *Aerobiologia (Bologna).* 19(3/4), 201–211. https://doi.org/10.1023/B:AERO.0000006571.23160.8a

CHAPTER 5

# The Environmental Microbiome

## A COMPUTATIONAL GENOMICS PERSPECTIVE

*Lenore Pipes*

## 5.1 PART 1: THE WORLD OF ENVIRONMENTAL MICROBIOMES

### 5.1.1 Introduction to the Environmental Microbiome

Microbial life is a ubiquitous feature of Earth's biosphere, which is characterized by immense phylogenetic and functional diversity. These microorganisms populate a vast range of habitats, from terrestrial and aquatic ecosystems to the tissues of other living organisms. This collective of microorganisms (including bacteria, archaea, fungi, viruses, and protists) and their collective genetic material is known as the microbiome (Lederberg and McCray, 2001). Microbiomes can be broadly categorized as either host-associated or environmental. While host-associated systems like the human microbiome are a major focus of research (Turnbaugh et al., 2007), the environmental microbiome represents the largest reservoir of microbial diversity (Whitman, Coleman, and Wiebe, 1998). This category encompasses the microbial communities found in terrestrial soils, aquatic systems, the atmosphere, and built environments.

The structure and function of environmental microbiomes are determined primarily by the physicochemical properties of their habitat (Fierer and Jackson, 2006). This principle distinguishes them from host-associated microbiomes, where community composition is strongly influenced by host-specific factors such as genetics and immune status (Ley et al., 2008). Nevertheless, these microbial domains are not discrete systems. A continuous interchange of microorganisms occurs between the external environment and its resident hosts. This microbial flux occurs via inhalation, ingestion, and contact with surfaces, immediately influencing the resident microbiome (Costello et al., 2012). The consequence of these environmental communities is far-reaching, as their collective metabolic activities drive the planetary-scale biogeochemical cycles that support ecosystem function and stability.

Environmental microorganisms function as the primary agents of global biogeochemical cycling (Blaser and Falkow, 2009). These key processes are responsible for the transformation and circulation of essential elements, including carbon, nitrogen, and sulfur, between the Earth's biotic and abiotic reservoirs (Figure 5.1).

The roles of environmental microbiomes are evident across a range of distinct ecosystems:

- **In Soil**: The soil microbiome is central to terrestrial ecosystems (Bardgett and Van Der Putten, 2014). As one of the most densely populated microbial habitats,


DOI: 10.1201/9781003637950-5

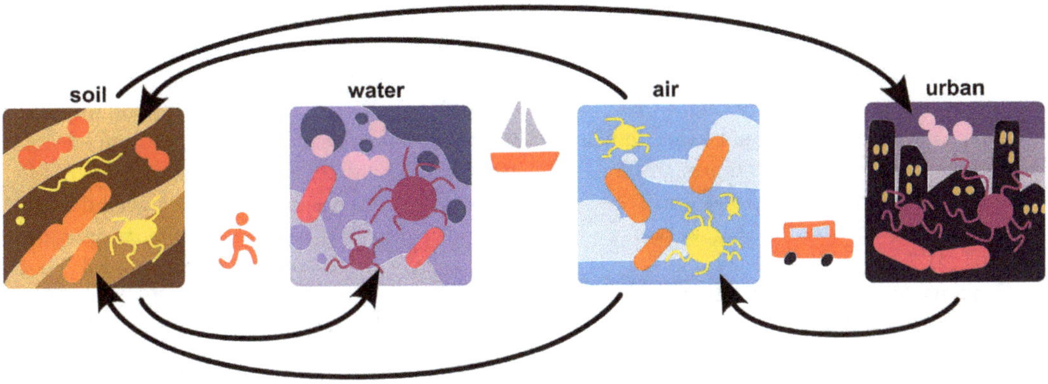


Figure 5.1 Environmental reservoirs and transmission pathways. Major habitats include soil, water, the aerobiome, and the built environment. Arrows show the exchange of taxa between these domains and human hosts, linking resident microbiota to their environmental sources.

soil can support up to $10^9$ microbial cells (Whitman, Coleman, and Wiebe, 1998), with a diversity encompassing thousands of distinct taxa (Torsvik, Goksøyr, and Daae, 1990). These microorganisms mediate key ecosystem processes, including the decomposition of organic matter, which increases nutrient bioavailability for plants (Falkowski, Fenchel, and Delong, 2008). They also form symbiotic relationships with plant roots which improve nutrient uptake and provide protection against pathogens, a function necessary for agriculture and food security (Berendsen, Pieterse, and Bakker, 2012). At a global scale, the soil microbiome plays a central role in regulating the carbon cycle by controlling the balance between carbon sequestration in soil organic matter and its respiratory release into the atmosphere (Schmidt et al., 2011).

- **In Water**: Aquatic microbiomes perform essential roles in marine and freshwater ecosystems, serving as the primary producers at the base of most aquatic food webs (Azam et al., 1983; Falkowski, Fenchel, and Delong, 2008). Photosynthetic microorganisms, such as cyanobacteria and eukaryotic phytoplankton, produce roughly half of Earth's oxygen and primary biomass (Field et al., 1998). Heterotrophic microbes drive the remineralization of dissolved and particulate organic matter, a process that recycles essential nutrients within the water column (Azam et al., 1983) and is critical for bioremediation. Given their role in these processes, changes in aquatic microbial composition act as indicators of environmental shifts like eutrophication, pollution, climate change, and salinization (Paerl and Otten, 2013).
- **In The Air**: The atmosphere, often called the aerobiome, is a vast, transient habitat that allows microorganisms to disperse globally (Fröhlich-Nowoisky et al., 2016). While microbial biomass density is significantly lower than in terrestrial or aquatic systems, the atmosphere contains a taxonomically diverse assemblage of bioaerosols, including bacteria, fungal spores, and viruses, that can be transported over intercontinental distances (Womack et al., 2010; Griffin, 2007). This atmospheric conveyance connects geographically distinct ecosystems and influences global biogeography and pathogen transmission (Smith et al., 2013). Beyond their role in dispersal, airborne microbes actively modulate meteorological processes. Specific bacterial taxa function as ice nucleating particles (INPs) and cloud condensation nuclei (CCN), linking microbial activity to cloud physics and the hydrological cycle (Morris et al., 2014; Christner et al., 2008).
- **In the Built Environment**: The built environment constitutes a distinct ecosystem that selects for specific microbial assemblages derived primarily from its

human occupants and the surrounding outdoor atmosphere (Gilbert and Stephens, 2018). Design choices like ventilation shape indoor microbes. Rooms with mechanical ventilation tend to accumulate communities like those on human skin, but natural ventilation allows for diversity closer to the outdoors (Kembel et al., 2012a). Human occupants act as the primary source of indoor bioaerosols, shedding vast quantities of biological particles, which accumulate on surfaces and in dust reservoirs (Hospodsky et al., 2012). These indoor microbial exposures play a major role in human health, shaping immune system development and affecting the prevalence of respiratory conditions like asthma and allergies (Stein et al., 2016; Ege et al., 2011).

Describing the environmental microbiome has become a key focus of contemporary biology. Characterizing the taxonomy, metabolism, and interactions of these communities supports ecosystem stewardship, sustainable agriculture, and the management of environmental health risks (Fierer, 2017; Stein et al., 2016). This chapter focuses on the computational methods driving modern microbiome science, specifically environmental DNA (eDNA) analysis. These bioinformatic frameworks process complex high-throughput sequencing (HTS) data, advancing our ability to monitor and interpret Earth's microscopic biosphere (Thomsen and Willerslev, 2015).

### 5.1.2 The Power of Environmental DNA (eDNA)

Past biodiversity assessments rested on physically identifying specimens through visual surveys or capture-based methods. However, high costs, logistical barriers, invasiveness, and taxonomic bias limit these protocols, especially for cryptic species or juvenile life stages that are hard to identify morphologically (Bohmann et al., 2014). The application of molecular diagnostics has introduced a non-invasive method for biological monitoring known as the analysis of eDNA. eDNA is defined as the total pool of genetic material, encompassing both cellular and extracellular nucleic acids, which is recovered directly from environmental matrices such as soil, water, or air. This allows for the detection of organisms in the absence of physical specimens (Ficetola et al., 2008; Rees et al., 2014).

The utility of eDNA analysis relies on the release of genetic material from organisms into their immediate environment via mechanisms such as epidermal exfoliation, excretion of metabolic waste, gamete release, and post-mortem decomposition (Thomsen and Willerslev, 2015). After release, the persistence of this extracellular DNA is governed by biogeochemical kinetics. In aquatic ecosystems, degradation rates are modulated by physicochemical variables including temperature, ultraviolet (UV) radiation, and pH, alongside microbial nuclease activity (Barnes et al., 2014; Strickler, Fremier, and Goldberg, 2015). Sedimentary environments with low temperatures, anoxia, or high mineral adsorption capacity can preserve DNA for long periods. This allows specific environments, such as permafrost or deep-sea sediments, to serve as natural archives for the recovery of ancient DNA and the reconstruction of past ecosystems (Willerslev et al., 2003; Pietramellara et al., 2009).

The analysis of eDNA provides a molecular profile of the biodiversity inhabiting a specific ecosystem. Through the sequestration and sequencing of nucleic acids from environmental matrices, such as water columns or sedimentary deposits, investigators can detect the presence of a broad array of taxonomic groups without the requirement for direct observation or physical handling of organisms (Taberlet et al., 2012a; Thomsen and Willerslev, 2015).

This methodology represents a shift in ecological monitoring for the following reasons:

- **Increased Sensitivity**: The high sensitivity of eDNA analysis allows for the detection of taxa present at low population densities, which exceeds the detection limits of conventional capture-based or visual surveys. Identifying invasions at this early stage increases the likelihood of successful eradication and control (Jerde, Mahon, et al., 2011; Dejean et al., 2012). Sensitive molecular tools reveal cryptic animals where physical sampling fails, clarifying the range of imperiled groups (Thomsen, Kielgast, et al., 2012).
- **Non-Invasiveness**: The collection of environmental samples is non-invasive, lessening the anthropogenic disturbance associated with traditional ecological surveys. Unlike capture-based techniques

such as electrofishing or netting, eDNA protocols eliminate the requirement for physical interaction with target organisms, which reduces physiological stress and the risk of accidental mortality (Rees et al., 2014). This characteristic makes the approach useful for the surveillance of threatened taxa or within fragile ecosystems in which physical disruption must be minimized (Bohmann et al., 2014). eDNA allows for frequent monitoring without the risk of behavioral avoidance or 'trap-shyness', ensuring reliable detection over time (Thomsen and Willerslev, 2015).

- **Comprehensive Biodiversity Assessment**: DNA metabarcoding combines HTS with eDNA analysis to identify diverse biological communities from a single sample. Amplifying these markers enables researchers to assess biodiversity at multiple levels, from microbial communities to macro-organisms (Taberlet et al., 2012b). This capacity provides a reconstruction of ecosystem structure that significantly exceeds the taxonomic breadth achievable through conventional, organism-centric survey methods (Ji et al., 2013; Thomsen and Willerslev, 2015).
- **Accessibility and Scalability**: The speed of eDNA sampling makes it easy to expand surveys across different locations and time periods. Using standard methods, scientists can acquire vast amounts of data far more quickly than with traditional specimen collection (Ji et al., 2013). This high-throughput capacity is necessary for generating the longitudinal datasets necessary to track ecosystem changes in response to global anthropogenic stressors, including climate change and chemical contamination (Bohmann et al., 2014).

eDNA analysis bypasses the logistical and taxonomic limits of traditional surveys, reshaping conservation biology and ecosystem monitoring (Thomsen and Willerslev, 2015; Bohmann et al., 2014). This approach provides a sensitive mechanism for biodiversity assessment, transitioning the focal point of observation from physical organismal capture to the analysis of persistent genetic traces (Rees et al., 2014; Taberlet et al., 2012a). The rest of this chapter outlines the data generation process and the computational frameworks used to interpret eDNA.

## 5.2 PART 2: FROM SAMPLE TO SEQUENCE: THE EDNA WORKFLOW

The translation of environmental samples into meaningful ecological insights requires a multi-stage workflow which connects field collection, molecular laboratory processing, and computational analysis. Although this chapter focuses on computational methodologies, an understanding of upstream experimental procedures is necessary for any genomicist working with eDNA data. Decisions about sampling, extraction, and library preparation can bias the data, skewing the taxonomic profile and affecting how sequences are interpreted (Zinger et al., 2019). Reliable analysis involves more than processing raw data; the bioinformatics pipeline must actively filter for sequencing errors, chimeras, index crossover, and contaminants found in negative controls (Deiner et al., 2017). The experimental and computational trajectory, from initial sample acquisition to final ecological inference, is outlined in Figure 5.2.

### 5.2.1 Field and Laboratory Methods

#### *5.2.1.1 Sample Collection and Preservation*

The initial acquisition of environmental samples is governed by the specific ecological hypothesis under investigation and the physicochemical nature of the target matrix. In aquatic ecosystems, sampling commonly includes the filtration of defined water volumes (typically 0.25 liters to several liters) through microporous membranes (typically 0.2–1.5 μm pore size) to entrap cellular and particle-associated DNA (Turner et al., 2014). Terrestrial sampling requires the collection of soil cores, where the spatial heterogeneity of the soil matrix requires randomization or stratification strategies to obtain representative biodiversity estimates (Fierer, 2017). The specific sampling parameters, including filtration volume, filter pore size, water depth, or soil horizon, are sources of technical variation that can alter detection probabilities and community diversity estimates (Goldberg et al., 2016). Following collection, immediate preservation is required to arrest DNA degradation driven by nuclease activity and microbial metabolism. While cryopreservation (flash-freezing) is

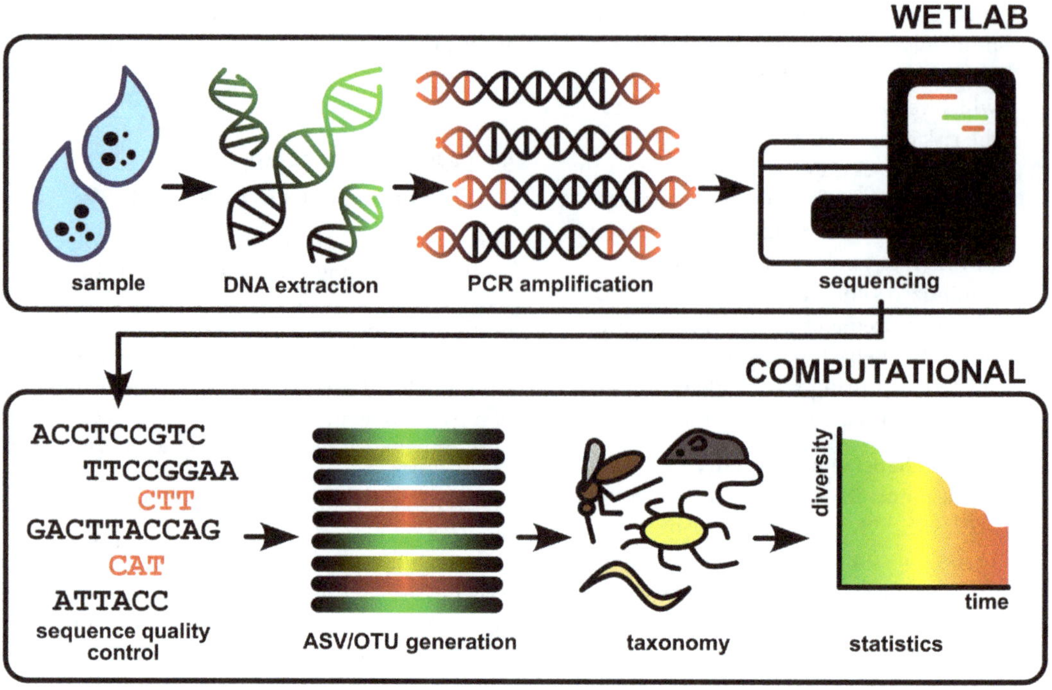


Figure 5.2 The eDNA workflow: from sample to ecological insight. This flowchart illustrates the end-to-end process of a typical eDNA study. It begins with sample collection in the field (i.e., water filtration), moves to laboratory procedures (DNA extraction, marker gene amplification via PCR, and high-throughput sequencing), and leads to the computational pipeline (sequence quality control, ASV/OTU generation, taxonomic assignment, and statistical analysis), which yields ecological interpretations such as biodiversity metrics and community composition plots.

often considered the standard, in situ chemical preservation using ethanol or lysis buffers provides a feasible alternative for remote fieldwork, albeit with potential trade-offs in DNA yield or extraction compatibility (Renshaw et al., 2015).

#### *5.2.1.2 DNA Extraction and Contamination Control*

The isolation of nucleic acids from environmental matrices, such as filtration membranes or sedimentary deposits, is a bottleneck in the analytical workflow. This step is commonly executed using commercial extraction chemistries or phenol-chloroform protocols formulated to separate DNA from enzymatic inhibitors, such as humic and fulvic acids, which are co-extracted from soil and aquatic samples. It is important to note that extraction protocols are rarely unbiased. Differential lysis efficiencies arising from variation in cell wall structure or mineral adsorption can systematically skew the recovered community composition, which favors organisms with labile membranes over those with recalcitrant cellular architectures (Zielińska et al., 2017).

Given the low standing stock of endogenous eDNA, these samples are sensitive to exogenous contamination from laboratory reagents, equipment, or operator shedding. Contamination management relies on a multi-tiered approach that uses field blanks, DNA capture controls, extraction blanks, and polymerase chain reaction (PCR) negative controls to monitor background DNA (Sepulveda et al., 2020). Sequencing these controls provides the baseline data needed for bioinformatic subtraction, a process that removes technical noise to reveal the actual biological community. The low concentration of endogenous genetic material in eDNA extracts makes them susceptible to contamination from exogenous DNA. This contaminant DNA can originate from both the physical materials used, such as laboratory reagents and sampling equipment, as well as through procedural cross-contamination from high-concentration samples (Salter et al., 2014). If left unresolved, such contamination can overwhelm the authentic environmental signal, leading to false positives and erroneous ecological inferences. The implementation of strict contamination mitigation measures, a practice largely inherited from the field of ancient DNA analysis, is a prerequisite for generating reliable eDNA data (Poinar and Cooper, 2000). These measures are implemented through a multi-tiered system of negative controls, including the following:

- **Field Negatives**: Field negative controls are designed to measure contamination that might occur while samples are being collected and transported, effectively monitoring the workflow before it even reaches the laboratory. These controls typically use a sterile substance, such as deionized water, that is certified to contain none of the DNA sequences being sought in the study, and they are handled in the field exactly like the biological samples (Dickie et al., 2018). These controls are subjected to the same handling procedures as the true samples, capturing any contaminant DNA introduced from sources such as airborne bioaerosols or residual genetic material on sampling equipment during collection and transport. Amplification in a negative control reveals a breach in protocol, acting as a benchmark to evaluate the accuracy of detections in the corresponding samples (Ficetola et al., 2015).
- **Extraction Blanks**: Extraction negative controls are implemented to identify and quantify contaminant DNA introduced during the laboratory processing phase. These controls, often referred to as extraction blanks, involve subjecting a sample-free matrix (such as a blank filter or purified water) to the entire DNA isolation and purification workflow. The resulting DNA represents the background contaminant profile, or 'kitome', originating from the extraction chemistry and laboratory environment (Eisenhofer et al., 2019). Characterizing this contaminant profile is necessary for distinguishing true low-level environmental signals from laboratory-derived artifacts, preventing the misinterpretation of results, particularly in low-biomass samples (Stinson, Keelan, and Payne, 2019).
- **PCR Negative Controls**: PCR negative controls, often known as no-template controls (NTCs), are used to find contamination introduced during library preparation and amplification. In these reactions, the DNA template is substituted with nuclease-free water to verify the sterility of the PCR master mix components, particularly Taq polymerases and primers, which are known vectors for bacterial DNA contamination (Salter et al., 2014). The inclusion of NTCs is important to distinguish whether amplified sequences originate from the environmental sample or from the reagents themselves, a distinction that is difficult when amplifying low-copy number targets where contaminant competition is most acute (Champlot et al., 2010; Goldberg et al., 2016).

Rather than just verifying experimental quality, these negative controls are included as essential data points within the sequencing results to aid in downstream analysis. To accurately define the background 'contaminome', these controls must be subjected to library preparation and HTS in parallel with biological samples (Kim et al., 2017). During downstream computational processing, these background sequences establish the statistical baseline necessary for *in silico* decontamination, which is a procedure that mathematically subtracts exogenous signals to resolve the authentic community composition (Davis et al., 2018). The omission of these controls from the bioinformatic workflow risks the inflation of diversity estimates and the reporting of erroneous ecological associations (Glassing et al., 2016). For this reason, the computational workflow must start by mapping the noise profile found in these experimental standards.

### 5.2.2 High-Throughput Sequencing of eDNA

Upon the isolation of eDNA, the analytical workflow progresses to the generation of digital sequence data via HTS technologies. This technology enables the read depth necessary to characterize the phylogenetic and functional diversity of microbial ecosystems (Sogin et al., 2006). The selection of a specific sequencing strategy serves as an analytical filter, which determines the resolution of taxonomic classification and the scope of functional inquiry attainable (Sharpton, 2014).

Contemporary environmental genomics relies on two methodologies to handle these objectives: targeted amplicon sequencing and whole-genome shotgun metagenomics.

### 5.2.2.1 Amplicon-Based Sequencing (Metabarcoding)

Targeted amplicon sequencing, designated as DNA metabarcoding, is the standard for taxonomic profiling within environmental microbiomes due to its high throughput and high throughput as well as its cost-efficiency. Rather than interrogating the total genomic content, this methodology relies on the selective amplification of specific, phylogenetically informative loci (Taberlet et al., 2012b). The utility of these marker genes is based on a specific structural duality: highly conserved flanking regions allow the annealing of broad-range universal primers, while the intervening hypervariable domains contain sufficient sequence divergence to distinguish taxa at the genus or species level (Cristescu, 2014). Methodologically, this workflow employs PCR to generate an enriched library of amplicons from the pool of extracted eDNA, which converts physical biological diversity into a comprehensive genomic profile. The selection of the marker gene depends on the specific taxonomic lineage under investigation, with the following loci serving as the standard community identifiers:

- **16S rRNA Gene**: This locus serves as the canonical molecular marker for the taxonomic profiling of prokaryotic domains (bacteria and archaea). Its phylogenetic utility is derived from a unique mosaic architecture, comprising highly conserved nucleotide sequences that allow broad-range primer annealing, interspersed with nine hypervariable regions (V1–V9) that provide the sequence divergence necessary for species discrimination (Chakravorty et al., 2007). Metabarcoding protocols typically target specific sub-regions, most notably the V4 or V3–V4 domains, which have been optimized to maximize the balance between amplification universality and taxonomic resolution (Klindworth et al., 2012).
- **18S rRNA Gene**: Functioning as the eukaryotic homolog to the 16S locus, the nuclear small subunit (18S) ribosomal RNA gene is the primary target for profiling the diversity of microbial eukaryotes, ranging from single-celled protists to microscopic metazoans. Metabarcoding protocols typically interrogate the V4 or V9 hypervariable regions, which have been validated to provide the compromise between primer universality spanning the eukaryotic domain and the phylogenetic resolution required for community analysis (Pawlowski et al., 2012). While useful for capturing high-level eukaryotic diversity, the evolutionary conservation of the 18S gene can limit species-level discrimination in certain lineages when compared to faster-evolving mitochondrial or spacer markers (Hadziavdic et al., 2014).
- **Internal Transcribed Spacer (ITS) Region**: Designated as the universal DNA barcode for the Kingdom Fungi, this nuclear locus provides the taxonomic granularity necessary to resolve mycobiomes (Schoch et al., 2012). The ITS region comprises two non-coding spacers (ITS1 and ITS2) flanking the conserved 5.8S rRNA gene; because these spacers are not incorporated into the mature ribosome, they tolerate high mutation rates that generate sequence divergence between closely related taxa. This evolutionary hypervariability makes possible species-level identification that is often not achievable with the more conserved 18S or 28S rRNA genes, though the variable amplicon lengths inherent to this region can complicate downstream alignment and clustering (Lindahl et al., 2013).
- **Cytochrome C Oxidase Subunit I (COI) Gene**: Established as the canonical barcode for the Kingdom Animalia, this mitochondrial locus provides the discrimination necessary to resolve metazoan diversity at the species level (Hebert et al., 2003). The utility of the COI gene is driven by the rapid rate of nucleotide substitution characteristic of the mitochondrial genome, which generates a distinct 'barcode gap' separating intraspecific variation from interspecific divergence. However, this hypervariability presents a technical challenge for

environmental metabarcoding; the lack of highly conserved flanking regions complicates the design of universal primers. As a result, researchers frequently use degenerate oligonucleotides to reduce taxonomic bias (Leray et al., 2013).

Targeted amplicon sequencing serves as a mechanism for resolving the taxonomic architecture of complex assemblages and addresses the question of community membership, yet it is subject to methodological limitations. The PCR amplification step is a primary source of technical variation. Specifically, differential primer binding efficiencies and stochastic amplification dynamics have the potential to alter the recovered community profiles and favor taxa with optimal primer complementarity (Elbrecht and Leese, 2015). The resulting sequence read counts function as semi-quantitative proxies rather than absolute measures of biological biomass. This requires the application of statistical frameworks designed for compositional data (Gloor et al., 2017). The restriction of this approach to phylogenetic marker loci precludes the direct interrogation of the community's metabolic potential and limits the capacity to infer functional attributes of the ecosystem (Ranjan et al., 2016).

#### 5.2.2.1.1 SHOTGUN METAGENOMICS

Shotgun metagenomics represents a departure from the scope of targeted amplicon sequencing. Rather than restricting analysis to a singular phylogenetic locus, this paradigm seeks to interrogate the total genomic content present within an environmental matrix (Handelsman et al., 1998). The methodology relies on the fragmentation of the pool of extracted DNA followed by the ligation of sequencing adapters. By eliminating the dependence on locus-specific PCR amplification, this strategy prevents the taxonomic biases associated with primer annealing efficiency and genomic copy number variation (Shah et al., 2011). The resulting data provides a comprehensive genomic profile that is not only taxonomic but functional and enables a simultaneous reconstruction of community composition and metabolic potential (Venter et al., 2004). This resolution offers several distinct benefits over amplicon-based protocols:

- **Taxonomic Profiling without Amplification Bias**: Direct sequencing avoids the randomness and selection bias found in PCR. While amplicon profiling is limited by primer binding, shotgun metagenomics reduces the distortions caused by uneven replication and genomic copy number variation. This yields a more accurate community structure, detecting lineages with divergent sequences that standard primers miss (Tremblay et al., 2015). Furthermore, this untargeted method detects a wider range of biological entities, including viruses that lack the conserved markers needed for targeted assays (Breitbart et al., 2002).
- **Functional Potential and Metabolic Reconstruction**: The sequencing of the total eDNA pool allows for the direct characterization of the community's functional repertoire and avoids the dependence on phylogenetic proxies to infer metabolic capacity. This method allows the quantification of gene families driving biogeochemical processes and allows the reconstruction of metabolic networks directly from environmental data (Sunagawa et al., 2015). The unbiased nature of shotgun sequencing supports the discovery of novel biocatalytic enzymes and biosynthetic gene clusters that differ from those found in cultured reference strains (Hess et al., 2011).
- **Genome Recovery and Assembly**: The deep sequencing enables the computational reconstruction of complete or near-complete microbial genomes directly from environmental matrices. Algorithms assemble overlapping sequence fragments into contiguous stretches while subsequent binning procedures group these contigs based on oligonucleotide composition and differential coverage patterns (Albertsen et al., 2013). The resulting metagenome-assembled genomes (MAGs) reveal the complete genomes of uncultivated lineages and allow for the prediction of metabolic capabilities that was previously restricted to cultured isolates (Parks, Rinke, et al., 2017).
- **Strain-Level Resolution**: The coverage of the genome afforded by shotgun sequencing permits the resolution of microbial diversity at the subspecies or strain level. While marker gene

surveys are typically limited to genus-level classification, the analysis of total community DNA enables the identification of single-nucleotide polymorphisms (SNPs) and gene content variation across the proteome. This profiling facilitates the discrimination of closely related genotypes (Luo et al., 2015). Researchers can use this data to trace individual strains within changing spatial and temporal landscapes, revealing micro-evolutionary dynamics (Simmons et al., 2008).

- **Multi-Kingdom and Viral Detection**: The nature of shotgun sequencing permits the simultaneous capture of genetic material across the biological spectrum and encompasses bacteria, archaea, eukaryotic microbes, and viral entities. This kingdom-agnostic resolution is useful for the characterization of the viral fraction, or virome, which lacks universal phylogenetic markers and is therefore invisible to standard amplicon surveys (Paez-Espino et al., 2016). The concurrent recovery of sequences from multiple domains allows for the description of cross-kingdom interactions as well as co-occurrence networks that drive ecosystem dynamics (Lima-Mendez et al., 2015).

The resolution afforded by shotgun metagenomics is counterbalanced by fiscal and computational prerequisites. To achieve sufficient coverage of complex microbial communities, this approach demands orders of magnitude more sequencing depth than amplicon-based surveys (Knight et al., 2018). This massive data generation creates a bioinformatic bottleneck. The assembly, annotation, and integration of fragmented genomic data require high-performance computing infrastructures and complex analysis pipelines that exceed the demands of marker gene analysis (Sboner et al., 2011). The analysis of these datasets constitutes a distinct discipline, the methodologies of which are detailed in the subsequent section.

## 5.3 PART 3: COMPUTATIONAL ANALYSIS OF ENVIRONMENTAL DNA DATA

The conversion of HTS data into ecological information depends on bioinformatic pipelines that perform essential quality control, removing non-biological sequences like adapters and primers to produce a reliable abundance table (Hugerth and Andersson, 2017). The structure of these pipelines is a critical experimental variable. Because different algorithms and parameter settings can alter community profiles, inconsistent processing can lead to spurious ecological associations (Pollock et al., 2018). We next describe the primary analytical workflows and statistical approaches needed to manage this data effectively.

### 5.3.1 Bioinformatic Pipelines for eDNA Metabarcoding

HTS platforms export data in the FASTQ format, a de facto standard that combines biological sequence data with per-base quality scores (Cock et al., 2009). These scores, known as Phred values, use a logarithmic scale to estimate the probability that a specific base call is incorrect (Ewing and Green, 1998). The primary task of the bioinformatic pipeline is to convert these raw, noisy reads into a Biological Observation Matrix (BIOM), a standardized format that maps observations (such as operational taxonomic units (OTUs)) to samples while handling the sparsity inherent in microbiome data (McDonald et al., 2012). To achieve this, researchers use a series of quality control steps to separate real biological signals from sequencing errors:

1. **Quality Filtering and Pre-Processing**: The integrity of downstream ecological inferences relies on the initial sanitization of raw sequencing data. High-throughput platforms generate dataset-specific noise profiles that include base-calling errors and technical artifacts which can artificially inflate estimates of alpha diversity if left unaddressed (Schirmer et al., 2015). Before further analysis, researchers must clean the sequence library using filtration and error-correction steps, such as discarding reads with high expected error rates, to isolate true biological variation from technical noise (Edgar and Flyvbjerg, 2015). This pre-processing phase typically requires the execution of the following computational procedures:

    a. **Adapter and Primer Trimming**: The construction of sequencing libraries requires the ligation of synthetic oligonucleotides such as adapters and amplification primers to the biological templates. These non-biological termini must be computationally excised prior to downstream analysis. Failure to remove these exogenous sequences results in their erroneous interpretation as biological signals and leads to the inflation of diversity estimates through the formation of spurious OTUs (Nelson et al., 2014). Algorithms based on semi-global alignment are used to identify and trim these technical artifacts accurately (Bolger et al., 2014).

b. **Quality Score Filtering**: The sequencing-by-synthesis chemistry inherent to Illumina platforms causes a systematic decay in base-calling accuracy that disproportionately affects the 3′ terminus of reads (Minoche, Dohm, and Himmelbauer, 2011). These algorithms typically calculate the cumulative expected error for each read and discard sequences that fail to meet a defined threshold, limiting downstream taxonomic classification to data that likely represents true biological variation (Edgar and Flyvbjerg, 2015).

c. **Paired-End Merging**: A standard method for contemporary amplicon sequencing is the paired-end strategy which interrogates both termini of a single DNA fragment to generate a set of mate-pair reads. These reads are intentionally generated with a region of overlap to allow their assembly into a single, contiguous sequence. Algorithms align these overlapping segments to reconstruct the complete intervening biological sequence. Merging the reads improves the accuracy of the base calls because the overlap region provides two independent observations of the same DNA. If the forward and reverse reads differ at a specific nucleotide, the software uses the quality scores to determine the correct base and produce a consensus (Zhang et al., 2013; Masella et al., 2012).

d. **PhiX Decontamination**: The calibration of Illumina sequencing instruments and the empirical estimation of run-specific error rates use the inclusion of a balanced, high-diversity control library for which the bacteriophage PhiX genome is the standard. A fraction of these control sequences can be erroneously assigned to environmental samples during the base-calling and demultiplexing process, an issue known as index hopping or bleed-through (Sinha et al., 2017). It is therefore a necessary pre-processing step to computationally identify and purge these PhiX-derived contaminants. To remove these contaminants, the pipeline aligns the raw reads against the PhiX reference genome with high stringency, preventing the software from confusing technical artifacts with real data (Bushnell, 2014).

e. **Complexity Filtering**: A subset of sequencing artifacts is characterized by low compositional complexity, such as homopolymeric tracts or tandemly repeated motifs. These simple sequences do not contain enough unique information to allow reliable classification, often leading to incorrect matches in the database (Altschul et al., 1994). Bioinformatic pipelines incorporate filters that calculate the sequence entropy or linguistic complexity of each read. These algorithms discard sequences that fall below a specified informational threshold to ensure that downstream analyses are based on phylogenetically informative data (Morgulis et al., 2006).

2. **Chimera Removal**: A class of molecular artifacts generated during PCR amplification are chimeric sequences. These spurious amplicons arise when a prematurely terminated DNA strand from one template functions as a primer on a heterologous template during a subsequent amplification cycle (Haas et al., 2011). The resulting hybrid molecule represents

a composite of two or more distinct biological entities. Its retention in the dataset leads to the erroneous inflation of taxonomic richness estimates. A critical step in the bioinformatic workflow is the application of algorithms for chimera detection. These methods operate either by comparing sequences to a reference database or through de novo models that identify putative chimeras as bimolecular combinations of more abundant parent sequences present within the sample (Edgar et al., 2011; Callahan et al., 2016).

3. **Feature Generation (OTUs versus Amplicon Sequence Variants (ASVs))**: Following the sanitization of the sequence library and the excision of chimeric artifacts, the analytical workflow progresses to the aggregation of individual reads into discrete, biologically meaningful units. This grouping step, broadly termed feature generation, constitutes the transformation of raw sequence data into a quantifiable community abundance matrix. This step is widely regarded as a decision point in the bioinformatic pipeline, as the algorithm selected dictates the taxonomic resolution and the definition of diversity for the entire study (Callahan et al., 2017). Historically, researchers have chosen between two primary approaches: the coarse-grained clustering of reads into OTUs and the fine-scale resolution of exact ASVs (Knight et al., 2018).

    a. **OTUs**: For much of the HTS era, the standard analytical paradigm has relied on the construction of OTUs. This approach functions as a data reduction strategy that aggregates sequencing reads into clusters based on a fixed dissimilarity threshold. A similarity cutoff of 97% has been employed for the 16S rRNA gene, a heuristic derived from early comparisons of DNA–DNA hybridization data intended to serve as a proxy for species-level differentiation (Stackebrandt and Goebel, 1994). Computational algorithms such as UCLUST (Edgar, 2010) and VSEARCH (Rognes et al., 2016) execute this process by collapsing sequences within this radius into a single representative centroid. This strategy streamlines the analysis and buffers against diversity inflation caused by sequencing errors, but it comes at the cost of limiting taxonomic resolution. By enforcing a fixed similarity boundary, OTU clustering often obscures genuine biological micro-diversity, merging ecologically distinct lineages that differ by less than the defined threshold (Tikhonov et al., 2015; Stackebrandt, 2006).

    b. **ASVs**: Modern analysis now favors the resolution of ASVs, a process based on the statistical inference of biological sequences rather than arbitrary clustering. Known as 'denoising', this approach uses probabilistic models of the sequencing error process to discriminate authentic biological variation from the stochastic noise introduced during PCR amplification and sequencing (Eren et al., 2013; Edgar, 2016). Software packages such as Deblur (Amir et al., 2017) and DADA2 (Callahan et al., 2016) execute this discrimination to resolve sequence variants with single-nucleotide precision. Unlike OTUs, which are emergent properties of a specific dataset, ASVs represent exact, error-corrected biological sequences that possess validity independent of the analysis context. This exactitude allows researchers to seamlessly merge data from multiple sources without needing to re-cluster the raw reads (Thompson et al., 2017).

Because ASVs provide high-resolution data and allow for easier comparison between studies, the field has shifted to using them as the primary unit of analysis (Prodan et al., 2020). Completing this phase produces two core data structures that form the basis for all subsequent ecological analysis:

1. **The Feature Table**: This abundance matrix serves as the core quantitative output of the pipeline. It is typically structured as a sparse contingency table where rows represent samples, columns represent features

(ASVs), and values indicate the frequency of observation. This data structure, often structured in the standardized BIOM format, provides the numerical basis for all subsequent alpha- and beta-diversity calculations (McDonald et al., 2012).

2. **The Representative Sequences**: Complementing the abundance table is a FASTA-formatted repository containing the specific nucleotide sequence associated with each ASV identifier. These sequences represent the biological reality of the features and are necessary for the subsequent computational steps of phylogenetic tree estimation and taxonomic classification (Bolyen et al., 2019).

Although calculating ASVs provides an error-corrected inventory of biological entities, the resulting nucleotide strings remain abstract in isolation. The translation of these feature sets into a coherent biological narrative requires the annotation of sequences using standard taxonomy. This interpretive step is important, as it enables researchers to infer the putative functional traits and place in the ecosystem by linking community members to the accumulated phenotypic knowledge associated with described taxa (Yarza et al., 2014). However, determining the correct identity of an environmental sequence presents technical and biological challenges, ranging from the resolution limits of the marker gene to the variable quality and completeness of reference databases (Glöckner et al., 2017). The following discussion details the algorithmic methods used to map these anonymous features against the known tree of life.

### 5.3.2 Taxonomic Assignment: Approaches and Challenges

Following the generation of a feature table, the analysis moves to the taxonomic annotation of the resolved ASVs. This phase aims to map each ASV to a hierarchical lineage ranging from phylum to species by querying it against a curated reference framework. The accuracy of this classification is important, as it provides the biological context necessary to interpret community ecology and infer putative metabolic functions. However, the fidelity of these assignments is linked to two factors: the comprehensiveness of the underlying reference database and the algorithmic strategy employed to traverse sequence space (Bokulich et al., 2018). Historically, researchers have relied on domain-specific resources such as SILVA for prokaryotic 16S rRNA or UNITE for fungal ITS markers (Nilsson et al., 2018). More recently, the development of unified resources like Eukaryome has broadened the scope of reference data, providing coverage for SSU, ITS, and LSU markers across the entire eukaryotic tree of life, including protists and metazoans (Tedersoo et al., 2024). Despite the availability of these databases, the approaches used for assignment remain highly heterogeneous. As illustrated in Figure 5.3, the past two and a half decades have witnessed a rapid increase in software packages designed for this task. The wide array of options indicates the absence of a single standardized consensus for taxonomic assignment, forcing researchers to choose among many competing computational paradigms.

While the bioinformatic landscape encompasses a diversity of software implementations, the underlying algorithmic logic predominant in marker-gene analysis generally coalesces into three different approaches: alignment-based similarity searching, composition-based classification, and phylogenetic placement. Table 5.1 provides a comparative overview of these strategies, delineating their core logic, strengths, and limitations. Alignment-based methods rely on direct global or local sequence comparison to identify the closest hits within a reference database. Composition-based approaches such as Naïve Bayes Classifiers (NBCs) analyze the frequency of subsequences (k-mers) to infer taxonomic probabilities without requiring alignment (Rosen et al., 2010). Finally, phylogenetic strategies, often designated as model-based approaches, utilize evolutionary models to place query sequences directly into a fixed reference tree topology (Matsen et al., 2010). Each strategy entails unique assumptions about computational demand and biological resolution, and the trade-offs associated with them are delineated in the following subsections.

#### *5.3.2.1 Alignment-Based Methods*

The standard strategy for taxonomic annotation relies on pairwise sequence alignment, a methodology epitomized by the basic local alignment search tool (BLAST) (Altschul et al., 1990). In this workflow, query ASVs are subjected to a homology search against a reference database to identify the highest-scoring local alignments.

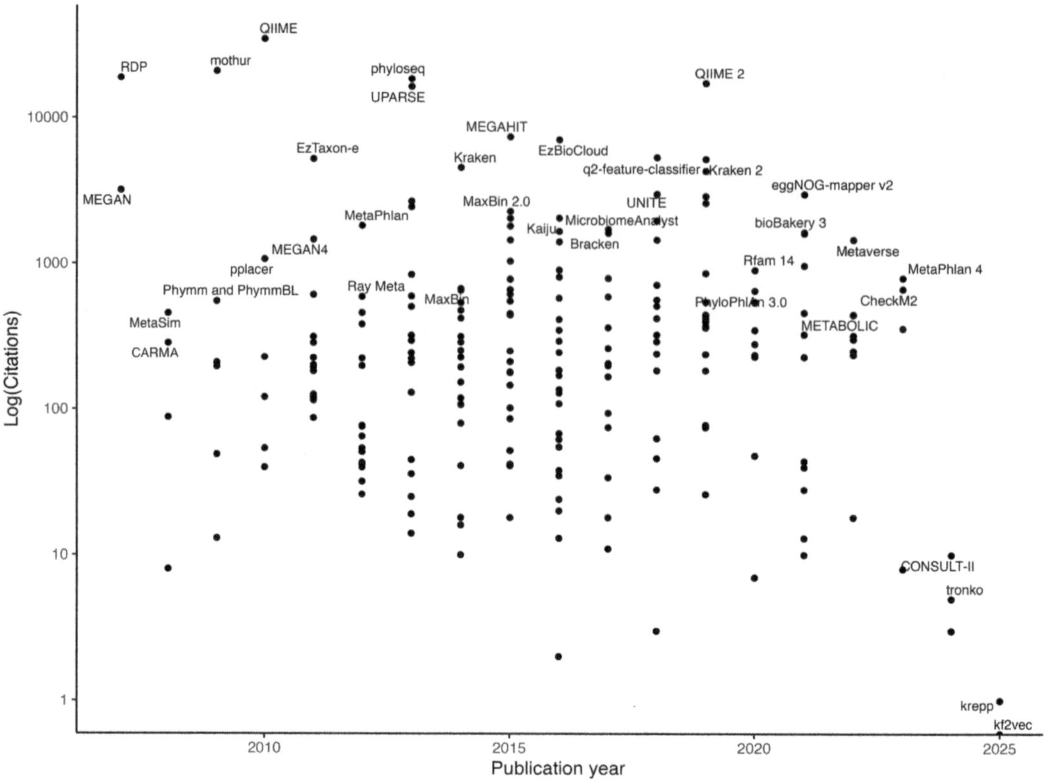


Figure 5.3 Citation frequency and release date for taxonomic classification software. The plot displays total citation counts on a logarithmic scale against the year of publication for 250 computational methods used in microbiome research. Prominent tools such as QIIME, mothur, and Kraken are labeled to indicate their relative impact and longevity in the field. The steady release of new software, represented by the density of points in recent years, indicates that the scientific community has not converged on a single standardized workflow. Instead, researchers must select from a growing list of computational options that vary in algorithmic approach and reference database compatibility.

The taxonomic identity is subsequently inferred from the annotation of the top-scoring hit or through a consensus of multiple top hits. Although resource-intensive, alignment-based methods generate a transparent metric of similarity (*i.e.*, bit score or percent identity) that is easily interpretable by researchers. However, depending on sequence similarity for classification is constrained by evolutionary reality. The application of universal identity thresholds (*i.e.*, defining species at ≥97% identity) is heuristic and biologically reductionist, as 16S rRNA gene evolutionary rates are non-uniform among distinct bacterial lineages (Rossi-Tamisier et al., 2015). A threshold that separates species in one genus may fail to distinguish distinct genera in another. Alignment algorithms are prone to misassignment when a query sequence is equidistant to multiple reference sequences from a large clade. When this occurs, standard alignment tools may arbitrarily select a single 'top hit' from the cluster, potentially assigning a specific species without statistical justification. Unlike phylogenetic placement methods, which resolve such ambiguity by placing the query on a basal node that reflects the evolutionary uncertainty rather than forcing a specific tip assignment. To address this risk within alignment approaches, pipelines frequently use a lowest common ancestor (LCA) algorithm. Instead of forcing an assignment to a specific leaf node, the LCA algorithm assigns the query to a higher, more conservative taxonomic rank (*i.e.*, family or order) when the top alignment hits share equal scores but possess conflicting taxonomic annotations.

### *5.3.2.2 Composition-Based Methods*

As an alternative to sequence alignment, composition-based strategies employ alignment-free algorithms to categorize sequences based on their intrinsic sequence structure. The most

TABLE 5.1

*Comparative analysis of predominant taxonomic classification strategies for marker- gene sequences.*

| | Alignment-based | Composition-based | Phylogenetic placement |
|---|---|---|---|
| **Core logic** | Identifies closest hits via global or local pairwise sequence alignment scores. | Calculates posterior probabilities based on shared subsequence frequencies (k-mers). | Inserts query sequences into a fixed reference phylogeny using evolutionary models. |
| **Primary strengths** | **Precision**: highly accurate for taxa explicitly present in the database.<br>**Interpretability**: direct sequence similarity (E-value, % identity) is intuitively understood. | **Speed and scalability**: extremely fast classification once trained; robust to minor sequencing errors.<br>**Adaptability**: can be retrained with ecological weights (*i.e.*, habitat-specific priors). | **Novelty handling**: superior for identifying novel taxa by placing them on deep branches rather than forcing assignment.<br>**Context**: provides evolutionary context beyond simple naming. |
| **Limitations** | **Sensitivity**: prone to false negatives for novel taxa; 'top-hit' bias can mask ambiguity.<br>**Speed**: traditional algorithms (*i.e.*, BLAST) are computationally prohibitive for massive datasets. | **Assumption violation**: assumes independence of k-mers (biologically incorrect).<br>**Resolution**: performance frequently degrades at the species level compared to exact alignment. | **Resource Intensity**: high memory (RAM) and CPU cost for large reference trees.<br>**Dependency**: strictly limited by the quality and topology of the underlying reference tree. |
| **Computational demand** | **Moderate to high**: scales linearly with database size; optimized tools (VSEARCH) mitigate this. | | |
| **Key software** | BLAST, VSEARCH, MEGAN | RDP classifier, QIIME 2 (sklearn), SINTAX | SEPP, pplacer, EPA-ng, Tronko |

Source: Extended from Richardson Bengtsson-Palme, and Johnson (2017).

common implementation of this paradigm is the NBC, exemplified by the Ribosomal Database Project (RDP) Classifier (Wang et al., 2007) and the scikit-learn classifier implemented in QIIME 2 (Bokulich et al., 2018). Rather than seeking a singular optimal alignment, these methods decompose both query and reference sequences into overlapping substrings of length *k* (typically 8-mers). The taxonomic assignment is subsequently derived by calculating the joint probability that the query sequence belongs to a specific genus or species given its k-mer composition (Lan et al., 2012). To assess reliability, these algorithms typically utilize a bootstrapping technique which involves randomly resampling the gene fragments to establish a confidence score for each taxonomic rank. This probabilistic framework delivers advantages in computational speed and tolerance for sequencing artifacts, as the presence of a few erroneous nucleotides (and thus erroneous k-mers) rarely disrupts the overall probability distribution enough to alter the classification. While the resulting confidence scores provide a measure of statistical uncertainty, this approach relies on the 'naïve' assumption that sequence features are statistically independent, which is a simplification of biological reality. These classifiers generate categorical labels with associated probabilities, but they do not provide the explicit phylogenetic topology or evolutionary distance metrics inherent to tree-based approaches.

#### 5.3.2.3 *Phylogenetic Placement*

To perform phylogenetic placement, researchers must integrate three distinct data structures. Instead of relying on pairwise similarity searching, this approach inserts query ASVs into a pre-computed reference phylogeny (Matsen et al., 2010). Methods such as pplacer and the evolutionary placement algorithm (EPA) apply maximum likelihood (ML) models to calculate the probability of a query sequence originating from specific branches within the fixed reference topology (Berger et al., 2011). By evaluating the likelihood weight of attachment at various nodes, these methods resolve the evolutionary source of the ASV. This process yields more than a simple taxonomic label, but a biologically meaningful position within the fixed phylogeny, preserving the phylogenetic relationships between novel environmental sequences and known reference taxa even in the absence of exact database matches (Janssen et al., 2018). In the past, using this approach was constrained by computational limits. However, recent algorithmic advancements, such as Tronko, have improved scalability, enabling rapid phylogenetic placement of millions of reads without compromising the accuracy advantages of the method (Pipes and Nielsen, 2024).

The operational execution of phylogenetic placement relies on the integration of three interdependent data structures. Unlike composition-based methods that analyze sequences in isolation, placement algorithms require a pre-computed evolutionary framework to calculate the likelihood of insertion events (Czech et al., 2022). The workflow requires the following inputs:

1. **A Reference Phylogeny**: A robustly inferred phylogenetic backbone tree that establishes the evolutionary relationships among known taxa. This tree defines the topology onto which query sequences will be placed.
2. **A Reference Multiple Sequence Alignment (MSA)**: The base alignment from which the backbone tree was reconstructed. This MSA provides the positional homology required to align the query sequences to the existing evolutionary model, often utilizing profile Hidden Markov Models (HMMs) to map short reads against full-length references without disrupting the existing alignment structure (Mirarab et al., 2012).
3. **The Query Sequences**: The collection of error-corrected query sequences derived from the environmental samples, which serve as the fragmentary input data to be placed within the reference tree.

The computational execution of this strategy is grounded in ML inference, seen in methods such as EPA-ng (Barbera et al., 2018). Unlike heuristic similarity searches, these methods examine the reference topology, treating each branch as a hypothetical insertion point for the query sequence. By optimizing branch lengths under a specified model of nucleotide substitution (*i.e.*, GTR + $\Gamma$), the algorithm calculates the conditional likelihood of observing the query sequence at every specific node. These values are subsequently normalized into likelihood weight ratios (LWRs), a metric that quantifies the confidence of assignment. This statistical approach allows the query to be assigned to the specific lineage yielding the highest probability mass, providing an estimation of evolutionary provenance rather than a simple best-hit match (Stark et al., 2010).

Using phylogenetic placement offers rigor by framing taxonomic assignment as an evolutionary inference problem. By mapping query sequences within a fixed topology, this method generates an explicit phylogenetic hypothesis for every feature. This approach excels at classifying novel organisms where similarity-based approaches often fail due to the absence of close relatives in reference databases. Instead of forcing an assignment to a specific taxon, phylogenetic placement maps divergent ASVs to basal nodes that represent ancestral relationships. For example, a sequence derived from an undescribed lineage within the *Lactobacillus* genus would be assigned to the ancestral branch of that clade to preserve taxonomic accuracy despite the lack of species-level resolution (Nguyen et al., 2014). Additionally, the reliability of these assignments is measured using ML models. This approach calculates the LWR to express the probability mass associated with the attachment of a sequence to specific branches. Although this approach requires greater computational resources than composition-based classifiers, it yields a detailed characterization of phylogenetic diversity. This precision is essential for the analysis of complex or under-sampled environments where the identification of novel lineages requires inference independent of exact database matches (Czech and Stamatakis, 2019).

### 5.3.3 Taxonomic Classification of Shotgun Metagenomic Data

Using WGS metagenomics calls for a different set of classification tactics than those used for amplicons. Unlike marker-gene surveys that target conserved sequences, each WGS study routinely yields a heterogeneous pool of billions of random genomic fragments. Processing this data creates a significant bottleneck, as exhaustive alignment algorithms against comprehensive whole-genome databases are impossible to run efficiently (Breitwieser, Lu, and Salzberg, 2017). Consequently, researchers have abandoned the exact alignment methods common in amplicon studies. The field now relies on heuristic tools designed for high speed. These methods typically fall into two categories: read-based profiling, which identifies fragments using exact k-mer matching or clade-specific markers (Sczyrba et al., 2017), and assembly-based taxonomy which classifies reconstructed MAGs through phylogenomic placement (Parks et al., 2018).

#### *5.3.3.1 Introduction: The Complexity of Untargeted Data*

The fundamental challenge of WGS lies in its indiscriminate nature. Unlike amplicon sequencing, which targets a specific gene, WGS captures DNA from bacteria, archaea, viruses, and single-celled eukaryotes all at once (Quince et al., 2017). This single dataset can include fragmented coding sequences, non-coding regions, viral elements, and host contamination (Sharpton, 2014). Consequently, reads from the same gene often do not overlap, making it impossible to align them directly to one another. Because these sequences lack a shared conserved locus, standard alignment tools cannot anchor them to a reference point. Instead, the analysis depends heavily on the quality of available genome catalogs. This is problematic because many environmental communities contain organisms that have never been cultured or sequenced. Researchers are often forced to map fragmented reads against reference databases that are incomplete, biased toward model organisms, and vast in size (Sharpton, 2014). To manage this complexity, bioinformatic pipelines have moved away from the alignment methods used in amplicon studies. Modern workflows generally follow one of two paths: read-based profiling, which prioritizes speed by using raw fragments, or assembly-based taxonomy, which attempts to reconstruct contiguous genomes prior to classification.

#### *5.3.3.2 Read-Based Taxonomic Profiling*

Read-based profiling involves mapping raw reads to reference sequences without assembling them first. By skipping assembly, this approach avoids heavy resource demands and the risk of creating chimeric artifacts, which permits a faster assessment of the community (Breitwieser et al., 2017). Different software packages use different algorithms and database scopes to achieve this. Most tools, however, rely on one of two approaches: matching exact k-mers against entire genomes or detecting specific diagnostic marker genes (Simon et al., 2019).

##### 5.3.3.2.1 EXACT K-MER MATCHING

Exact k-mer matching represents a heuristic alternative to alignment-based profiling, designed to handle the throughput of modern sequencing platforms. Rather than aligning full reads, these algorithms decompose query sequences and reference genomes into short, fixed-length substrings (k-mers), usually between 30 and 35 base pairs. To minimize search times, the reference database is built into a high-efficiency index using structures like hash tables or Bloom filters. Assignment follows LCA logic: when a k-mer matches multiple reference genomes (*i.e.*, conserved across *Escherichia* and *Salmonella*), the read is assigned to the specific taxonomic rank that includes all targets. By doing so, the method avoids making false species calls based on conserved DNA, defaulting to the genus or family level. Widely used tools like Kraken2 and Centrifuge apply this approach to process millions of reads per minute (Wood et al., 2019; Kim et al., 2016). However, indexing a large database like NCBI's RefSeq database demands heavy computational resources, specifically random access memory (RAM) amounts that go beyond the capacity of standard workstations. Furthermore, because these approaches depend on exact matches, they are sensitive to errors. A single nucleotide difference will cause a match failure, which can result in false negatives when analyzing divergent environmental strains (Nasko et al., 2018).

##### 5.3.3.2.2 MARKER-GENE PROFILING

Marker-gene profiling offers an efficient alternative to whole-genome alignment by limiting the analysis to a selected group of phylogenetically informative loci. Unlike amplicon sequencing, which uses PCR to amplify a single gene, this strategy finds shotgun reads that map to gene families diagnostic for specific lineages. By using databases of clade-specific markers, algorithms can classify taxonomy and estimate abundance much faster than alignment against entire genomes. Software such as MetaPhlAn demonstrates this approach, using a catalog of millions of marker sequences to distinguish subspecies and strains in complex mixtures (Blanco-Míguez et al., 2023). Similarly, the mOTUs profiler utilizes phylogenetic marker genes to quantify taxa that correspond to microbial OTUs, enabling the profiling of known and unknown species without the need for reference genomes (Milanese et al., 2019). Although accurate, this method ignores reads that do not match the specific marker sets, leaving the majority of the metagenomic data unclassified.

### *5.3.3.3 Assembly-Based Taxonomy (the Genome-Centric Approach)*

The transition from read-based profiling to assembly-based taxonomy represents a shift from viewing the microbiome as a composite of disconnected gene fragments to a structured community of distinct organisms. While read-based approaches are efficient for quantification, they often disconnect phylogenetic identity from metabolic potential. To address this, modern environmental genomics uses a genome-centric strategy to reconstruct complete or near-complete genomes directly from sequencing data. This process permits the characterization of the 'uncultured majority' of microorganisms that cannot be grown in the lab, revealing their evolutionary history and ecosystem roles (Sharon and Banfield, 2013). By going beyond single-gene proxies, this method uses the full genomic context to improve microbial systematics (Albertsen et al., 2013). The following subsections describe how to reconstruct these genomes and place them in an evolutionary context.

#### 5.3.3.3.1 METAGENOME-ASSEMBLED GENOMES (MAGS)

Although read-based profiling provides a rapid list of the community in a sample, it treats the microbiome as an unstructured mix of DNA sequences. Consequently, researchers cannot easily link metabolic pathways to specific organisms. To resolve this limitation, the field has adopted genome-resolved metagenomics, a method designed to recover whole genomes from environmental data (Tully et al., 2018). This approach requires the computational reconstruction of MAGs whereby short reads are assembled into contiguous sequences and subsequently organized into genomic bins based on oligonucleotide composition and differential coverage patterns (Kang et al., 2019). Following the quality assessment of these bins for completeness and contamination using lineage-specific marker sets, as implemented in programs such as CheckM, the analysis transitions from the classification of millions of anonymous reads to the taxonomic assignment of discrete, reconstructed organisms (Parks et al., 2015; Bowers et al., 2017).

#### 5.3.3.3.2 PHYLOGENETIC PLACEMENT

The taxonomic assignment of MAGs is based on phylogenomics, which provides higher evolutionary resolution than single-gene markers. While the 16S rRNA gene often lacks the variation needed for species-level discrimination, phylogenomic strategies target a wide range of conserved marker proteins. This usually involves 120 markers for bacteria and 122 for archaea, a selection designed to maximize signal and minimize errors from lateral gene transfer (Parks et al., 2018; Chaumeil et al., 2019). Software PhyloPhlAn manages this process by finding and translating these markers, then merging them into a concatenated sequence (Segata et al., 2013). This aggregation of sequence data helps against topological incongruence arising from individual gene histories and allows the placement of the query genome within a fixed reference phylogeny using ML inference (Chaumeil et al., 2019).

#### 5.3.3.3.3 THE GENOME TAXONOMY DATABASE (GTDB) FRAMEWORK

Genome-based classification has reshaped the field, leading to the establishment of the Genome Taxonomy Database (GTDB). Historically, microbial taxonomy lacked consistency and relied on a mixture of phenotypic traits and 16S rRNA gene thresholds that often resulted in polyphyletic groups and inconsistent ranking criteria (Hugenholtz et al., 2021). The GTDB solves these issues by establishing a standardized taxonomy based on genome phylogeny and relative evolutionary divergence (RED). Using this metric means that taxonomic ranks such as phylum, class, and order represent comparable evolutionary depths across all lineages (Parks et al., 2020). For environmental microbiologists, this database is useful as it provides a systematic nomenclature for the vast majority of uncultured diversity that lacks physical type strains. Tools such as GTDB-Tk allow for the placement of environmental MAGs into this global reference phylogeny with accuracy (Chaumeil et al., 2022).

#### *5.3.3.4 Summary of Trade-Offs*

The distinction between read-based and assembly-based strategies constitutes a strategic decision in the design of metagenomic studies. Read-based profiling serves as a computationally efficient mechanism for establishing a taxonomic profile and is especially useful for the detection of low-abundance taxa that frequently lack sufficient coverage for de novo assembly (Breitwieser et al., 2017). These methods provide a broad overview of community diversity. In contrast, assembly-based taxonomy demands more sequencing data and processing power, but it reveals the complete genomic structure of the ecosystem. This resolution is required to describe novel environmental lineages. By recovering whole genomes, researchers can link phylogenetic identity directly to metabolic potential and evolutionary history without needing to culture the organisms (Quince et al., 2017).

### 5.3.4 Statistical Analysis of Environmental Microbiome Data

Once raw sequencing reads are grouped into a feature table, the focus of the study moves from identifying individual sequences to analyzing entire microbial communities. This shift requires statistical methods designed for data that is sparse, contains thousands of variables, and represents relative proportions rather than absolute counts (Gloor et al., 2017). The analysis usually starts by measuring richness and evenness within samples (alpha diversity), then compares the turnover of species between sites (beta diversity), and finally identifies specific taxa that differ between groups (differential abundance analysis). Using multivariate tools from numerical ecology, researchers can measure changes in community structure and test how environmental factors shape microbial distribution (Anderson, 2001). The goal is to convert complex abundance matrices into interpretable ecological models that explain the assembly and function of the environmental microbiome (Knight et al., 2018).

#### *5.3.4.1 Alpha Diversity: Within-Sample Diversity*

Alpha diversity provides a measure of the ecological complexity of a single biological sample, characterizing the community structure within a defined local scale (Whittaker, 1960). This concept is divided into two distinct components: richness and evenness. Richness denotes the aggregate count of distinct taxa, such as ASVs or species, observed within the dataset. However, because HTS is a sampling process that may not detect the entirety of the underlying community, reliance on raw counts of observed features can result in downward bias. To address this, non-parametric estimators such as the Chao1 index are often used to extrapolate the theoretical lower bound of true richness by using the frequency of rare, single-occurrence taxa (Chao, 1984). In parallel, evenness describes the equitability of abundance distributions among the detected features. A community is considered even if taxa are present in comparable proportions, whereas dominance by a few highly abundant lineages indicates unevenness. To capture the relationship between these two dimensions, integrative metrics like the Shannon index are utilized to calculate the uncertainty of predicting the identity of a randomly selected entity, providing a value that increases with both the number of unique taxa and the equality of their distribution (Spellerberg and Fedor, 2003). These metrics serve as bioindicators for assessing ecosystem state. For example, they have the potential to determine whether anthropogenic stressors such as hydrocarbon contamination cause a significant decline in microbial diversity relative to undisturbed reference sites (Atlas, 1991).

#### *5.3.4.2 Beta Diversity: Between-Sample Diversity*

Beta diversity quantifies the compositional differentiation between biological samples, addressing the extent to which microbial communities vary across environmental gradients or experimental conditions (Anderson et al., 2011). The analytical workflow begins with the computation of a pairwise distance or dissimilarity matrix, a process that reduces the feature table into a set of values representing the ecological distance between every possible pair of samples. The selection of the specific metric used to populate this matrix is a decision that determines how community differences are weighted. Common non-phylogenetic metrics handle data in different ways: the Jaccard index measures overlap using only presence/absence data, while the Bray–Curtis dissimilarity uses relative abundance to weigh the differences between communities (Bray and Curtis, 1957). However, these taxonomy-independent metrics treat all features as equally distinct, ignoring the evolutionary relatedness among taxa. To address this, phylogenetic metrics such as UniFrac have been developed to incorporate the branch lengths of the underlying phylogenetic tree into the distance calculation. This approach allows researchers to distinguish between communities that differ by closely related strains versus those separated by deep evolutionary divergences (Lozupone and Knight, 2005). Specific variants of this metric, such as unweighted and weighted UniFrac, allow investigators to tune the analysis to emphasize either the presence of rare lineages or the shifts in dominant taxa, respectively (Lozupone et al., 2007). Following the calculation of pairwise distances, the high-dimensional structure of the data is projected into a lower-dimensional space to allow visualization and interpretation. This dimensionality reduction is commonly achieved through ordination techniques such as principal coordinates analysis (PCoA) or non-metric multidimensional scaling (NMDS). These statistical techniques generate scatter plots wherein each point represents the entire microbial community of a single sample, and the spatial distance between points approximates their ecological dissimilarity (Ramette, 2007). Samples that cluster together in the ordination space exhibit similar compositional profiles, while dispersed points approximate ecological differences. To statistically validate these groupings in these visualizations, non-parametric tests such as permutational multivariate analysis of variance (PERMANOVA) are applied directly to the distance matrices. These tests determine whether the observed clustering between predefined experimental groups (such as the distinction between polluted and undisturbed sites) constitutes a significant shift in community composition rather than random variation (McArdle and Anderson, 2001).

#### *5.3.4.3 Differential Abundance Analysis*

While beta diversity measures provide a macroscopic view of community dissimilarity, they do not identify the specific taxonomic features driving these shifts. To identify the specific organisms causing these changes, differential abundance analysis is used to detect taxa that shows significant shifts in prevalence between experimental conditions. Analyzing this data is statistically challenging because microbiome datasets are compositional. Since sequencing depth varies by sample, an increase in the relative abundance of one taxon causes a decrease in others, inducing correlations if not properly normalized (Mandal et al., 2015). Environmental datasets are also highly sparse, meaning most taxa appear as zeros in any single sample. Consequently, standard parametric tests such as the Student's t-test are often inappropriate for these data structures. To address these constraints, approaches such as Analysis of Compositions of Microbiomes with Bias Correction (ANCOM-BC) and adaptations of negative binomial models like DESeq2 have been developed. These methods use log-ratio transformations and probabilistic modeling to distinguish genuine biological variation from technical sampling bias (Lin and Peddada, 2020; Love et al., 2014). Ultimately, this process identifies the specific indicator species and putative biomarkers that define the ecosystem (Segata et al., 2011).

## 5.4 PART 4: APPLICATIONS, CHALLENGES, AND FUTURE DIRECTIONS

Using the computational strategies outlined above, environmental genomics is shifting from a purely academic discipline to one that actively informs policy and management. By converting sequence data into biodiversity metrics, these techniques offer a way to solve specific problems in ecosystem stewardship, conservation biology, and public health. This capacity has advanced biomonitoring by allowing for the discovery of elusive taxa and the assessment of ecosystem health with a sensitivity and scalability that exceeds traditional morphological surveys by orders of magnitude (Deiner et al., 2017). Continued success, however, requires that the community solve specific technical limitations. This final section reviews the real-world uses of environmental microbiome research while examining the statistical and computational hurdles that must be addressed to realize the potential of planetary-scale genomic monitoring (Ruppert, Kline, and Rahman, 2019).

### 5.4.1 Applications of Environmental Microbiome Research

Recovering DNA from the environment allows ecologists to answer questions that were previously impossible to address. Since researchers no longer need to physically handle specimens, they can track biodiversity with minimal intrusion, a distinct advantage over conventional surveying (Kelly et al., 2014). As highlighted by Nicole S Webster (2025), incorporating eco-evolutionary genomics into sustainability strategies allows for the mapping of microbial diversity across ecosystems and the forecasting of how communities will react to environmental stress. These technologies have been adopted across disciplines ranging from the real-time detection of biological invasions and pathogens to the reconstruction of ancient floral and faunal assemblages (Lodge et al., 2012; Rawlence et al., 2014). The following subsections explore domains where environmental genomics has revealed ecosystem dynamics, historical baselines, and biosecurity risks.

#### 5.4.1.1 Ecosystem Monitoring and Conservation

Environmental microbiomes function as sensitive bioindicators of ecosystem integrity because microbial community composition shifts in response to physicochemical perturbations. This capacity to respond is leveraged to monitor the impacts of anthropogenic stressors such as chemical pollution and land-use intensification. For instance, the metagenomic analysis of riverine microbial assemblages has been shown to describe the biological impact of agricultural runoff with a resolution that captures persistent disturbances often missed by traditional chemical monitoring (Beattie et al., 2020). In marine systems, the integration of eDNA with machine learning software allows for the assessment of sediment quality status, providing a high-throughput alternative to macrobenthic surveys for monitoring organic enrichment and industrial pollution (Cordier et al., 2017). The characterization of the coral microbiome provides early warning signals for dysbiosis associated with thermal stress, while shifts in planktonic communities act as sentinels for global phenomena such as ocean acidification (Bourne et al., 2016). Data from these surveys are becoming a standard part of modern conservation planning and environmental management.

#### 5.4.1.2 Invasive Species Detection

The deployment of eDNA analysis has proven effective for the surveillance of non-native invasive species in both aquatic and terrestrial ecosystems. By identifying genetic material shed into the environment, this methodology allows the detection of taxa at population densities below the detection thresholds of conventional capture-based surveys (Jerde et al., 2011). This capacity to provide an early warning signal is useful for biosecurity protocols because eradication and containment strategies are most efficacious when initiated during the nascent stages of invasion before populations become established. The utility of this approach has been validated through the monitoring of high-profile invaders, such as the tracking of Asian carp encroachment into North American river systems (Jerde et al., 2013) and the mapping of American bullfrog expansion in Europe (Dejean et al., 2012). This sensitivity extends to the surveillance of microscopic biological threats, including the pathogenic chytrid fungus *Batrachochytrium dendrobatidis* that threatens global amphibian populations (Schmidt et al., 2013). eDNA monitoring provides a scalable and non-invasive strategy for screening vulnerable habitats to yield the distribution data necessary for targeted management interventions.

#### 5.4.1.3 Paleogenomics: Reconstructing Past Ecosystems

The application of environmental genomics has expanded temporally to encompass the reconstruction of biological history through the analysis of sedimentary ancient DNA (sedaDNA). Under specific physicochemical conditions characterized by low temperatures or favorable mineral adsorption, genetic material can exhibit recalcitrance to degradation over geological timescales. Such preservation occurs across various habitats such as permafrost, deep lake sediments, cave deposits, marine archives, and glacial ice (Willerslev et al., 2003; Gelabert et al., 2025). These natural archives function as stratified molecular biobanks where sediment accumulation occurs sequentially over millennia. The vertical extraction of DNA allows researchers to access a time-resolved history of local biodiversity that is independent of the macro-fossil record and its inherent taphonomic biases (Pedersen et al., 2016; Lindahl et al., 2025). This paleogenomic approach has advanced Quaternary ecology by providing the resolution necessary to map ecosystem dynamics in response to past climate oscillations. Computational analysis of sedaDNA has played a key role in tracking the population dynamics and extinction chronologies of Pleistocene megafauna as it reveals the persistence of carnivores and hominins in refugia long after their disappearance from the morphological record (Slon et al., 2017; Gelabert et al., 2025). Methodological comparisons reveal that although metabarcoding is still effective for Holocene reconstructions, shotgun metagenomics offers superior fidelity for deep-time samples by minimizing the PCR-induced biases associated with ancient and fragmented DNA templates (Holman et al., 2025; Kjær et al., 2022). By establishing pre-anthropogenic baselines of ecosystem structure and genetic diversity, paleogenomics provides the long-term context necessary for distinguishing natural ecological variability from modern human impact (Capo et al., 2021; Lindahl et al., 2025).

#### 5.4.1.4 *Public Health and Biosecurity*

The integration of environmental genomics into epidemiology is central to the One Health goal of connecting environmental, animal, and human health. This approach treats the environment as a dynamic reservoir and transmission route for infectious diseases. Consequently, molecular surveillance is being used in various water systems to detect biological threats before outbreaks occur. Moving past standard fecal indicators, these efforts now include detecting pathogens directly in potable water reserves, recreational aquatic zones, and agricultural irrigation networks (Leifels et al., 2022). High-sensitivity eDNA assays can now be used to track the spatiotemporal distribution of specific pathogens including opportunistic bacteria such as *Mycobacterium* spp., enteric viruses, and pathogenic protozoa like *Acanthamoeba* more accurately than culture-based methodologies (Brumfield et al., 2020). Surveillance has also extended to the aerobiome where it has emerged as a frontier for biosecurity. Systematic reviews of filter-based sampling methodologies demonstrate that viral RNA from respiratory pathogens can be reliably detected and quantified in aerosols collected from indoor healthcare and sociosanitary settings (Alfaro-Perez et al., 2025). These non-invasive bioaerosol monitoring techniques are increasingly applied at the human–animal interface to track zoonotic threats. Recent implementations include the detection of novel influenza A viruses in air samples from swine farms and live bird markets which allows for the early identification of spillover risks without the need for individual animal swabbing (Badra et al., 2025).

An application of this approach is the characterization of the environmental resistome or the collection of antibiotic resistance genes (ARGs) present within a microbial community. Anthropogenic environments such as wastewater treatment plants and hospital effluents act as evolutionary hotspots that facilitate the proliferation and horizontal transfer of resistance determinants between environmental bacteria and human pathogens (Larsson and Flach, 2022). Metagenomics resolves these ARGs and their genetic context by mapping them to mobile genetic elements. Surveilling urban sewage provides a scalable and non-invasive mechanism for global antimicrobial resistance monitoring as it provides an aggregate readout of community health and resistance trends that complements individual clinical diagnostics (Hendriksen et al., 2019; Aarestrup and Woolhouse, 2020).

### 5.4.2 Current Challenges and Future Perspectives

Environmental genomics has changed the observational scope of ecology by permitting the detection of biodiversity across large spatiotemporal scales. Despite this capacity, the field remains characterized by a lack of methodological standardization and analytical challenges that constrain the ecological interpretation of sequence data (Cristescu and Hebert, 2018). The transition from simple species lists to ecosystem models requires the resolution of biases associated with molecular quantification and taxonomic resolution. Future progress relies on the development of improved statistical methods that integrate genomic data with autonomous monitoring systems to describe the functional dynamics of the environmental microbiome (Zinger et al., 2019; Burian et al., 2021).

#### 5.4.2.1 *From Presence to Abundance: The Quantitative Challenge*

The inference of absolute biological abundance from compositional sequence data is a major analytical challenge in environmental genomics. While HTS generates large datasets, the resulting read counts function as relative probabilities rather than direct measures of organismal density (Gloor et al., 2017). This difference arises from biological variables including interspecific variation in extracellular DNA shedding rates and the heterogeneity of marker gene copy numbers across taxonomic lineages (Kembel et al., 2012b). The physical transport and differential degradation kinetics of genetic material within environmental matrices can spatially decouple the DNA signal from its source population (Barnes et al., 2014). Technical biases are superimposed upon these biological factors as primer mismatch and stochastic amplification efficiency during PCR can systematically distort community profiles (McLaren et al., 2019). To resolve these limitations, researchers are adopting orthogonal methodologies such as droplet digital PCR (ddPCR) which enables the absolute quantification of target molecules without the need for standard curves (Doi et al., 2015). For broader community analysis, the integration of internal DNA standards or synthetic spike-ins allows for the transformation of relative abundance data into absolute quantitative estimates, allowing more accurate ecological modeling (Ushio et al., 2017).

### 5.4.2.2 *The Expanding Reference Database*

The accuracy of taxonomic annotation is linked to the scope and curation quality of the underlying reference repositories (Martí et al., 2025). While centralized reference databases have expanded, the vast proportion of global microbial diversity remains uncultured and genomically uncharacterized. This reference deficit creates a taxonomic void where a fraction of environmental ASVs fails to map to any known lineage. This issue is especially severe within underexplored biomes such as deep-subsurface aquifers or tropical soils where the dissociation between sequence data and described taxonomy is most diverged (Anantharaman et al., 2016). Resolving this problem requires the expansion of reference databases through cultivation-independent methodologies. Technologies such as single-cell genomics and the recovery of MAGs provide the necessary data to populate the reference phylogeny with unrepresented lineages and reveal the functional capacity of the uncultivated majority of organisms (Almeida et al., 2021; Pasolli et al., 2019).

### 5.4.2.3 *Integration of Multi-Omics Data*

While metabarcoding and shotgun metagenomics provide a snapshot of taxonomic composition and functional potential, they do not explicitly capture the physiological state or metabolic activity of the community. Understanding how the ecosystem functions in real time requires moving beyond the static assessment of genetic potential to the direct measurement of biological expression. The field is adopting a multi-omics strategy that synthesizes genomic data with complementary molecular layers to construct an integrated model of microbial activity. This framework typically incorporates metatranscriptomics to profile the pool of expressed RNA transcripts and identify the genes actively transcribed under specific environmental conditions (Bashiardes et al., 2016). To resolve the translational landscape, metaproteomics is used to identify and quantify the catalytic proteins that constitute the functional machinery of the ecosystem (Hultman et al., 2015). Finally, metabolomics provides the phenotypic readout by measuring the small-molecule end products of cellular metabolism which links upstream genetic regulation to downstream biogeochemical outcomes (Patti et al., 2012). The synthesis of these heterogeneous datasets presents an analytical challenge but offers the capability to integrate community structure with ecosystem function (Narayanasamy et al., 2016).

### 5.4.2.4 *Toward Real-Time and Autonomous Monitoring*

The future of environmental monitoring relies on the deployment of autonomous sensing infrastructures capable of delivering near real-time assessments of biological data (Yamahara et al., 2025). Recent technological advances have allowed the development of robotic sampling platforms that function as autonomous genomic observatories in remote or hazardous locations. The deployment of uncrewed surface vessels equipped with electromechanical filtration units permits the high-frequency collection of eDNA across oceanic transects without direct human intervention (Preston et al., 2024). These autonomous systems are increasingly integrated into national early detection systems such as the READI-Net initiative, which leverages networked eDNA robots to identify invasive species dynamics and pathogen outbreaks with a temporal resolution that manual surveying cannot achieve (Jones et al., 2024). The coupling of these sampling technologies with field-deployable third-generation sequencing platforms has advanced. Recent evaluations of the latest Nanopore chemistries demonstrate the capacity for rapid genomic surveillance of wastewater, which enables the characterization of viral variants with accuracy equivalent to laboratory standards (Xu et al., 2024). The synthesis of these robotic samplers with on-site molecular processing remains a developmental frontier. Current autonomous workflows are constrained by the time required for sample retrieval and extraction, which results in an observational lag that must be accounted for when designing rapid response protocols (Yamahara et al., 2025).

The analysis of the environmental microbiome constitutes a discipline that is reshaping the observational boundaries of ecological science. By resolving persistent quantitative biases and integrating multi-omics layers into unified analytical frameworks, researchers are moving beyond descriptive inventories toward predictive ecosystem modeling. Combining autonomous robotic sampling with emerging real-time sequencing technologies will eventually establish a responsive global sensor web capable of detecting biosecurity threats and biodiversity shifts with high temporal resolution (Preston et al., 2024; Badra et al., 2025). As these frontiers advance, they will provide the scaffolding to understand the functional complexity of the planetary microbiome and guide global ecosystem management.

## 5.5 CONCLUDING SUMMARY

HTS along with advancements in computational genomics has caused a transformation in the ecological sciences. The observational scope has shifted from the physical capture of macro-organisms to the molecular interrogation of community-level genetic assemblages. This chapter has outlined the steps required to translate genetic material from the environment into ecological insights. We have followed the computational workflow from the initial quality control of sequence data through the transition from heuristic clustering to the statistical resolution of ASVs. This shift toward exact sequence inference represents a shift where the field now values accuracy and consistency over computational speed. We have also emphasized the need for evolutionarily aware methodologies such as phylogenetic placement to interpret the unclassified biodiversity. These types of analyses allow researchers to move beyond simple taxonomic labeling to the reconstruction of evolutionary relationships for novel lineages that lack representation in reference databases. The applications of these computational strategies are as diverse as the environments they interrogate. In the domain of conservation biology, these tools have become biosensors for monitoring ecosystem health and detecting the encroachment of invasive species before they become established populations (Deiner et al., 2017). In the context of public health, the molecular surveillance of wastewater and bioaerosols provides a scalable mechanism for tracking infectious disease dynamics and the proliferation of antimicrobial resistance genes within the built environment (Pruden et al., 2013). Moreover, the application of these methods to sedimentary archives has advanced our understanding of deep time by permitting the reconstruction of past ecosystems and the description of long-term biotic responses to climatic forcing (Armbrecht et al., 2022). Yet, the discipline faces analytical challenges that define the current frontier of research. The resolution of biases associated with biomass quantification and the closure of the taxonomic void within reference databases remain primary objectives. Future work will require the integration of metagenomics with metatranscriptomics and metabolomics in order to bridge the gap between genetic potential and functional realization. This allows for the direct assessment of metabolic activity and ecosystem process rates (Makiola et al., 2020). Finally, the projected convergence of autonomous robotic sampling with real-time sequencing technologies signals a new era of planetary monitoring. These responsive sensor webs will provide the temporal resolution necessary to detect and manage rapid environmental changes. Ultimately, these computational tools translate the complexity of the microscopic biosphere into actionable knowledge. This provides the empirical foundation for interpreting living systems and informs sustainable stewardship of Earth's microbial reservoirs.

## ACKNOWLEDGMENTS

The author wishes to thank several individuals and institutions whose support was invaluable during the preparation of this chapter. The illustrations presented herein would not have been possible without the exceptional talent of Petra Korlevic. I also thank the current members of the Pipes Lab (Stephanie Murray and Alexandra Lamtyugina) at the University of Hawai'i at Mānoa for fostering a stimulating and supportive intellectual environment. Funding for the Pipes Lab is provided by NIH grant R00GM144747.

## REFERENCES

Aarestrup, Frank M., and Mark E. J. Woolhouse. 2020. "Using Sewage for Surveillance of Antimicrobial Resistance." *Science* 367 (6478): 630–32. https://doi.org/10.1126/science.aba3432.

Albertsen, Mads, Philip Hugenholtz, Adam Skarshewski, Kåre L. Nielsen, Gene W. Tyson, and Per H. Nielsen. 2013. "Genome Sequences of Rare, Uncultured Bacteria Obtained by Differential Coverage Binning of Multiple Metagenomes." *Nature Biotechnology* 31 (6): 533–38.

Alfaro-Perez, C., M. Barberá-Riera, R. de Llanos, and J. M. Delgado-Saborit. 2025. "Systematic Review and Meta-Analysis of Methodological Approaches for Characterising Airborne SARS-CoV-2 RNA for Environmental Surveillance." *Npj Climate and Atmospheric Science* 8 (1): 310.

Almeida, Alexandre, Stephen Nayfach, Miguel Boland, et al. 2021. "A Unified Catalog of 204,938 Reference Genomes from the Human Gut Microbiome." *Nature Biotechnology* 39 (1): 105–14.

Altschul, Stephen F., Mark S. Boguski, Warren Gish, and John C. Wootton. 1994. "Issues in Searching Molecular Sequence Databases." *Nature Genetics* 6 (2): 119–29.

Altschul, Stephen F., Warren Gish, Webb Miller, Eugene W. Myers, and David J. Lipman. 1990. "Basic Local Alignment Search Tool." *Journal of Molecular Biology* 215 (3): 403–10. https://doi.org/10.1016/S0022-2836(05)80360-2.

Amir, Amnon, Daniel McDonald, Jose A. Navas-Molina, et al. 2017. "Deblur Rapidly Resolves Single-Nucleotide Community Sequence Patterns." *mSystems* 2 (2):. e00191-16. https://doi.org/10.1128/msystems.00191-16.

Anantharaman, Karthik, Christopher T. Brown, Laura A. Hug, et al. 2016. "Thousands of Microbial Genomes Shed Light on Interconnected Biogeochemical Processes in an Aquifer System." *Nature Communications* 7 (1): 13219.

Anderson, Marti J. 2001. "A New Method for Non-Parametric Multivariate Analysis of Variance." *Austral Ecology* 26 (1): 32–46. https://doi.org/10.1111/j.1442-9993.2001.01070.pp.x.

Anderson, Marti J., Thomas O. Crist, Jonathan M. Chase, et al. 2011. "Navigating the Multiple Meanings of β Diversity: A Roadmap for the Practicing Ecologist." *Ecology Letters* 14 (1): 19–28. https://doi.org/10.1111/j.1461-0248.2010.01552.x.

Armbrecht, Linda, Michael E. Weber, Maureen E. Raymo, et al. 2022. "Ancient Marine Sediment DNA Reveals Diatom Transition in Antarctica." *Nature Communications* 13 (1): 5787.

Atlas, Ronald M. 1991. "Microbial Hydrocarbon Degradation—Bioremediation of Oil Spills." *Journal of Chemical Technology & Biotechnology* 52 (2): 149–56.

Azam, Farooq, Tom Fenchel, John G. Field, et al. 1983. "The Ecological Role of Water-Column Microbes in the Sea." *Marine Ecology Progress Series. Oldendorf* 10 (3): 257–63.

Badra, Rebecca, Wenqing Zhang, John S. L. Tam, et al. 2025. "A Systematic Review of New, Enhanced Surveillance Systems and Methodologies for Zoonotic Influenza Viruses in Animals and Human–Animal Interface." *Influenza and Other Respiratory Viruses* 19 (11): e70178. https://doi.org/10.1111/irv.70178.

Barbera, Pierre, Alexey M. Kozlov, Lucas Czech, et al. 2018. "EPA-Ng: Massively Parallel Evolutionary Placement of Genetic Sequences." *Systematic Biology* 68 (2): 365–69. https://doi.org/10.1093/sysbio/syy054.

Bardgett, Richard D., and Wim H. Van Der Putten. 2014. "Belowground Biodiversity and Ecosystem Functioning." *Nature* 515 (7528): 505–11.

Barnes, Matthew A., Cameron R. Turner, Christopher L. Jerde, Mark A. Renshaw, W. Lindsay Chadderton, and David M. Lodge. 2014. "Environmental Conditions Influence eDNA Persistence in Aquatic Systems." *Environmental Science & Technology* 48 (3): 1819–27.

Bashiardes, Stavros, Gili Zilberman-Schapira, and Eran Elinav. 2016. "Use of Metatranscriptomics in Microbiome Research." *Bioinformatics and Biology Insights* 10: BBI.S34610. https://doi.org/10.4137/BBI.S34610.

Beattie, Rachelle E., Aditya Bandla, Sanjay Swarup, and Krassimira R. Hristova. 2020. "Freshwater Sediment Microbial Communities Are Not Resilient to Disturbance from Agricultural Land Runoff." *Frontiers in Microbiology* 11. 539921. https://doi.org/10.3389/fmicb.2020.539921.

Berendsen, Roeland L., Corné M. J. Pieterse, and Peter A. H. M. Bakker. 2012. "The Rhizosphere Microbiome and Plant Health." *Trends in Plant Science* 17 (8): 478–86.

Berger, Simon A., Denis Krompass, and Alexandros Stamatakis. 2011. "Performance, Accuracy, and Web Server for Evolutionary Placement of Short Sequence Reads Under Maximum Likelihood." *Systematic Biology* 60 (3): 291–302. https://doi.org/10.1093/sysbio/syr010.

Blanco-Míguez, Aitor, Francesco Beghini, Fabio Cumbo, et al. 2023. "Extending and Improving Metagenomic Taxonomic Profiling with Uncharacterized Species Using MetaPhlAn 4." *Nature Biotechnology* 41 (11): 1633–44.

Blaser, Martin J., and Stanley Falkow. 2009. "What Are the Consequences of the Disappearing Human Microbiota?" *Nature Reviews Microbiology* 7 (12): 887–94.

Bohmann, Kristine, Alice Evans, M. Thomas P. Gilbert, et al. 2014. "Environmental DNA for Wildlife Biology and Biodiversity Monitoring." *Trends in Ecology & Evolution* 29 (6): 358–67.

Bokulich, Nicholas A., Benjamin D. Kaehler, Jai Ram Rideout, et al. 2018. "Optimizing Taxonomic Classification of Marker-Gene Amplicon Sequences with QIIME 2's Q2-Feature-Classifier Plugin." *Microbiome* 6 (1): 90.

Bolger, Anthony M., Marc Lohse, and Bjoern Usadel. 2014. "Trimmomatic: A Flexible Trimmer for Illumina Sequence Data." *Bioinformatics* 30 (15): 2114–20. https://doi.org/10.1093/bioinformatics/btu170.

Bolyen, Evan, Jai Ram Rideout, Matthew R. Dillon, et al. 2019. "Reproducible, Interactive, Scalable and Extensible Microbiome Data Science Using QIIME 2." *Nature Biotechnology* 37 (8): 852–57.

Bourne, David G., Kathleen M. Morrow, and Nicole S. Webster. 2016. "Insights into the Coral Microbiome: Underpinning the Health and Resilience of Reef Ecosystems." *Journal Article. Annual Review of Microbiology* 70: 317–40. https://doi.org/10.1146/annurev-micro-102215-095440.

Bowers, Robert M., Nikos C. Kyrpides, Ramunas Stepanauskas, et al. 2017. "Minimum Information about a Single Amplified Genome (MISAG) and a Metagenome-Assembled Genome (MIMAG) of Bacteria and Archaea." *Nature Biotechnology* 35 (8): 725–31.

Bray, J. Roger, and John T. Curtis. 1957. "An Ordination of the Upland Forest Communities of Southern Wisconsin." *Ecological Monographs* 27 (4): 326–49.

Breitbart, Mya, Peter Salamon, Bjarne Andresen, et al. 2002. "Genomic Analysis of Uncultured Marine Viral Communities." *Proceedings of the National Academy of Sciences* 99 (22): 14250–55. https://doi.org/10.1073/pnas.202488399.

Breitwieser, Florian P., Jennifer Lu, and Steven L. Salzberg. 2017. "A Review of Methods and Databases for Metagenomic Classification and Assembly." *Briefings in Bioinformatics* 20 (4): 1125–36. https://doi.org/10.1093/bib/bbx120.

Brumfield, Kyle D., Nur A. Hasan, Menu B. Leddy, et al. 2020. "A Comparative Analysis of Drinking Water Employing Metagenomics." *PLOS ONE* 15 (4): 1–27. https://doi.org/10.1371/journal.pone.0231210.

Burian, Alfred, Quentin Mauvisseau, Mark Bulling, Sami Domisch, Song Qian, and Michael Sweet. 2021. "Improving the Reliability of eDNA Data Interpretation." *Molecular Ecology Resources* 21 (5): 1422–33. https://doi.org/10.1111/1755-0998.13367.

Bushnell, Brian. 2014. *BBMap: A Fast, Accurate, Splice-Aware Aligner*. No. LBNL-7065E. Ernest Orlando Lawrence Berkeley National Laboratory, Berkeley, CA.

Callahan, Benjamin J., Paul J. McMurdie, and Susan P. Holmes. 2017. "Exact Sequence Variants Should Replace Operational Taxonomic Units in Marker-Gene Data Analysis." *The ISME Journal* 11 (12): 2639–43. https://doi.org/10.1038/ismej.2017.119.

Callahan, Benjamin J., Paul J. McMurdie, Michael J. Rosen, Andrew W. Han, Amy Jo A. Johnson, and Susan P. Holmes. 2016. "DADA2: High-Resolution Sample Inference from Illumina Amplicon Data." *Nature Methods* 13 (7): 581–83.

Capo, Eric, Charline Giguet-Covex, Alexandra Rouillard, et al. 2021. "Lake Sedimentary DNA Research on Past Terrestrial and Aquatic Biodiversity: Overview and Recommendations." *Quaternary* 4 (1): 6. https://doi.org/10.3390/quat4010006.

Chakravorty, Soumitesh, Danica Helb, Michele Burday, Nancy Connell, and David Alland. 2007. "A Detailed Analysis of 16S Ribosomal RNA Gene Segments for the Diagnosis of Pathogenic Bacteria." *Journal of Microbiological Methods* 69 (2): 330–39. https://doi.org/10.1016/j.mimet.2007.02.005.

Champlot, Sophie, Camille Berthelot, Mélanie Pruvost, E. Andrew Bennett, Thierry Grange, and Eva-Maria Geigl. 2010. "An Efficient Multistrategy DNA Decontamination Procedure of PCR Reagents for Hypersensitive PCR Applications." *PLOS ONE* 5 (9): 1–15. https://doi.org/10.1371/journal.pone.0013042.

Chao, Anne. 1984. "Nonparametric Estimation of the Number of Classes in a Population." *Scandinavian Journal of Statistics*, 11, 265–70.

Chaumeil, Pierre-Alain, Aaron J. Mussig, Philip Hugenholtz, and Donovan H. Parks. 2019. "GTDB-Tk: A Toolkit to Classify Genomes with the Genome Taxonomy Database." *Bioinformatics* 36 (6): 1925–27. https://doi.org/10.1093/bioinformatics/btz848.

Chaumeil, Pierre-Alain, Aaron J. Mussig, Philip Hugenholtz, and Donovan H. Parks. 2022. "GTDB-Tk V2: Memory Friendly Classification with the Genome Taxonomy Database." *Bioinformatics* 38 (23): 5315–16. https://doi.org/10.1093/bioinformatics/btac672.

Christner, Brent C., Cindy E. Morris, Christine M. Foreman, Rongman Cai, and David C. Sands. 2008. "Ubiquity of Biological Ice Nucleators in Snowfall." *Science* 319 (5867): 1214–14. https://doi.org/10.1126/science.1149757.

Cock, Peter J. A., Christopher J. Fields, Naohisa Goto, Michael L. Heuer, and Peter M. Rice. 2009. "The Sanger FASTQ File Format for Sequences with Quality Scores, and the Solexa/Illumina FASTQ Variants." *Nucleic Acids Research* 38 (6): 1767–71. https://doi.org/10.1093/nar/gkp1137.

Cordier, Tristan, Philippe Esling, Franck Lejzerowicz, et al. 2017. "Predicting the Ecological Quality Status of Marine Environments from eDNA Metabarcoding Data Using Supervised Machine Learning." *Environmental Science & Technology* 51 (16): 9118–26. https://doi.org/10.1021/acs.est.7b01518.

Costello, Elizabeth K., Keaton Stagaman, Les Dethlefsen, Brendan J. M. Bohannan, and David A. Relman. 2012. "The Application of Ecological Theory Toward an Understanding of the Human Microbiome." *Science* 336 (6086): 1255–62. https://doi.org/10.1126/science.1224203.

Cristescu, Melania E. 2014. "From Barcoding Single Individuals to Metabarcoding Biological Communities: Towards an Integrative Approach to the Study of Global Biodiversity." *Trends in Ecology & Evolution* 29 (10): 566–71.

Cristescu, Melania E., and Paul D. N. Hebert. 2018. "Uses and Misuses of Environmental DNA in Biodiversity Science and Conservation." *Journal Article. Annual Review of Ecology, Evolution, and Systematics* 49: 209–30. https://doi.org/10.1146/annurev-ecolsys-110617-062306.

Czech, Lucas, and Alexandros Stamatakis. 2019. "Scalable Methods for Analyzing and Visualizing Phylogenetic Placement of Metagenomic Samples." *PLOS ONE* 14 (5): 1–50. https://doi.org/10.1371/journal.pone.0217050.

Czech, Lucas, Alexandros Stamatakis, Micah Dunthorn, and Pierre Barbera. 2022. "Metagenomic Analysis Using Phylogenetic Placement—A Review of the First Decade." *Frontiers in Bioinformatics* 2. 871393. https://doi.org/10.3389/fbinf.2022.871393.

Davis, Nicole M., Diana M. Proctor, Susan P. Holmes, David A. Relman, and Benjamin J. Callahan. 2018. "Simple Statistical Identification and Removal of Contaminant Sequences in Marker-Gene and Metagenomics Data." *Microbiome* 6 (1): 226.

Deiner, Kristy, Holly M. Bik, Elvira Mächler, et al. 2017. "Environmental DNA Metabarcoding: Transforming How We Survey Animal and Plant Communities." *Molecular Ecology* 26 (21): 5872–95. https://doi.org/10.1111/mec.14350.

Dejean, Tony, Alice Valentini, Christian Miquel, Pierre Taberlet, Eva Bellemain, and Claude Miaud. 2012. "Improved Detection of an Alien Invasive Species Through Environmental DNA Barcoding: The Example of the American Bullfrog Lithobates Catesbeianus." *Journal of Applied Ecology* 49 (4): 953–59. https://doi.org/10.1111/j.1365-2664.2012.02171.x.

Dickie, Ian A., Stephane Boyer, Hannah L. Buckley, et al. 2018. "Towards Robust and Repeatable Sampling Methods in eDNA-Based Studies." *Molecular Ecology Resources* 18 (5): 940–52. https://doi.org/10.1111/1755-0998.12907.

Doi, Hideyuki, Kimiko Uchii, Teruhiko Takahara, Saeko Matsuhashi, Hiroki Yamanaka, and Toshifumi Minamoto. 2015. "Use of Droplet Digital PCR for Estimation of Fish Abundance and Biomass in Environmental DNA Surveys." *PLOS ONE* 10 (3): 1–11. https://doi.org/10.1371/journal.pone.0122763.

Edgar, Robert C. 2010. "Search and Clustering Orders of Magnitude Faster Than BLAST." *Bioinformatics* 26 (19): 2460–61.

Edgar, Robert C. 2016. "UNOISE2: Improved Error-Correction for Illumina 16S and ITS Amplicon Sequencing." *bioRxiv*, ahead of print. https://doi.org/10.1101/081257.

Edgar, Robert C., and Henrik Flyvbjerg. 2015. "Error Filtering, Pair Assembly and Error Correction for Next-Generation Sequencing Reads." *Bioinformatics* 31 (21): 3476–82. https://doi.org/10.1093/bioinformatics/btv401.

Edgar, Robert C., Brian J. Haas, Jose C. Clemente, Christopher Quince, and Rob Knight. 2011. "UCHIME Improves Sensitivity and Speed of Chimera Detection." *Bioinformatics* 27 (16): 2194–200. https://doi.org/10.1093/bioinformatics/btr381.

Ege, Markus J., Melanie Mayer, Anne-Cécile Normand, et al. 2011. "Exposure to Environmental Microorganisms and Childhood Asthma." *New England Journal of Medicine* 364 (8): 701–9. https://doi.org/10.1056/NEJMoa1007302.

Eisenhofer, Raphael, Jeremiah J. Minich, Clarisse Marotz, Alan Cooper, Rob Knight, and Laura S. Weyrich. 2019. "Contamination in Low Microbial Biomass Microbiome Studies: Issues and Recommendations." *Trends in Microbiology* 27 (2): 105–17.

Elbrecht, Vasco, and Florian Leese. 2015. "Can DNA-Based Ecosystem Assessments Quantify Species Abundance? Testing Primer Bias and Biomass—Sequence Relationships with an Innovative Metabarcoding Protocol." *PloS One* 10 (7): e0130324.

Eren, A. Murat, Loïs Maignien, Woo Jun Sul, et al. 2013. "Oligotyping: Differentiating Between Closely Related Microbial Taxa Using 16S rRNA Gene Data." *Methods in Ecology and Evolution* 4 (12): 1111–19. https://doi.org/10.1111/2041-210X.12114.

Ewing, Brent, and Phil Green. 1998. "Base-Calling of Automated Sequencer Traces Using Phred. II. Error Probabilities." *Genome Research* 8 (3): 186–94.

Falkowski, Paul G., Tom Fenchel, and Edward F. Delong. 2008. "The Microbial Engines That Drive Earth's Biogeochemical Cycles." *Science* 320 (5879): 1034–39. https://doi.org/10.1126/science.1153213.

Ficetola, Gentile F., Johan Pansu, Aurélie Bonin, et al. 2015. "Replication Levels, False Presences and the Estimation of the Presence/Absence from eDNA Metabarcoding Data." *Molecular Ecology Resources* 15 (3): 543–56. https://doi.org/10.1111/1755-0998.12338.

Ficetola, Gentile Francesco, Claude Miaud, François Pompanon, and Pierre Taberlet. 2008. "Species Detection Using Environmental DNA from Water Samples." *Biology Letters* 4 (4): 423–25.

Field, Christopher B., Michael J. Behrenfeld, James T. Randerson, and Paul Falkowski. 1998. "Primary Production of the Biosphere: Integrating Terrestrial and Oceanic Components." *Science* 281 (5374): 237–40. https://doi.org/10.1126/science.281.5374.237.

Fierer, Noah. 2017. "Embracing the Unknown: Disentangling the Complexities of the Soil Microbiome." *Nature Reviews Microbiology* 15 (10): 579–90.

Fierer, Noah, and Robert B. Jackson. 2006. "The Diversity and Biogeography of Soil Bacterial Communities." *Proceedings of the National Academy of Sciences* 103 (3): 626–31. https://doi.org/10.1073/pnas.0507535103.

Fröhlich-Nowoisky, Janine, Christopher J. Kampf, Bettina Weber, et al. 2016. "Bioaerosols in the Earth System: Climate, Health, and Ecosystem Interactions." *Atmospheric Research* 182: 346–76. https://doi.org/10.1016/j.atmosres.2016.07.018.

Gelabert, Pere, Victoria Oberreiter, Lawrence Guy Straus, et al. 2025. "A Sedimentary Ancient DNA Perspective on Human and Carnivore Persistence Through the Late Pleistocene in El Mirón Cave, Spain." *Nature Communications* 16 (1): 107.

Gilbert, Jack A., and Brent Stephens. 2018. "Microbiology of the Built Environment." *Nature Reviews Microbiology* 16 (11): 661–70.

Glassing, Angela, Scot E. Dowd, Susan Galandiuk, Brian Davis, and Rodrick J. Chiodini. 2016. "Inherent Bacterial DNA Contamination of Extraction and Sequencing Reagents May Affect Interpretation of Microbiota in Low Bacterial Biomass Samples." *Gut Pathogens* 8 (1): 24.

Glöckner, Frank Oliver, Pelin Yilmaz, Christian Quast, et al. 2017. "25 Years of Serving the Community with Ribosomal RNA Gene Reference Databases and Tools." *Journal of Biotechnology* 261: 169–76. https://doi.org/10.1016/j.jbiotec.2017.06.1198.

Gloor, Gregory B., Jean M. Macklaim, Vera Pawlowsky-Glahn, and Juan J. Egozcue. 2017. "Microbiome Datasets Are Compositional: And This Is Not Optional." *Frontiers in Microbiology* 8, 2224. https://doi.org/10.3389/fmicb.2017.02224.

Goldberg, Caren S., Cameron R. Turner, Kristy Deiner, et al. 2016. "Critical Considerations for the Application of Environmental DNA Methods to Detect Aquatic Species." *Methods in Ecology and Evolution* 7 (11): 1299–307. https://doi.org/10.1111/2041-210X.12595.

Griffin, Dale W. 2007. "Atmospheric Movement of Microorganisms in Clouds of Desert Dust and Implications for Human Health." *Clinical Microbiology Reviews* 20 (3): 459–77. https://doi.org/10.1128/cmr.00039-06.

Haas, Brian J, Dirk Gevers, Ashlee M Earl, et al. 2011. "Chimeric 16S rRNA Sequence Formation and Detection in Sanger and 454-Pyrosequenced PCR Amplicons." *Genome Research* 21 (3): 494–504.

Hadziavdic, Kenan, Katrine Lekang, Anders Lanzen, Inge Jonassen, Eric M. Thompson, and Christofer Troedsson. 2014. "Characterization of the 18S rRNA Gene for Designing Universal Eukaryote Specific Primers." *PloS One* 9 (2): e87624.

Handelsman, Jo, Michelle R. Rondon, Sean F. Brady, Jon Clardy, and Robert M. Goodman. 1998. "Molecular Biological Access to the Chemistry of Unknown Soil Microbes: A New Frontier for Natural Products." *Chemistry & Biology* 5 (10): R245–49.

Hebert, Paul D. N., Alina Cywinska, Shelley L. Ball, and Jeremy R. DeWaard. 2003. "Biological Identifications Through DNA Barcodes." *Proceedings of the Royal Society of London. Series B: Biological Sciences* 270 (1512): 313–21.

Hendriksen, Rene S., Patrick Munk, Patrick Njage, et al. 2019. "Global Monitoring of Antimicrobial Resistance Based on Metagenomics Analyses of Urban Sewage." *Nature Communications* 10 (1): 1124.

Hess, Matthias, Alexander Sczyrba, Rob Egan, et al. 2011. "Metagenomic Discovery of Biomass-Degrading Genes and Genomes from Cow Rumen." *Science* 331 (6016): 463–67. https://doi.org/10.1126/science.1200387.

Holman, Luke E., Giulia Zampirolo, Richard Gyllencreutz, et al. 2025. "Navigating Past Oceans: Comparing Metabarcoding and Metagenomics of Marine Ancient Sediment Environmental DNA." *Molecular Ecology Resources* 25 (6): e14086. https://doi.org/10.1111/1755-0998.14086.

Hospodsky, Denina, Jing Qian, William W. Nazaroff, et al. 2012. "Human Occupancy as a Source of Indoor Airborne Bacteria." *PLOS ONE* 7 (4): 1–10. https://doi.org/10.1371/journal.pone.0034867.

Hugenholtz, Philip, Maria Chuvochina, Aharon Oren, Donovan H. Parks, and Rochelle M. Soo. 2021. "Prokaryotic Taxonomy and Nomenclature in the Age of Big Sequence Data." *The ISME Journal* 15 (7): 1879–92. https://doi.org/10.1038/s41396-021-00941-x.

Hugerth, Luisa W., and Anders F. Andersson. 2017. "Analysing Microbial Community Composition Through Amplicon Sequencing: From Sampling to Hypothesis Testing." *Frontiers in Microbiology* 8.1561, https://doi.org/10.3389/fmicb.2017.01561.

Hultman, Jenni, Mark P. Waldrop, Rachel Mackelprang, et al. 2015. "Multi-Omics of Permafrost, Active Layer and Thermokarst Bog Soil Microbiomes." *Nature* 521 (7551): 208–12.

Janssen, Stefan, Daniel McDonald, Antonio Gonzalez, et al. 2018. "Phylogenetic Placement of Exact Amplicon Sequences Improves Associations with Clinical Information." *mSystems* 3 (3): e00021-18. https://doi.org/10.1128/msystems.00021-18.

Jerde, Christopher L., W. Lindsay Chadderton, Andrew R. Mahon, et al. 2013. "Detection of Asian Carp DNA as Part of a Great Lakes Basin-Wide Surveillance Program." *Canadian Journal of Fisheries and Aquatic Sciences* 70 (4): 522–26. https://doi.org/10.1139/cjfas-2012-0478.

Jerde, Christopher L., Andrew R. Mahon, W. Lindsay Chadderton, and David M. Lodge. 2011. "'Sight-Unseen' Detection of Rare Aquatic Species Using Environmental DNA." *Conservation Letters* 4 (2): 150–57. https://doi.org/10.1111/j.1755-263X.2010.00158.x.

Ji, Yinqiu, Louise Ashton, Scott M. Pedley, et al. 2013. "Reliable, Verifiable and Efficient Monitoring of Biodiversity via Metabarcoding." *Ecology Letters* 16 (10): 1245–57. https://doi.org/10.1111/ele.12162.

Jones, Devin N., Ben C. Augustine, Patrick Hutchins, et al. 2024. "Autonomous Samplers and Environmental DNA Metabarcoding: Sampling Day and Primer Choice Have Greatest Impact on Fish Detection Probabilities." *Metabarcoding and Metagenomics* 8: e122375.

Jones, Devin N., Kaylin R. Clements, and Adam J. Sepulveda. 2024. "A Workshop to Advance Invasive Species Early Detection Capacity of the Rapid Environmental DNA Assessment and Deployment Initiative & Network (READI-Net)." *Management of Biological Invasions* 15 (1), 159–167.

Kang, Dongwan D., Feng Li, Edward Kirton, et al. 2019. "MetaBAT 2: An Adaptive Binning Algorithm for Robust and Efficient Genome Reconstruction from Metagenome Assemblies." *PeerJ* 7: e7359.

Kelly, Ryan P., Jesse A. Port, Kevan M. Yamahara, et al. 2014. "Harnessing DNA to Improve Environmental Management." *Science* 344 (6191): 1455–56. https://doi.org/10.1126/science.1251156.

Kembel, Steven W., Evan Jones, Jeff Kline, et al. 2012a. "Architectural Design Influences the Diversity and Structure of the Built Environment Microbiome." *The ISME Journal* 6 (8): 1469–79. https://doi.org/10.1038/ismej.2011.211.

Kembel, Steven W., Martin Wu, Jonathan A. Eisen, and Jessica L. Green. 2012b. "Incorporating 16S Gene Copy Number Information Improves Estimates of Microbial Diversity and Abundance." *PLOS Computational Biology* 8 (10): 1–11. https://doi.org/10.1371/journal.pcbi.1002743.

Kim, Daehwan, Li Song, Florian P. Breitwieser, and Steven L. Salzberg. 2016. "Centrifuge: Rapid and Sensitive Classification of Metagenomic Sequences." *Genome Research* 26 (12): 1721–29.

Kim, Dorothy, Casey E. Hofstaedter, Chunyu Zhao, et al. 2017. "Optimizing Methods and Dodging Pitfalls in Microbiome Research." *Microbiome* 5 (1): 52.

Kjær, Kurt H., Mikkel Winther Pedersen, Bianca De Sanctis, et al. 2022. "A 2-Million-Year-Old Ecosystem in Greenland Uncovered by Environmental DNA." *Nature* 612 (7939): 283–91.

Klindworth, Anna, Elmar Pruesse, Timmy Schweer, et al. 2012. "Evaluation of General 16S Ribosomal RNA Gene PCR Primers for Classical and Next-Generation Sequencing-Based Diversity Studies." *Nucleic Acids Research* 41 (1): e1–1. https://doi.org/10.1093/nar/gks808.

Knight, Rob, Alison Vrbanac, Bryn C. Taylor, et al. 2018. "Best Practices for Analysing Microbiomes." *Nature Reviews Microbiology* 16 (7): 410–22.

Lan, Yemin, Qiong Wang, James R. Cole, and Gail L. Rosen. 2012. "Using the RDP Classifier to Predict Taxonomic Novelty and Reduce the Search Space for Finding Novel Organisms." *PLOS ONE* 7 (March): 1–15. https://doi.org/10.1371/journal.pone.0032491.

Larsson, D. G. Joakim, and Carl-Fredrik Flach. 2022. "Antibiotic Resistance in the Environment." *Nature Reviews Microbiology* 20 (5): 257–69.

Lederberg, Joshua, and Alexa T. McCray. 2001. "Ome SweetOmics–a Genealogical Treasury of Words." *The Scientist* 15 (7): 8–8.

Leifels, Mats, Omar Khalilur Rahman, I-Ching Sam, et al. 2022. "The One Health Perspective to Improve Environmental Surveillance of Zoonotic Viruses: Lessons from COVID-19 and Outlook Beyond." *ISME Communications* 2 (1): 107. https://doi.org/10.1038/s43705-022-00191-8.

Leray, Matthieu, Joy Y. Yang, Christopher P. Meyer, et al. 2013. "A New Versatile Primer Set Targeting a Short Fragment of the Mitochondrial COI Region for Metabarcoding Metazoan Diversity: Application for Characterizing Coral Reef Fish Gut Contents." *Frontiers in Zoology* 10 (1): 34.

Ley, Ruth E., Micah Hamady, Catherine Lozupone, et al. 2008. "Evolution of Mammals and Their Gut Microbes." *Science* 320 (5883): 1647–51. https://doi.org/10.1126/science.1155725.

Lima-Mendez, Gipsi, Karoline Faust, Nicolas Henry, et al. 2015. "Determinants of Community Structure in the Global Plankton Interactome." *Science* 348 (6237): 1262073. https://doi.org/10.1126/science.1262073.

Lin, Huang, and Shyamal Das Peddada. 2020. "Analysis of Compositions of Microbiomes with Bias Correction." *Nature Communications* 11 (1): 3514.

Lindahl, Amanda, Laura S. Epp, Sanne Boessenkool, et al. 2025. "Palaeogenomic Inference of Biodiversity Dynamics Across Quaternary Timescales." *Nature Reviews Biodiversity*, 1, 233–247.

Lindahl, Björn D., R. Henrik Nilsson, Leho Tedersoo, et al. 2013. "Fungal Community Analysis by High-Throughput Sequencing of Amplified Markers – a User's Guide." *New Phytologist* 199 (1): 288–99. https://doi.org/10.1111/nph.12243.

Lodge, David M., Cameron R. Turner, Christopher L. Jerde, et al. 2012. "Conservation in a Cup of Water: Estimating Biodiversity and Population Abundance from Environmental DNA." *Molecular Ecology* 21 (11): 2555–58. https://doi.org/10.1111/j.1365-294X.2012.05600.x.

Love, Michael I., Wolfgang Huber, and Simon Anders. 2014. "Moderated Estimation of Fold Change and Dispersion for RNA-Seq Data with DESeq2." *Genome Biology* 15 (12): 550.

Lozupone, Catherine, and Rob Knight. 2005. "UniFrac: A New Phylogenetic Method for Comparing Microbial Communities." *Applied and Environmental Microbiology* 71 (12): 8228–35. https://doi.org/10.1128/AEM.71.12.8228-8235.2005.

Lozupone, Catherine A., Micah Hamady, Scott T. Kelley, and Rob Knight. 2007. "Quantitative and Qualitative β Diversity Measures Lead to Different Insights into Factors That Structure Microbial Communities." *Applied and Environmental Microbiology* 73 (5): 1576–85. https://doi.org/10.1128/AEM.01996-06.

Luo, Chengwei, Rob Knight, Heli Siljander, Mikael Knip, Ramnik J. Xavier, and Dirk Gevers. 2015. "ConStrains Identifies Microbial Strains in Metagenomic Datasets." *Nature Biotechnology* 33 (10): 1045–52.

Makiola, Andreas, Zacchaeus G. Compson, Donald J. Baird, et al. 2020. "Key Questions for Next-Generation Biomonitoring." *Frontiers in Environmental Science* 7. 197. https://doi.org/10.3389/fenvs.2019.00197.

Mandal, Siddhartha, Will Van Treuren, Richard A. White, Merete Eggesbø, Rob Knight, and Shyamal D. Peddada. 2015. "Analysis of Composition of Microbiomes: A Novel Method for Studying Microbial Composition." *Microbial Ecology in Health and Disease* 26 (1): 27663.

Martí, Jose Manuel, Car Reen Kok, James B. Thissen, et al. 2025. "Addressing the Dynamic Nature of Reference Data: A New Nucleotide Database for Robust Metagenomic Classification." *mSystems* 10 (4): e01239-24. https://doi.org/10.1128/msystems.01239-24.

Masella, Andre P., Andrea K. Bartram, Jakub M. Truszkowski, Daniel G. Brown, and Josh D. Neufeld. 2012. "PANDAseq: Paired-End Assembler for Illumina Sequences." *BMC Bioinformatics* 13 (1): 31.

Matsen, Frederick A., Robin B. Kodner, and E. Virginia Armbrust. 2010. "Pplacer: Linear Time Maximum-Likelihood and Bayesian Phylogenetic Placement of Sequences onto a Fixed Reference Tree." *BMC Bioinformatics* 11 (1): 538.

McArdle, Brian H., and Marti J. Anderson. 2001. "Fitting Multivariate Models to Community Data: A Comment on Distance-Based Redundancy Analysis." *Ecology* 82 (1): 290–97.

McDonald, Daniel, Jose C. Clemente, Justin Kuczynski, et al. 2012. "The Biological Observation Matrix (BIOM) Format or: How i Learned to Stop Worrying and Love the Ome-Ome." *Gigascience* 1 (1): 7.

McLaren, Michael R., Amy D. Willis, and Benjamin J. Callahan. 2019. "Consistent and Correctable Bias in Metagenomic Sequencing Experiments." *eLife* 8 (September): e46923. https://doi.org/10.7554/eLife.46923.

Milanese, Alessio, Daniel R. Mende, Lucas Paoli, et al. 2019. "Microbial Abundance, Activity and Population Genomic Profiling with mOTUs2." *Nature Communications* 10 (1): 1014.

Minoche, André E, Juliane C. Dohm, and Heinz Himmelbauer. 2011. "Evaluation of Genomic High-Throughput Sequencing Data Generated on Illumina HiSeq and Genome Analyzer Systems." *Genome Biology* 12 (11): R112.

Mirarab, S., N. Nguyen, and T. Warnow. 2012. SEPP: SATé-enabled phylogenetic placement. *Pacific Symposium on Biocomputing* 247–58. https://doi.org/10.1142/9789814366496_0024.

Morgulis, Aleksandr, E. Michael Gertz, Alejandro A. Schäffer, and Richa Agarwala. 2006. "A Fast and Symmetric DUST Implementation to Mask Low-Complexity DNA Sequences." *Journal of Computational Biology* 13 (5): 1028–40. https://doi.org/10.1089/cmb.2006.13.1028.

Morris, Cindy E., Franz Conen, J. Alex Huffman, Vaughan Phillips, Ulrich Pöschl, and David C. Sands. 2014. "Bioprecipitation: A Feedback Cycle Linking Earth History, Ecosystem Dynamics and Land Use Through Biological Ice Nucleators in the Atmosphere." *Global Change Biology* 20 (2): 341–51. https://doi.org/10.1111/gcb.12447.

Narayanasamy, Shaman, Yohan Jarosz, Emilie E. L. Muller, et al. 2016. "IMP: A Pipeline for Reproducible Reference-Independent Integrated Metagenomic and Metatranscriptomic Analyses." *Genome Biology* 17 (1): 260.

Nasko, Daniel J., Sergey Koren, Adam M. Phillippy, and Todd J. Treangen. 2018. "RefSeq Database Growth Influences the Accuracy of k-Mer-Based Lowest Common Ancestor Species Identification." *Genome Biology* 19 (1): 165.

Nelson, Michael C., Hilary G. Morrison, Jacquelynn Benjamino, Sharon L. Grim, and Joerg Graf. 2014. "Analysis, Optimization and Verification of Illumina-Generated 16S rRNA Gene Amplicon Surveys." *PLOS ONE* 9 (4): 1–14. https://doi.org/10.1371/journal.pone.0094249.

Nguyen, Nam-phuong, Siavash Mirarab, Bo Liu, Mihai Pop, and Tandy Warnow. 2014. "TIPP: Taxonomic Identification and Phylogenetic Profiling." *Bioinformatics* 30 (24): 3548–55. https://doi.org/10.1093/bioinformatics/btu721.

Nilsson, Rolf Henrik, Karl-Henrik Larsson, Andy F. S. Taylor, et al. 2018. "The UNITE Database for Molecular Identification of Fungi: Handling Dark Taxa and Parallel Taxonomic Classifications." *Nucleic Acids Research* 47 (D1): D259–64. https://doi.org/10.1093/nar/gky1022.

Paerl, Hans W., and Timothy G. Otten. 2013. "Harmful Cyanobacterial Blooms: Causes, Consequences, and Controls." *Microbial Ecology* 65 (4): 995–1010.

Paez-Espino, David, Emiley A. Eloe-Fadrosh, Georgios A. Pavlopoulos, et al. 2016. "Uncovering Earth's Virome." *Nature* 536 (7617): 425–30.

Parks, Donovan H., Maria Chuvochina, Pierre-Alain Chaumeil, Christian Rinke, Aaron J. Mussig, and Philip Hugenholtz. 2020. "A Complete Domain-to-Species Taxonomy for Bacteria and Archaea." *Nature Biotechnology* 38 (9): 1079–86.

Parks, Donovan H., Maria Chuvochina, David W. Waite, et al. 2018. "A Standardized Bacterial Taxonomy Based on Genome Phylogeny Substantially Revises the Tree of Life." *Nature Biotechnology* 36 (10): 996–1004.

Parks, Donovan H., Michael Imelfort, Connor T. Skennerton, Philip Hugenholtz, and Gene W. Tyson. 2015. "CheckM: Assessing the Quality of Microbial Genomes Recovered from Isolates, Single Cells, and Metagenomes." *Genome Research* 25 (7): 1043–55.

Parks, Donovan H., Christian Rinke, Maria Chuvochina, et al. 2017. "Recovery of Nearly 8,000 Metagenome-Assembled Genomes Substantially Expands the Tree of Life." *Nature Microbiology* 2 (11): 1533–42.

Pasolli, Edoardo, Francesco Asnicar, Serena Manara, et al. 2019. "Extensive Unexplored Human Microbiome Diversity Revealed by over 150,000 Genomes from Metagenomes Spanning Age, Geography, and Lifestyle." *Cell* 176 (3): 649–62.

Patti, G. J., O. Yanes, and G. Siuzdak. 2012. "Innovation: Metabolomics: The Apogee of the Omics Trilogy." *Nature Reviews Molecular Cell Biology*, 13 (4): 263–9.

Pawlowski, Jan, Stéphane Audic, Sina Adl, et al. 2012. "CBOL Protist Working Group: Barcoding Eukaryotic Richness Beyond the Animal, Plant, and Fungal Kingdoms." *PLOS Biology* 10 (11): 1–5. https://doi.org/10.1371/journal.pbio.1001419.

Pedersen, Mikkel W., Anthony Ruter, Charles Schweger, et al. 2016. "Postglacial Viability and Colonization in North America's Ice-Free Corridor." *Nature* 537 (7618): 45–49.

Pietramellara, Giacomo, Judith Ascher, Federica Borgogni, Maria Teresa Ceccherini, Giulia Guerri, and Paolo Nannipieri. 2009. "Extracellular DNA in Soil and Sediment: Fate and Ecological Relevance." *Biology and Fertility of Soils* 45 (3): 219–35.

Pipes, Lenore, and Rasmus Nielsen. 2024. "A Rapid Phylogeny-Based Method for Accurate Community Profiling of Large-Scale Metabarcoding Datasets." *eLife* 13 (August): e85794. https://doi.org/10.7554/eLife.85794.

Poinar, Hendrik N., and A. Cooper. 2000. "Ancient DNA: Do It Right or Not at All." *Science* 5482 (1139): 416.

Pollock, Jolinda, Laura Glendinning, Trong Wisedchanwet, and Mick Watson. 2018. "The Madness of Microbiome: Attempting to Find Consensus "Best Practice" for 16S Microbiome Studies." *Applied and Environmental Microbiology* 84 (7): e02627–17. https://doi.org/10.1128/AEM.02627-17.

Preston, Christina, Kevan Yamahara, Douglas Pargett, et al. 2024. "Autonomous eDNA Collection Using an Uncrewed Surface Vessel over a 4200-Km Transect of the Eastern Pacific Ocean." *Environmental DNA* 6 (1): e468. https://doi.org/10.1002/edn3.468.

Prodan, Andrei, Valentina Tremaroli, Harald Brolin, Aeilko H. Zwinderman, Max Nieuwdorp, and Evgeni Levin. 2020. "Comparing Bioinformatic Pipelines for Microbial 16S rRNA Amplicon Sequencing." *PLOS ONE* 15 (1): 1–19. https://doi.org/10.1371/journal.pone.0227434.

Pruden, Amy, D. G. Joakim Larsson, Alejandro Amézquita, et al. 2013. "Management Options for Reducing the Release of Antibiotics and Antibiotic Resistance Genes to the Environment." *Environmental Health Perspectives* 121 (8): 878–85. https://doi.org/10.1289/ehp.1206446.

Quince, Christopher, Alan W. Walker, Jared T. Simpson, Nicholas J. Loman, and Nicola Segata. 2017. "Shotgun Metagenomics, from Sampling to Analysis." *Nature Biotechnology* 35 (9): 833–44.

Ramette, Alban. 2007. "Multivariate Analyses in Microbial Ecology." *FEMS Microbiology Ecology* 62 (2): 142–60. https://doi.org/10.1111/j.1574-6941.2007.00375.x.

Ranjan, Ravi, Asha Rani, Ahmed Metwally, Halvor S. McGee, and David L. Perkins. 2016. "Analysis of the Microbiome: Advantages of Whole Genome Shotgun Versus 16S Amplicon Sequencing." *Biochemical and Biophysical Research Communications* 469 (4): 967–77. https://doi.org/10.1016/j.bbrc.2015.12.083.

Rawlence, Nicolas J., David J. Lowe, Jamie R. Wood, et al. 2014. "Using Palaeoenvironmental DNA to Reconstruct Past Environments: Progress and Prospects." *Journal of Quaternary Science* 29 (7): 610–26. https://doi.org/10.1002/jqs.2740.

Rees, Helen C., Ben C. Maddison, David J. Middleditch, James R. M. Patmore, and Kevin C. Gough. 2014. "REVIEW: The Detection of Aquatic Animal Species Using Environmental DNA – a Review of eDNA as a Survey Tool in Ecology." *Journal of Applied Ecology* 51 (5): 1450–59. https://doi.org/10.1111/1365-2664.12306.

Renshaw, Mark A., Brett P. Olds, Christopher L. Jerde, Margaret M. McVeigh, and David M. Lodge. 2015. "The Room Temperature Preservation of Filtered Environmental DNA Samples and Assimilation into a Phenol–Chloroform–Isoamyl Alcohol DNA Extraction." *Molecular Ecology Resources* 15 (1): 168–76. https://doi.org/10.1111/1755-0998.12281.

Richardson, Rodney T., Johan Bengtsson-Palme, and Reed M. Johnson. 2017. "Evaluating and Optimizing the Performance of Software Commonly Used for the Taxonomic Classification of DNA Metabarcoding Sequence Data." *Molecular Ecology Resources* 17 (4): 760–69. https://doi.org/10.1111/1755-0998.12628.

Rognes, Torbjørn, Tomáš Flouri, Ben Nichols, Christopher Quince, and Frédéric Mahé. 2016. "VSEARCH: A Versatile Open Source Tool for Metagenomics." *PeerJ* 4: e2584.

Rosen, Gail L., Erin R. Reichenberger, and Aaron M. Rosenfeld. 2010. "NBC: The Naïve Bayes Classification Tool Webserver for Taxonomic Classification of Metagenomic Reads." *Bioinformatics* 27 (1): 127–29. https://doi.org/10.1093/bioinformatics/btq619.

Rossi-Tamisier, Morgane, Samia Benamar, Didier Raoult, and Pierre-Edouard Fournier. 2015. "Cautionary Tale of Using 16S rRNA Gene Sequence Similarity Values in Identification of Human-Associated Bacterial Species." *International Journal of Systematic and Evolutionary Microbiology* 65 (6): 1929–34.

Ruppert, Krista M., Richard J. Kline, and Md Saydur Rahman. 2019. "Past, Present, and Future Perspectives of Environmental DNA (eDNA) Metabarcoding: A Systematic Review in Methods, Monitoring, and Applications of Global eDNA." *Global Ecology and Conservation* 17: e00547. https://doi.org/10.1016/j.gecco.2019.e00547.

Salter, Susannah J., Michael J. Cox, Elena M. Turek, et al. 2014. "Reagent and Laboratory Contamination Can Critically Impact Sequence-Based Microbiome Analyses." *BMC Biology* 12 (1): 87.

Sboner, Andrea, Xinmeng Jasmine Mu, Dov Greenbaum, Raymond K. Auerbach, and Mark B. Gerstein. 2011. "The Real Cost of Sequencing: Higher Than You Think!" *Genome Biology* 12 (8): 125.

Schirmer, Melanie, Umer Z. Ijaz, Rosalinda D'Amore, Neil Hall, William T. Sloan, and Christopher Quince. 2015. "Insight into Biases and Sequencing Errors for Amplicon Sequencing with the Illumina MiSeq Platform." *Nucleic Acids Research* 43 (6): e37. https://doi.org/10.1093/nar/gku1341.

Schmidt, Benedikt R., Marc Kéry, Sylvain Ursenbacher, Oliver J. Hyman, and James P. Collins. 2013. "Site Occupancy Models in the Analysis of Environmental DNA Presence/Absence Surveys: A Case Study of an Emerging Amphibian Pathogen." *Methods in Ecology and Evolution* 4 (7): 646–53. https://doi.org/10.1111/2041-210X.12052.

Schmidt, Michael W. I., Margaret S. Torn, Samuel Abiven, et al. 2011. "Persistence of Soil Organic Matter as an Ecosystem Property." *Nature* 478 (7367): 49–56.

Schoch, Conrad L., Keith A. Seifert, Sabine Huhndorf, et al. 2012. "Nuclear Ribosomal Internal Transcribed Spacer (ITS) Region as a Universal DNA Barcode Marker for fungi." *Proceedings of the National Academy of Sciences* 109 (16): 6241–46. https://doi.org/10.1073/pnas.1117018109.

Sczyrba, Alexander, Peter Hofmann, Peter Belmann, et al. 2017. "Critical Assessment of Metagenome Interpretation—A Benchmark of Metagenomics Software." *Nature Methods* 14 (11): 1063–71.

Segata, Nicola, Daniela Börnigen, Xochitl C. Morgan, and Curtis Huttenhower. 2013. "PhyloPhlAn Is a New Method for Improved Phylogenetic and Taxonomic Placement of Microbes." *Nature Communications* 4 (1): 2304.

Segata, Nicola, Jacques Izard, Levi Waldron, et al. 2011. "Metagenomic Biomarker Discovery and Explanation." *Genome Biology* 12 (6): R60.

Sepulveda, Adam J., Patrick R. Hutchins, Meghan Forstchen, Madeline N. Mckeefry, and Anna M. Swigris. 2020. "The Elephant in the Lab (and Field): Contamination in Aquatic Environmental DNA Studies." *Frontiers in Ecology and Evolution* 8: 609973.

Shah, Neethu, Haixu Tang, Thomas G. Doak, and Yuzhen Ye. 2011. "Comparing Bacterial Communities Inferred from 16S rRNA gene sequencing and Shotgun Metagenomics." *Pacific Symposium on Biocomputing* 165–76. https://doi.org/10.1142/9789814335058_0018.

Sharon, Itai, and Jillian F. Banfield. 2013. "Genomes from Metagenomics." *Science* 342 (6162): 1057–58. https://doi.org/10.1126/science.1247023.

Sharpton, Thomas J. 2014. "An Introduction to the Analysis of Shotgun Metagenomic Data." *Frontiers in Plant Science* 5: 209.

Simmons, Sheri L., Genevieve DiBartolo, Vincent J. Denef, Daniela S. Aliaga Goltsman, Michael P. Thelen, and Jillian F. Banfield. 2008. "Population Genomic Analysis of Strain Variation in Leptospirillum Group II Bacteria Involved in Acid Mine Drainage Formation." *PLoS Biology* 6 (7): e177.

Simon, H. Ye, Katherine J. Siddle, Daniel J. Park, and Pardis C. Sabeti. 2019. "Benchmarking Metagenomics Tools for Taxonomic Classification." *Cell* 178 (4): 779–94.

Sinha, Rashmi, Galeb Abu-Ali, Emily Vogtmann, et al. 2017. "Assessment of Variation in Microbial Community Amplicon Sequencing by the Microbiome Quality Control (MBQC) Project Consortium." *Nature Biotechnology* 35 (11): 1077–86.

Slon, Viviane, Charlotte Hopfe, Clemens L. Weiß, et al. 2017. "Neandertal and Denisovan DNA from Pleistocene Sediments." *Science* 356 (6338): 605–8. https://doi.org/10.1126/science.aam9695.

Smith, David J., Hilkka J. Timonen, Daniel A. Jaffe, et al. 2013. "Intercontinental Dispersal of Bacteria and Archaea by Transpacific Winds." *Applied and Environmental Microbiology* 79 (4): 1134–39. https://doi.org/10.1128/AEM.03029-12.

Sogin, Mitchell L., Hilary G. Morrison, Julie A. Huber, et al. 2006. "Microbial Diversity in the Deep Sea and the Underexplored 'Rare Biosphere'." *Proceedings of the National Academy of Sciences* 103 (32): 12115–20. https://doi.org/10.1073/pnas.0605127103.

Spellerberg, Ian F., and Peter J. Fedor. 2003. "A Tribute to Claude Shannon (1916–2001) and a Plea for More Rigorous Use of Species Richness, Species Diversity and the 'Shannon–Wiener'index." *Global Ecology and Biogeography* 12 (3): 177–79.

Stackebrandt, Erko. 2006. "Taxonomic Parameters Revisited: Tarnished Gold Standards." *Microbial Today* 33: 152.

Stackebrandt, E., and B. M. Goebel. 1994. "Taxonomic Note: A Place for DNA-DNA Reassociation and 16S rRNA Sequence Analysis in the Present Species Definition in Bacteriology." *International Journal of Systematic and Evolutionary Microbiology* 44 (4): 846–49.

Stark, Manuel, Simon A. Berger, Alexandros Stamatakis, and Christian von Mering. 2010. "MLTreeMap-Accurate Maximum Likelihood Placement of Environmental DNA Sequences into Taxonomic and Functional Reference Phylogenies." *BMC Genomics* 11 (1): 461.

Stein, Michelle M., Cara L. Hrusch, Justyna Gozdz, et al. 2016. "Innate Immunity and Asthma Risk in Amish and Hutterite Farm Children." *New England Journal of Medicine* 375 (5): 411–21. https://doi.org/10.1056/NEJMoa1508749.

Stinson, L. F., J. A. Keelan, and M. S. Payne. 2019. "Identification and Removal of Contaminating Microbial DNA from PCR Reagents: Impact on Low-biomass Microbiome Analyses." *Letters in Applied Microbiology* 68 (1): 2–8. https://doi.org/10.1111/lam.13091.

Strickler, Katherine M., Alexander K. Fremier, and Caren S. Goldberg. 2015. "Quantifying Effects of UV-b, Temperature, and pH on eDNA Degradation in Aquatic Microcosms." *Biological Conservation* 183: 85–92. https://doi.org/10.1016/j.biocon.2014.11.038.

Sunagawa, Shinichi, Luis Pedro Coelho, Samuel Chaffron, et al. 2015. "Structure and Function of the Global Ocean Microbiome." *Science* 348 (6237): 1261359. https://doi.org/10.1126/science.1261359.

Taberlet, Pierre, Eric Coissac, Mehrdad Hajibabaei, and Loren H. Rieseberg. 2012a. "Environmental DNA." *Molecular Ecology* 21 (8): 1789–93. https://doi.org/10.1111/j.1365-294X.2012.05542.x.

Taberlet, Pierre, Eric Coissac, Francois Pompanon, Christian Brochmann, and Eske Willerslev. 2012. "Towards Next-Generation Biodiversity Assessment Using DNA Metabarcoding." *Molecular Ecology* 21 (8): 2045–50. https://doi.org/10.1111/j.1365-294X.2012.05470.x.

Tedersoo, Leho, Mahdieh S. Hosseyni Moghaddam, Vladimir Mikryukov, et al. 2024. "EUKARYOME: The rRNA Gene Reference Database for Identification of All Eukaryotes." *Database* 2024 (June): baae043. https://doi.org/10.1093/database/baae043.

Thompson, Luke R., Jon G. Sanders, Daniel McDonald, et al. 2017. "A Communal Catalogue Reveals Earth's Multiscale Microbial Diversity." *Nature* 551 (7681): 457–63.

Thomsen, Philip Francis, Jos OS Kielgast, Lars L. Iversen, et al. 2012. "Monitoring Endangered Freshwater Biodiversity Using Environmental DNA." *Molecular Ecology* 21 (11): 2565–73. https://doi.org/10.1111/j.1365-294X.2011.05418.x.

Thomsen, Philip Francis, and Eske Willerslev. 2015. "Environmental DNA – an Emerging Tool in Conservation for Monitoring Past and Present Biodiversity." *Biological Conservation* 183: 4–18. https://doi.org/10.1016/j.biocon.2014.11.019.

Tikhonov, Mikhail, Robert. W. Leach, and Ned S. Wingreen. 2015. "Interpreting 16S Metagenomic Data Without Clustering to Achieve Sub-OTU Resolution." *The ISME Journal* 9 (1): 68–80.

Torsvik, V., J. Goksøyr, and F. L. Daae. 1990. "High Diversity in DNA of Soil Bacteria." *Applied and Environmental Microbiology* 56 (3): 782–87. https://doi.org/10.1128/aem.56.3.782-787.1990.

Tremblay, Julien, Kanwar Singh, Alison Fern, et al. 2015. "Primer and Platform Effects on 16S rRNA Tag Sequencing." *Frontiers in Microbiology* 6. 771. https://doi.org/10.3389/fmicb.2015.00771.

Tully, Benjamin J., Elaina D. Graham, and John F. Heidelberg. 2018. "The Reconstruction of 2,631 Draft Metagenome-Assembled Genomes from the Global Oceans." *Scientific Data* 5 (1): 1–8.

Turnbaugh, Peter J., Ruth E. Ley, Micah Hamady, Claire M. Fraser-Liggett, Rob Knight, and Jeffrey I. Gordon. 2007. "The Human Microbiome Project." *Nature* 449 (7164): 804–10.

Turner, Cameron R., Matthew A. Barnes, Charles C. Y. Xu, Stuart E. Jones, Christopher L. Jerde, and David M. Lodge. 2014. "Particle Size Distribution and Optimal Capture of Aqueous Macrobial eDNA." *Methods in Ecology and Evolution* 5 (7): 676–84. https://doi.org/10.1111/2041-210X.12206.

Ushio, Masayuki, Hiroaki Murakami, Reiji Masuda, et al. 2017. "Quantitative Monitoring of Multispecies Fish Environmental DNA Using High-Throughput Sequencing." BioRxiv, 113472.

Venter, J. Craig, Karin Remington, John F. Heidelberg, et al. 2004. "Environmental Genome Shotgun Sequencing of the Sargasso Sea." *Science* 304 (5667): 66–74. https://doi.org/10.1126/science.1093857.

Wang, Qiong, George M. Garrity, James M. Tiedje, and James R. Cole. 2007. "Naïve Bayesian Classifier for Rapid Assignment of rRNA Sequences into the New Bacterial Taxonomy." *Applied and Environmental Microbiology* 73 (16): 5261–67. https://doi.org/10.1128/AEM.00062-07.

Webster, Nicole S. 2025. "Microorganisms as Architects of a Sustainable Future." *Nature Reviews Genetics*, 27 (1): 5–6.

Whitman, William B., David C. Coleman, and William J. Wiebe. 1998. "Prokaryotes: The Unseen Majority." *Proceedings of the National Academy of Sciences* 95 (12): 6578–83. https://doi.org/10.1073/pnas.95.12.6578.

Whittaker, Robert Harding. 1960. "Vegetation of the Siskiyou Mountains, Oregon and California." *Ecological Monographs* 30 (3): 279–338.

Willerslev, Eske, Anders J. Hansen, Jonas Binladen, et al. 2003. "Diverse Plant and Animal Genetic Records from Holocene and Pleistocene Sediments." *Science* 300 (5620): 791–95. https://doi.org/10.1126/science.1084114.

Womack, Ann M., Brendan J. M. Bohannan, and Jessica L. Green. 2010. "Biodiversity and Biogeography of the Atmosphere." *Philosophical Transactions of the Royal Society B: Biological Sciences* 365 (1558): 3645–53.

Wood, Derrick E., Jennifer Lu, and Ben Langmead. 2019. "Improved Metagenomic Analysis with Kraken 2." *Genome Biology* 20 (1): 257.

Xu, Xiaoqing, Yu Deng, Jiahui Ding, et al. 2024. "High-Resolution and Real-Time Wastewater Viral Surveillance by Nanopore Sequencing." *Water Research* 256: 121623. https://doi.org/10.1016/j.watres.2024.121623.

Yamahara, Kevan M., Elizabeth A. Allan, Julie Robidart, et al. 2025. "A State-of-the-Art Review of Aquatic eDNA Sampling Technologies and Instrumentation: Advancements, Challenges, and Future Prospects." *Environmental DNA* 7 (4): e70170. https://doi.org/10.1002/edn3.70170.

Yarza, Pablo, Pelin Yilmaz, Elmar Pruesse, et al. 2014. "Uniting the Classification of Cultured and Uncultured Bacteria and Archaea Using 16S rRNA Gene Sequences." *Nature Reviews Microbiology* 12 (9): 635–45.

Zhang, Jiajie, Kassian Kobert, Tomáš Flouri, and Alexandros Stamatakis. 2013. "PEAR: A Fast and Accurate Illumina Paired-End reAd mergeR." *Bioinformatics* 30 (5): 614–20. https://doi.org/10.1093/bioinformatics/btt593.

Zielińska, Sylwia, Piotr Radkowski, Aleksandra Blendowska, Agnieszka Ludwig-Gałęzowska, Joanna M. Łoś, and Marcin Łoś. 2017. "The Choice of the DNA Extraction Method May Influence the Outcome of the Soil Microbial Community Structure Analysis." *MicrobiologyOpen* 6 (4): e00453. https://doi.org/10.1002/mbo3.453.

Zinger, Lucie, Aurélie Bonin, Inger G. Alsos, et al. 2019. "DNA Metabarcoding—Need for Robust Experimental Designs to Draw Sound Ecological Conclusions." *Molecular Ecology* 28 (8): 1857–62. https://doi.org/10.1111/mec.15060.

CHAPTER 6

# The Secret Life of Ecosystems

## UNCOVERING THE ROLES OF MICROBES IN SHAPING ENVIRONMENTAL PROCESSES

*Hina H. Radadiya, Rajesh K. Patel, Charmy R. Kothari, and Ramesh K. Kothari*

### 6.1 INTRODUCTION

Microorganisms are the primordial and numerically dominant components of the biosphere, present since ≥3.5 Ga and responsible for most of Earth's biochemical transformations (Schopf, 2006). Their collective genomic and metabolic repertoire far exceeds that of macroorganisms: a single gram of soil can contain thousands of taxa and up to $10^{10}$ cells, many of which remain uncultured and functionally uncharacterized (Locey & Lennon, 2016). Rather than functioning as isolated biological entities, these microorganisms operate as dynamic, interacting assemblages whose metabolic activities directly regulate ecosystem-scale fluxes of energy and matter.

This extraordinary diversity confers unparalleled metabolic versatility, ranging from autotrophic carbon fixation and chemosynthesis to nitrogen fixation, nitrification/denitrification, sulfur redox cycling, and phosphorus mobilization, enabling microbes to mediate the fluxes of energy and nutrients that determine ecosystem productivity, stability, and Earth-system feedbacks (Falkowski et al., 2008; Muyzer & Stams, 2008; Richardson & Simpson, 2011).

Importantly, the ecological significance of microbial processes lies not only in their presence but also in the magnitude and direction of the fluxes they regulate under specific environmental constraints. Beyond elemental cycling, microbial interactions, including symbiosis, competition, syntrophy, predation, and viral lysis, reorganize biomass and biochemical pools across spatial and temporal scales. These interactions introduce nonlinearities that complicate attempts to predict ecosystem responses based solely on taxonomic composition or gene abundance. As a result, microbial communities often exert disproportionate influence on ecosystem function relative to their biomass, particularly in systems subject to strong environmental gradients or disturbance.

Despite substantial progress driven by cultivation-independent methods, major gaps remain in our mechanistic understanding of how microbial genetic potential scales to ecosystem process rates and resilience under environmental change. Advances in high-throughput sequencing, genome-resolved metagenomics, single-cell genomics, stable-isotope probing (SIP), and multi-omics have revealed trait distributions and metabolic pathways at unprecedented resolution.

However, translating these molecular insights into quantitative predictions of biogeochemical fluxes remains a central challenge in environmental microbiology.

From an environmental monitoring perspective, this disconnect has practical implications. Monitoring programs increasingly use microbial

DOI: 10.1201/9781003637950-6

indicators as proxies for ecosystem function, yet the relationship between microbial community attributes and ecosystem-level outcomes is often assumed rather than empirically constrained. This assumption risks oversimplifying microbial regulation of ecosystem processes, particularly in heterogeneous environments where localized microbial activity may be decoupled from bulk measurements of fluxes or concentrations.

Accordingly, this chapter synthesizes current mechanistic understanding of microbial drivers of ecosystem processes while explicitly addressing the limitations of linking molecular data to environmental outcomes. We (i) summarize taxonomic and functional diversity among bacteria, archaea, fungi, protists, and viruses, along with their ecological niches; (ii) review microbial controls on carbon, nitrogen, sulfur, and phosphorus cycling; (iii) examine soil and aquatic microbial ecology, with emphasis on rhizosphere dynamics, decomposition, and the biological carbon pump; (iv) highlight extremophile systems as models of physiological adaptation and ecosystem initiation; (v) evaluate microbe–climate feedbacks and greenhouse gas dynamics; (vi) address symbiotic, pathogenic, and anthropogenic influences, including pollution and antimicrobial resistance; and (vii) assess emerging methodological frameworks, including genome-resolved approaches, SIP, and trait-based modeling.

Our objective is not to provide an exhaustive catalog of microbial processes but to identify regulatory control points where microbial activity most strongly shapes measurable environmental signals. By emphasizing these control points, we aim to inform the design of monitoring strategies and conceptual models that more accurately capture microbial contributions to ecosystem function amid ongoing environmental change.

## 6.2 MICROBIAL DIVERSITY AND ECOLOGICAL NICHES

### 6.2.1 Taxonomic Groups: Bacteria, Archaea, Fungi, Protists, and Viruses

Microbial diversity spans all domains of life and extends to acellular entities such as viruses, collectively forming the most functionally versatile component of Earth's biosphere, as first quantified by Whitman et al. (1998). Bacteria exhibit the broadest taxonomic and metabolic diversity, occupying nearly every ecological niche. They drive processes ranging from oxygenic and anoxygenic carbon fixation to the degradation of recalcitrant organic matter, underscoring their central role in Earth-system functioning (Falkowski et al., 2008). Cyanobacteria fuel global primary production, particularly in aquatic ecosystems (Partensky et al., 1999), while actinobacteria specialize in decomposing complex polymers and producing antibiotics, siderophores, and extracellular enzymes that mediate competitive and cooperative interactions within microbial communities (Hibbing et al., 2010; Madigan et al., 2021). Archaea, once thought restricted to extreme environments, are now recognized as abundant in marine pelagic zones, soils, and animal microbiomes. Molecular surveys have revealed their widespread distribution and ecological relevance beyond extreme habitats (Offre et al., 2013). Their ecological roles include methanogenesis in anoxic habitats, ammonia oxidation by Thaumarchaeota in oligotrophic waters, and hydrocarbon degradation in deep-sea sediments. The dominance of archaeal ammonia monooxygenase in global nitrification, particularly in nutrient-poor marine systems, was demonstrated by Stahl and de la Torre (2012) and further synthesized by Offre et al. (2013), underscoring their biogeochemical significance.

Fungi are essential eukaryotic decomposers and symbionts. Saprotrophic Ascomycota and Basidiomycota degrade lignin, cellulose, and other complex biopolymers that many bacteria cannot efficiently process, a functional specialization documented by Treseder and Lennon (2015). Symbiotic groups—most notably arbuscular mycorrhizal fungi (AMF) and ectomycorrhizal fungi (ECM)—enhance nutrient acquisition, regulate plant community composition, and contribute to long-term soil carbon storage, as established through physiological and ecological syntheses by Smith and Read (2008) and confirmed by large-scale analyses of mycorrhizal function (van der Heijden et al., 2015).

Protists function primarily as microbial predators and nutrient recyclers. Amoebae, ciliates, and flagellates exert strong top-down control on bacterial and algal populations, accelerating nutrient regeneration in soil and aquatic systems, a role demonstrated through global soil and aquatic surveys by Geisen et al. (2018). Mixotrophic protists integrate photosynthesis with heterotrophy, effectively linking primary production and microbial grazing within the same trophic unit, a strategy shown to be widespread and ecologically significant by Flynn et al. (2019).

Viruses are the most abundant biological entities on Earth, with an estimated global population exceeding $10^{31}$ particles, as quantified by Suttle (2007). Through host lysis, viruses release cellular contents into dissolved organic matter pools, fueling microbial loops via the "viral shunt" and diverting carbon from higher trophic transfer, a process conceptualized and reviewed by Weitz and Wilhelm (2012). Viral-mediated horizontal gene transfer profoundly reshapes microbial genome evolution and ecosystem functional potential, as demonstrated by marine viromics and comparative genomic studies summarized by Brum and Sullivan (2015). Together, these microbial groups form a multilayered network of interacting taxa whose metabolic flexibility and evolutionary adaptability support ecosystem productivity, resilience, and biogeochemical continuity.

### 6.2.2 Adaptations to Extreme Environments

Microorganisms exhibit remarkable physiological and biochemical adaptability, enabling survival under conditions that approach or exceed the limits tolerated by most forms of life. Thermophilic microorganisms, such as *Thermus aquaticus* and members of the genus *Sulfolobus*, thrive at temperatures above 80°C by employing thermostable enzymes, specialized chaperone systems, and membrane architectures that preserve structural integrity under intense thermal stress, as experimentally demonstrated by Stetter (1999). These adaptations show that metabolic activity can be maintained under conditions once considered incompatible with life.

Halophilic microorganisms persist in hypersaline environments by accumulating compatible solutes or, alternatively, by adopting a "salt-in" strategy in which intracellular proteins are structurally adapted to function at high ionic strength. Physiological analyses of extreme halophiles have shown that certain taxa, including *Halobacterium salinarum*, exploit bacteriorhodopsin-based phototrophy to generate energy under extreme salinity (Oren, 2011). Acidophiles, exemplified by *Acidithiobacillus ferrooxidans*, maintain cytoplasmic pH homeostasis and utilize acid-stable enzymes that permit growth at pH values below 3, a mechanism elucidated through studies of acid mine drainage systems (Baker-Austin & Dopson, 2007).

Cold-adapted psychrophiles inhabit polar and alpine environments, where metabolic activity persists at subzero temperatures through the production of antifreeze proteins, cryoprotectants, and unsaturated membrane lipids, as reviewed by D'Amico et al. (2006). In the deep ocean, barophilic microorganisms tolerate hydrostatic pressures exceeding 100 MPa by modifying protein structure and membrane composition. Radiophiles such as *Deinococcus radiodurans* possess highly efficient DNA repair systems that enable survival under extreme ionizing radiation, as revealed by genome-based analyses (Makarova et al., 2001). Collectively, extremophiles delineate the biochemical boundaries of life and provide insight into early Earth conditions and potential extraterrestrial habitats.

### 6.2.3 Microbial Community Interactions: Symbiosis, Competition, and Cooperation

Microbial communities are structured not only by the physiological traits of individual taxa but also by a complex web of interactions that govern collective behavior and emergent ecosystem properties. Symbiotic relationships span mutualism, commensalism, and parasitism, often involving tightly coordinated metabolic exchanges. Comparative genomic and evolutionary studies have shown that rhizobia–legume nitrogen fixation, mycorrhizal nutrient exchange between fungi and plants, and insect–endosymbiont associations supply essential nutrients to hosts with restricted diets (Moran et al., 2008). Such interactions can stabilize ecosystem processes by buffering partners against environmental variability. Competition is equally pervasive, particularly in nutrient-limited environments. Microorganisms compete through siderophore-mediated iron acquisition, antibiotic and bacteriocin production, and spatial strategies such as biofilm formation. Experimental and conceptual syntheses have shown that quorum sensing enables coordinated population-level responses, regulating behaviors that enhance collective fitness, including secondary metabolite production and biofilm architecture (Hibbing et al., 2010; Nadell et al., 2009). These competitive dynamics strongly shape community composition and resource partitioning. In contrast, cooperation emerges when metabolic interdependence confers a net energetic advantage. Syntrophic partnerships, such as those between fermentative bacteria and methanogenic archaea, enable energetically unfavorable reactions to proceed by maintaining low concentrations of inhibitory intermediates, a principle formalized by Morris et al. (2012). Biofilms further exemplify cooperative organization, often exhibiting division of labor among subpopulations that specialize in complementary metabolic roles. While such cooperation enhances community resilience, it also complicates efforts to attribute ecosystem processes to individual taxa.

Together, symbiosis, competition, and cooperation generate highly structured microbial networks whose collective dynamics govern ecosystem function. Recognizing the primacy of these interactions is essential for understanding how microbial communities sustain biogeochemical processes and adapt to environmental change.

## 6.3 MICROBES AND BIOGEOCHEMICAL CYCLES

Microorganisms are the primary biological agents driving the transformation and redistribution of elements in Earth's biogeochemical cycles. Early syntheses of Earth-system functioning showed that microbial metabolism underpins the cycling of major elements at the planetary scale (Falkowski et al., 2008). Through diverse metabolic pathways, microbes regulate the movement of carbon, nitrogen, sulfur, and phosphorus between organic and inorganic pools, thereby shaping ecosystem productivity, nutrient availability, and long-term Earth-system dynamics. Because microbial metabolism responds rapidly to environmental conditions, these processes vary strongly across space and time, producing heterogeneous biogeochemical fluxes across ecosystems, as emphasized in global biogeochemical frameworks (Schimel & Schaeffer, 2012).

### 6.3.1 Microbial Mediation of the Carbon Cycle

Microbial processes govern both carbon fixation and organic matter decomposition, placing microorganisms at the center of the global carbon cycle. Photosynthetic microbes, including cyanobacteria and microalgae, fix $CO_2$ into organic biomass via the Calvin–Benson–Bassham cycle and alternative pathways, as shown by comparative biochemical analyses of carbon fixation strategies (Berg et al., 2010). In marine environments, small picocyanobacteria such as *Prochlorococcus* and *Synechococcus* account for a substantial fraction of primary production, particularly in nutrient-poor regions, a conclusion supported by ocean-scale observations and physiological studies (Partensky et al., 1999).

Heterotrophic bacteria and fungi remineralize organic carbon via aerobic respiration, returning $CO_2$ to the atmosphere. Under anoxic conditions, fermentative pathways and anaerobic respiration dominate, culminating in methanogenesis by archaea that produce methane in wetlands, sediments, and animal digestive systems, as demonstrated by physiological and isotopic studies of methanogenic pathways (Thauer et al., 2008). Methane emissions are partially constrained by methanotrophic microorganisms, which oxidize $CH_4$ under both oxic and anoxic conditions, a process documented in marine and sedimentary systems by Knittel and Boetius (2009). The balance among these processes determines whether ecosystems act as net sources or sinks of greenhouse gases.

Microbial activity also contributes to long-term carbon storage. In the oceans, microbes underpin the biological carbon pump by facilitating the export of organic matter to depth, a process established by biogeochemical and sediment-trap studies (Ducklow et al., 2001). In soils, microbial residues and mineral associations stabilize organic carbon over extended timescales, as shown by soil fractionation and mineral-protection studies (Lehmann & Kleber, 2015).

### 6.3.2 Microbial Roles in the Nitrogen Cycle

The nitrogen cycle depends almost entirely on microbial transformations because atmospheric $N_2$ is biologically inaccessible to most organisms. Diazotrophic bacteria and archaea introduce reactive nitrogen into ecosystems through biological nitrogen fixation, thereby supporting primary productivity in terrestrial and aquatic environments, as quantified in global nitrogen budgets (Raymond et al., 2004). Following fixation, nitrogen undergoes successive microbial transformations. Nitrification converts ammonia to nitrate under aerobic conditions and is mediated by ammonia-oxidizing bacteria and archaea, together with nitrite-oxidizing bacteria. This process has been characterized through physiological and molecular studies (Kowalchuk & Stephen, 2001). Under oxygen-limited conditions, denitrification reduces nitrate to gaseous nitrogen species, returning nitrogen to the atmosphere and contributing to nitrous oxide emissions, as established in classical enzymatic and ecological analyses (Zumft, 1997).

In addition, anaerobic ammonium oxidation (anammox) provides an alternative pathway for nitrogen loss in oxygen-depleted marine zones and sediments, a process first identified and experimentally validated by Strous et al. (1999). Through these coupled processes, microbial nitrogen cycling regulates nutrient availability, ecosystem productivity, and greenhouse gas fluxes, while remaining highly sensitive to environmental change, as emphasized in ecosystem-scale syntheses (Galloway et al., 2004).

### 6.3.3 Microbial Contributions to the Sulfur Cycle

Microbial sulfur cycling is driven by redox transformations across oxic and anoxic environments. In anoxic sediments, sulfate-reducing bacteria oxidize organic matter while reducing sulfate to sulfide, constituting a major pathway of carbon mineralization, as shown by sediment biogeochemistry and cultivation studies (Muyzer & Stams, 2008). The resulting reduced sulfur compounds are then oxidized by sulfur-oxidizing bacteria in oxic or microoxic environments, often coupled to carbon fixation, a process observed in both marine and terrestrial systems (Friedrich et al., 2001).

These transformations link sulfur cycling to other elemental cycles, particularly carbon and nitrogen. In marine systems, microbial sulfur metabolism also influences atmospheric chemistry by producing volatile sulfur compounds that affect cloud formation and climate processes. This climate connection was first articulated by Charlson et al. (1987) and has been supported by subsequent observational studies.

### 6.3.4 Microbial Roles in the Phosphorus Cycle

Unlike other major biogeochemical cycles, phosphorus lacks a gaseous phase and is often limited by low solubility and strong mineral associations. Microorganisms play a central role in mobilizing phosphorus by mineralizing organic phosphorus compounds and solubilizing inorganic phosphate through organic acid production, as demonstrated in soil and aquatic systems (Richardson & Simpson, 2011; Rodríguez & Fraga, 1999). Some microbial taxa temporarily store excess phosphorus as intracellular polyphosphate, buffering fluctuations in environmental availability and influencing longer-term phosphorus retention, as shown in microbial physiological and ecological studies (Hupfer and Lewandowski~~et al.~~, 2008). Through these processes, microbes regulate phosphorus availability and contribute to nutrient limitation patterns across ecosystems.

## 6.4 MICROBES IN SOIL ECOSYSTEMS

Soils are among the most biologically complex habitats on Earth, hosting dense and diverse assemblages of bacteria, archaea, fungi, protists, and viruses. Early global syntheses identified soils as major reservoirs of microbial diversity and biomass (Whitman et al., 1998). This belowground diversity underpins key ecosystem functions, including organic matter decomposition, nutrient cycling, soil structure formation, and long-term carbon storage, as shown in ecosystem-scale soil studies (Bardgett & van der Putten, 2014). Rather than acting independently, soil microorganisms form tightly coupled networks whose collective activity governs soil fertility and ecosystem resilience, a concept emphasized in integrative soil ecology frameworks (Fierer, 2017).

### 6.4.1 Rhizosphere Processes and Plant–Microbe Interactions

The rhizosphere is a highly dynamic interface shaped by the release of root exudates, including sugars, amino acids, and organic acids, which selectively stimulate microbial growth, as originally conceptualized by Hiltner and later refined through experimental studies (Bais et al., 2006). These inputs create steep chemical gradients that structure microbial communities and intensify biogeochemical activity near plant roots (Philippot et al., 2013). Within this zone, plant-associated microorganisms influence nutrient acquisition, root development, and plant stress tolerance, as shown in rhizosphere ecology studies (Berendsen et al., 2012). Nitrogen-fixing bacteria, phosphate-solubilizing microorganisms, and siderophore-producing taxa enhance nutrient availability, while other rhizosphere inhabitants suppress pathogens through competition or the production of antimicrobial compounds, as documented in plant–microbe interaction research (Lugtenberg & Kamilova, 2009). Beyond nutrient supply, many root-associated microbes modulate plant physiology by producing phytohormones or altering stress signaling pathways (Vacheron et al., 2013). As a result, plant performance emerges from integrated plant–microbe interactions rather than from plant traits alone, a conclusion supported by experimental and ecological syntheses (Friesen et al., 2011).

### 6.4.2 Mycorrhizal Associations and Soil Structure

Mycorrhizal fungi form some of the most widespread and influential symbioses in terrestrial ecosystems, as established by classical and contemporary mycorrhizal research (Smith & Read, 2008). By extending hyphal networks beyond the root depletion zone, these fungi increase access to phosphorus, nitrogen, and micronutrients, particularly in nutrient-poor soils, as demonstrated by physiological and isotopic studies (van der Heijden et al., 2015). In return, plants allocate a substantial fraction of photosynthetically fixed carbon to their fungal partners (Hobbie & Hobbie, 2006). Beyond nutrient exchange, mycorrhizal fungi also influence soil physical properties. Fungal hyphae and extracellular compounds contribute to soil aggregation, enhancing water retention and stabilizing organic matter, as shown in soil structural studies (Rillig & Mummey, 2006). Extensive mycorrhizal networks can connect multiple plants, facilitating resource redistribution and influencing plant community dynamics (Simard et al., 2012). These effects position mycorrhizal fungi as key regulators of both nutrient cycling and soil structure.

### 6.4.3 Decomposition, Carbon Turnover, and Soil Organic Matter Formation

Soil microorganisms drive the decomposition of plant litter, root residues, and microbial necromass through a succession of enzymatic processes, as shown in litter decomposition experiments and soil enzyme studies (Schimel & Schaeffer, 2012). Bacteria dominate the rapid turnover of labile substrates, whereas fungi specialize in breaking down structurally complex compounds such as lignin and cellulose (Treseder & Lennon, 2015). Actinobacteria further contribute to the degradation of aromatic compounds and the production of secondary metabolites that shape microbial interactions (Goodfellow & Williams, 1983). As decomposition proceeds, some organic material is converted into more stable soil organic matter. Microbial residues interact with mineral surfaces and soil aggregates, reducing decomposer access and promoting long-term carbon persistence, as shown by mineral-protection and microbial necromass studies (Lehmann & Kleber, 2015). At the same time, microbial respiration releases carbon dioxide and, under anoxic conditions, methane, linking soil processes to atmospheric carbon dynamics (Conrad, 2009). The balance between carbon stabilization and mineralization therefore depends strongly on microbial activity and environmental conditions.

## 6.5 MICROBES IN AQUATIC ECOSYSTEMS

Aquatic ecosystems, encompassing marine and freshwater environments, are fundamentally structured by microbial life, as established in early marine microbial ecology syntheses (Azam et al., 1983). In these systems, microorganisms dominate in abundance, metabolic diversity, and biogeochemical influence, forming the base of food webs and regulating elemental cycling across spatial scales from micrometers to entire ocean basins (Falkowski et al., 2008). Unlike many terrestrial environments, where plant biomass provides an obvious structural framework, aquatic ecosystems rely heavily on microbial processes to sustain productivity and regulate chemical gradients. Microbial communities in aquatic systems are shaped by physical stratification, light availability, nutrient supply, and hydrological connectivity. Vertical gradients in oxygen, temperature, and nutrient concentrations generate distinct ecological niches, often producing sharp transitions in microbial community composition over small spatial scales, as demonstrated in stratified water columns (Wright et al., 2012). Consequently, aquatic microbial processes are highly dynamic, responding rapidly to seasonal mixing, stratification, and episodic disturbances (Simon et al., 2002).

### 6.5.1 Marine Microbial Communities and Primary Production

Marine ecosystems account for a substantial share of global primary production, much of it driven by microscopic photosynthetic organisms, as quantified in global ocean productivity assessments (Field et al., 1998). Phytoplankton, including diatoms, dinoflagellates, and cyanobacteria, convert dissolved inorganic carbon into organic matter that fuels marine food webs. Despite their small size, these organisms collectively fix carbon comparable to that of terrestrial vegetation, underscoring the outsized role of microbes in ocean productivity (Falkowski et al., 1998). Among marine primary producers, picocyanobacteria such as *Prochlorococcus* and *Synechococcus* are particularly significant in oligotrophic regions. Their streamlined genomes and efficient light-harvesting strategies enable them to thrive under low-nutrient conditions, where larger phytoplankton are competitively disadvantaged, as shown by genomic and physiological studies (Partensky et al., 1999). These taxa dominate vast expanses of the open ocean, illustrating how microbial adaptation shapes ecosystem-scale productivity patterns. Below the sunlit surface waters, microbial primary production continues via chemolithoautotrophic pathways. Ammonia-oxidizing archaea and bacteria fix carbon while oxidizing reduced nitrogen compounds, contributing to carbon cycling in mesopelagic and deeper waters (Herndl et al., 2005). In hydrothermal vent systems, chemosynthetic microbes oxidize sulfur, hydrogen, and methane to support dense animal communities entirely independent of photosynthesis, a phenomenon first documented in deep-sea vent ecosystems (Jannasch & Mottl, 1985). These environments highlight the metabolic breadth of aquatic microbes and their ability to sustain ecosystems under extreme conditions.

### 6.5.2 Microbial Regulation of Oxygen and Carbon Fluxes

Aquatic microorganisms play a central role in regulating oxygen production and carbon sequestration. Photosynthetic microbes account for approximately half of global oxygen generation, directly linking microbial activity to atmospheric composition (Falkowski, 2012). At the same time, heterotrophic bacteria consume phytoplankton-produced organic matter, respiring carbon back to $CO_2$ and controlling the efficiency of organic carbon transfer to higher trophic levels or export to depth (Azam & Malfatti, 2007). The biological carbon pump is one of the most important mechanisms by which aquatic microbes influence the global carbon cycle. Organic matter produced in surface waters is transformed, aggregated, and partially remineralized as it sinks through the water column, a process documented by sediment-trap and biogeochemical studies (Ducklow et al., 2001). Microbial degradation during this descent determines how much carbon is respired to $CO_2$ versus how much is sequestered in deep waters or sediments over long timescales (Boyd & Trull, 2007). Small changes in microbial community composition or activity can therefore have disproportionate effects on oceanic carbon storage. Viruses exert additional control over carbon cycling through viral lysis. By infecting and lysing microbial cells, viruses redirect organic carbon from particulate to dissolved pools, stimulating microbial loop activity while reducing the efficiency of carbon transfer to higher trophic levels. This "viral shunt" alters both the magnitude and pathway of carbon fluxes, as conceptualized and reviewed by Wilhelm and Suttle (1999) and later synthesized by Weitz and Wilhelm (2012), emphasizing that microbial mortality processes are integral components of aquatic biogeochemistry.

### 6.5.3 Freshwater Microbial Communities and Nutrient Cycling

Freshwater ecosystems, including lakes, rivers, wetlands, and reservoirs, host diverse microbial communities that regulate nutrient dynamics and organic matter turnover (Wetzel, 2001). Although freshwater systems occupy a relatively small fraction of Earth's surface, they exert a disproportionate influence on biogeochemical cycling, particularly carbon and nitrogen fluxes (Cole et al., 2007). Primary production in freshwater environments is driven by eukaryotic algae and cyanobacteria. Under nutrient-enriched conditions, certain cyanobacterial taxa can form dense blooms that restructure food webs and alter ecosystem functioning, as documented in eutrophication studies (Paerl & Otten, 2013). These blooms often reflect shifts in nutrient availability, temperature, and hydrological stability, underscoring the sensitivity of freshwater microbial communities to environmental change.

Heterotrophic bacteria play a central role in processing organic matter from both internal primary production and terrestrial inputs. In lakes and rivers, large fractions of organic carbon originate in surrounding catchments, and microbial degradation of this material links terrestrial and aquatic biogeochemical cycles (Tranvik et al., 2009). Wetlands and stratified lakes further support anaerobic microbial processes, including methanogenesis and anaerobic methane oxidation, which strongly influence greenhouse gas emissions (Segers, 1998). Nitrogen cycling in freshwater systems is mediated by the coordinated activity of nitrifying, denitrifying, and anammox microorganisms. The balance among these processes determines nitrogen availability and the release of nitrogenous gases to the atmosphere, as shown by freshwater nitrogen budgets (Seitzinger et al., 2006). Because freshwater systems often experience strong temporal variability in oxygen and nutrient concentrations, microbial nitrogen transformations can shift rapidly, producing highly variable biogeochemical outcomes.

### 6.5.4 Coupling between Physical Processes and Microbial Dynamics

A defining feature of aquatic ecosystems is the tight coupling between physical processes and microbial activity. Seasonal stratification and mixing events redistribute nutrients and oxygen, restructuring microbial communities and altering metabolic pathways, as shown by long-term lake and ocean observations (Sommer et al., 2012). In stratified lakes and oceans, sharp chemical gradients form at oxic–anoxic interfaces, supporting specialized microbial assemblages that exploit redox transitions (Lam & Kuypers, 2011). Hydrological connectivity further links microbial processes across spatial scales. Rivers transport microorganisms and organic matter from terrestrial environments to lakes and oceans, while coastal systems integrate marine and freshwater microbial communities, a linkage emphasized in river–ocean continuum studies (Bauer et al., 2013). These connections underscore that aquatic microbial ecology cannot be understood in isolation but must be viewed within the broader context of landscape and oceanographic processes.

## 6.6 EXTREMOPHILES AS ENVIRONMENTAL PIONEERS

Extremophilic microorganisms inhabit environments with physicochemical conditions that exceed the tolerance limits of most life forms, including extremes of temperature, salinity, pH, pressure, and radiation, as comprehensively reviewed in early extremophile research (Rothschild & Mancinelli, 2001). These organisms are not marginal components of the biosphere; rather, they play central roles in driving energy flow and elemental transformations in environments where biological activity would otherwise be severely constrained. In many extreme settings, microbial metabolism represents the dominant biological interface between geochemical energy sources and ecosystem-level processes, a concept emphasized in geomicrobiological syntheses (Nealson & Ghiorse, 2001). Research on extremophiles has substantially expanded understanding of the limits of life and demonstrated that microbial systems can remain metabolically active under conditions previously assumed to preclude sustained biological function (Stetter, 1999). Importantly, extremophiles often act as ecological pioneers, initiating biological colonization and shaping environmental chemistry in newly formed or highly constrained habitats (Cockell & Raven, 2004).

### 6.6.1 Hydrothermal Vents and Deep-Sea Environments

Hydrothermal vent ecosystems exemplify environments sustained entirely by microbial metabolism in the absence of sunlight, as first demonstrated in deep-sea vent studies (Jannasch & Mottl, 1985). Chemolithoautotrophic bacteria and archaea exploit reduced compounds such as hydrogen sulfide, hydrogen, and methane to fuel primary production on the deep seafloor (Karl et al., 1980). These microbial processes form the energetic foundation of vent ecosystems and support dense invertebrate communities through obligate symbiotic associations, as documented in studies of vent fauna (Cavanaugh et al., 1981). Beyond vent systems, deep-sea sediments harbor microbial communities adapted to high hydrostatic pressure, low temperature, and limited energy availability (Jørgensen & Boetius, 2007). A particularly important process in these environments is anaerobic methane oxidation, carried out by consortia of methanotrophic archaea and sulfate-reducing bacteria, as demonstrated by geochemical and molecular analyses (Boetius et al., 2000). This metabolism consumes methane before it can escape into the water column, representing a major biological control on methane release from marine sediments and underscoring the climatic relevance of deep-sea microbial activity (Reeburgh, 2007).

### 6.6.2 Cold and Cryogenic Environments

Polar regions and permanently cold environments host diverse microbial assemblages adapted to subzero temperatures, freeze–thaw cycles, and prolonged low-light periods, as reviewed in polar microbiology studies (D'Amico et al., 2006). Psychrophilic and psychrotolerant microorganisms maintain metabolic activity through structural adaptations that preserve enzyme flexibility and membrane fluidity at low temperatures (Feller & Gerday, 2003). These traits enable biochemical reactions to proceed under conditions that strongly constrain metabolic rates in mesophilic organisms. In permafrost soils, microbial activity plays a key role in regulating long-term carbon storage and release. Organic matter preserved under frozen conditions can remain stable for extended periods, but warming-induced thaw rapidly reactivates microbial decomposition, increasing emissions of carbon dioxide and methane, as demonstrated in permafrost incubation and field studies (Schuur et al., 2015). These responses highlight the sensitivity of cold-region carbon reservoirs to microbial reactivation and environmental change.

### 6.6.3 Acidic, Alkaline, and Hypersaline Systems

Microbial life also persists under extreme chemical conditions. Acidophilic microorganisms inhabit environments such as acid mine drainage and volcanic soils, where they maintain internal pH homeostasis despite highly acidic external conditions, as shown by physiological and genomic studies of acidophiles (Baker-Austin & Dopson, 2007). These organisms often derive energy from oxidizing iron and sulfur compounds, driving geochemical transformations that shape mineral stability and metal mobility (Johnson & Hallberg, 2003). At the opposite end of the pH spectrum, alkaliphilic microorganisms dominate soda lakes and alkaline soils, relying on specialized ion transport systems and enzyme adaptations to sustain growth (Horikoshi, 2006). Hypersaline environments, including salt lakes and evaporative basins, support communities of halophilic bacteria and archaea that counteract osmotic stress through compatible solutes or high intracellular salt concentrations (Oren, 2011). In some halophiles, light-driven energy generation supplements metabolic pathways, illustrating the diversity of strategies that support life under extreme salinity.

### 6.6.4 Extremophiles as Drivers of Ecosystem Initiation

Extremophiles often serve as pioneer organisms in newly formed or severely disturbed environments, including volcanic deposits, retreating glaciers, and mining-impacted landscapes, as observed in primary succession studies (Edwards et al., 2014). Through mineral weathering, organic matter production, and nutrient mobilization, early microbial colonizers modify physicochemical conditions and facilitate subsequent biological succession (Gadd, 2010). These initial microbial processes often exert a long-lasting influence on ecosystem development, well before plants or animals become established. Beyond their ecological significance, extremophiles provide important models for understanding early Earth environments and the potential for life beyond Earth (Cockell et al., 2016). Their metabolic pathways and survival strategies demonstrate how biological systems can persist under severe constraints, offering insight into the resilience and adaptability of life.

## 6.7 MICROBES AND CLIMATE REGULATION

Microorganisms profoundly influence Earth's climate system by controlling greenhouse gas production and consumption, carbon sequestration, and feedbacks that link ecosystem function to atmospheric composition, as synthesized in Earth-system and biogeochemical frameworks (Falkowski et al., 2008; Schimel & Schaeffer, 2012). Microbial metabolism regulates the fluxes of carbon dioxide ($CO_2$), methane ($CH_4$), and nitrous oxide ( $N_2O$) and shapes the stability of terrestrial and marine carbon reservoirs (Conrad, 2009). Because microbial activity responds sensitively to environmental conditions such as temperature, moisture, oxygen availability, and nutrient supply, microbial–climate interactions are inherently dynamic and strongly context-dependent (Allison et al., 2010). Climate change, in turn, alters microbial community composition and function by modifying the physical and chemical constraints on metabolism (Crowther et al., 2016). These reciprocal interactions create feedback loops that can either amplify or dampen climate forcing, positioning microorganisms as central regulators within the Earth system rather than passive responders to environmental change (Davidson & Janssens, 2006).

### 6.7.1 Microbial Control of Greenhouse Gas Fluxes

Microorganisms drive both the production and consumption of major biogenic greenhouse gases. Carbon dioxide fluxes are governed by the balance between microbial carbon fixation and heterotrophic respiration, a relationship demonstrated across terrestrial and aquatic ecosystems (Raich & Schlesinger, 1992). Photosynthetic microbes, including cyanobacteria and microalgae, fix $CO_2$ into organic biomass, whereas heterotrophic bacteria and fungi decompose organic matter and release $CO_2$ through respiration. The temperature sensitivity of microbial respiration is a critical determinant of carbon balance, particularly in soils, where warming can accelerate decomposition and increase atmospheric $CO_2$ concentrations (Davidson & Janssens, 2006).

Methane dynamics are governed almost entirely by microbial processes. Methanogenic archaea produce methane under anoxic conditions in wetlands, rice paddies, sediments, and animal digestive systems, as documented in physiological and ecosystem-scale studies (Conrad, 2009). These organisms use substrates such as hydrogen, acetate, and methylated compounds to generate energy, producing $CH_4$ as a metabolic end product (Thauer et al., 2008). Methane emissions are partially offset by methanotrophic microorganisms, which oxidize methane under both aerobic and anaerobic conditions (Knittel & Boetius, 2009). The relative activities of methanogens and methanotrophs determine whether ecosystems act as net methane sources or sinks, making $CH_4$ fluxes particularly sensitive to changes in hydrology, oxygen availability, and temperature (Segers, 1998). Nitrous oxide is produced primarily during microbial nitrogen transformations, especially nitrification and denitrification. Ammonia-oxidizing bacteria and archaea can generate $N_2O$ as a by-product of nitrification, while incomplete denitrification under low-oxygen conditions can lead to substantial $N_2O$ emissions, as established by enzymatic and ecological analyses (Zumft, 1997). Because $N_2O$ has a high global warming potential and a long atmospheric lifetime, even small shifts in microbial nitrogen cycling can have disproportionate climatic effects (Ravishankara et al., 2009).

### 6.7.2 Microbial Carbon Sequestration in Soils and Oceans

Beyond regulating gaseous emissions, microorganisms play key roles in long-term carbon sequestration. In terrestrial ecosystems, microbial residues contribute significantly to soil organic matter pools. After decomposition, microbial necromass can associate with mineral surfaces or be physically protected within soil aggregates, reducing its susceptibility to further degradation, as shown by mineral-association studies (Lehmann & Kleber, 2015). These stabilization mechanisms allow carbon to persist in soils over decadal to centennial timescales, making microbial processes central to terrestrial carbon storage (Schimel & Schaeffer, 2012). In marine systems, microorganisms underpin the biological carbon pump, which transfers carbon from surface waters to the deep ocean. Phytoplankton fix $CO_2$ into organic matter that is then transformed, aggregated, and partially remineralized as it sinks through the water column (Ducklow et al., 2001). Heterotrophic bacteria regulate the fraction of organic carbon that is respired versus exported to depth, while viral lysis redirects carbon into dissolved pools, altering sequestration efficiency (Azam & Malfatti, 2007; Weitz & Wilhelm, 2012). Small changes in microbial community structure or activity can therefore influence the magnitude and persistence of oceanic carbon storage (Boyd & Trull, 2007).

### 6.7.3 Climate Change Effects on Microbial Communities

Climate-driven changes in temperature, moisture regimes, and ocean chemistry are reshaping microbial communities across ecosystems. Warming generally accelerates microbial metabolic rates, increasing respiration and decomposition, particularly in carbon-rich soils (Allison et al., 2010). In high-latitude regions, thawing permafrost reactivates dormant microbial communities, leading to rapid mineralization of previously frozen organic matter and enhanced emissions of $CO_2$ and $CH_4$, as shown in permafrost studies (Schuur et al., 2015). These processes constitute strong positive feedbacks between microbial activity and climate warming. In aquatic systems, ocean warming and acidification alter microbial community composition and metabolic pathways. Changes in stratification affect nutrient availability and oxygen distribution, reshaping microbial niches and influencing carbon export efficiency (Boyd et al., 2015). Acidification can modify microbial enzyme activity and nutrient cycling, with implications for both primary production and remineralization (Liu et al., 2010). Together, these shifts demonstrate that microbial responses to climate change are complex and ecosystem-specific.

### 6.7.4 Microbial Feedbacks in the Earth System

Microbial processes operate at the intersection of biology, chemistry, and physics, linking local metabolic activity to global climate dynamics. Feedbacks arise when climate-driven changes in microbial activity alter greenhouse gas fluxes, which in turn influence climate forcing (Davidson & Janssens, 2006). These feedbacks can be positive, amplifying warming through increased emissions, or negative, dampening change through enhanced sequestration or gas consumption. Despite growing recognition of microbial importance, microbial processes remain underrepresented in many climate models, as noted in Earth-system modeling assessments (Wieder et al., 2015). This limitation reflects both the complexity of microbial systems and the difficulty of scaling microbial activity from local measurements to global predictions. Integrating microbial traits, functional diversity, and environmental constraints into Earth-system models remains a key challenge for advancing predictive understanding of climate–biosphere interactions.

## 6.8 SYMBIOTIC AND PATHOGENIC ROLES OF MICROORGANISMS

Microorganisms influence ecosystem structure and function not only through biogeochemical transformations but also through a broad spectrum of symbiotic and pathogenic interactions, as emphasized in ecological syntheses of host–microbe interactions (Douglas, 2010; Friesen et al., 2011). These relationships shape host physiology, population dynamics, and community structure, thereby exerting indirect yet often profound effects on ecosystem processes. Symbiosis and pathogenesis should therefore be regarded as integral components of ecosystem function rather than secondary biological phenomena.

Mutualistic interactions between microbes and plants are among the most influential symbioses in terrestrial ecosystems. Nitrogen-fixing bacteria provide bioavailable nitrogen to host plants in exchange for carbon, a process established by physiological and ecological studies of rhizobia–legume associations (Oldroyd & Dixon, 2014). Mycorrhizal fungi enhance nutrient and water uptake, influence plant competitive dynamics, and contribute to soil carbon stabilization, as demonstrated in syntheses of mycorrhizal ecology (Smith & Read, 2008; van der Heijden et al., 2015). Endophytic bacteria and fungi further modulate plant performance by affecting hormone signaling, stress tolerance, and resistance to pathogens (Rodriguez et al., 2009). Collectively, these interactions extend plant functional capacity beyond the limits imposed by plant physiology alone, embedding microbial metabolism within plant ecological strategies.

Similar dependencies occur in animal systems. Gut-associated microbial communities enable the digestion of complex or otherwise inaccessible substrates, as seen in cellulose degradation in ruminants and lignocellulose processing in termites, processes documented in animal–microbe symbiosis studies (Flint et al., 2008; Brune, 2014). In marine environments, many invertebrates rely on microbial symbionts for nutrition, particularly in chemosynthetic systems where microbial metabolism replaces photosynthesis as the primary energy source (Dubilier et al., 2008). These host–microbe associations expand the ecological niches available to animals and shape the structure of entire communities.

Pathogenic interactions occupy the opposite end of the symbiotic spectrum and play important regulatory roles in ecosystems. Microbial pathogens can reduce host fitness, alter population sizes, and drive shifts in species composition. Plant pathogens influence vegetation patterns and successional dynamics, while animal pathogens have caused widespread declines in wildlife populations, as documented in disease ecology studies (Anderson et al., 2004; Fisher et al., 2012). Such impacts illustrate how disease can act as a selective force with ecosystem-level consequences.

Importantly, the boundary between mutualism and pathogenesis is often fluid. Environmental conditions, host status, and community context can shift interactions along this continuum, transforming beneficial associations into harmful ones or vice versa (Bronstein, 2009). Recognizing this context dependence is essential to understanding how microbial interactions contribute to ecosystem resilience or vulnerability under environmental change.

## 6.9 HUMAN INFLUENCE ON MICROBIAL ECOLOGY

Human activities increasingly shape microbial communities across terrestrial, aquatic, and built environments. Land-use change, agriculture, industrialization, urbanization, pollution, and climate change modify the physical and chemical conditions that govern microbial metabolism, selection, and evolution, as synthesized in global change microbiology frameworks (Fierer, 2017). As a result, microbial community composition and function are being altered across spatial scales, from individual habitats to the global biosphere.

In agricultural systems, management practices exert strong selective pressures on soil microbial communities. Intensive fertilizer use alters nutrient availability and can promote microbial pathways linked to nitrogen loss and greenhouse gas emissions, as shown in agroecosystem studies (Galloway et al., 2008). Pesticides and other agrochemicals may suppress sensitive taxa while favoring resistant or opportunistic microorganisms (Imfeld & Vuilleumier, 2012). In contrast, practices such as organic amendments, crop diversification, and reduced tillage tend to support greater microbial diversity and functional stability, with implications for nutrient cycling and soil structure (Lori et al., 2017).

Industrial activity and environmental pollution further reshape microbial ecosystems. Hydrocarbons, heavy metals, and synthetic compounds impose strong selection on microorganisms capable of detoxification or degradation, a pattern documented in contaminated environments (Gadd, 2010). While these traits underpin bioremediation potential, they also alter community structure and can disrupt established biogeochemical processes. In aquatic environments, wastewater inputs introduce nutrients, organic matter, and contaminants that modify microbial metabolism, often leading to eutrophication, oxygen depletion, and altered nutrient cycling (Diaz & Rosenberg, 2008).

A particularly consequential human influence is the environmental release of antibiotics and antimicrobial compounds. These inputs select for resistant microorganisms and facilitate the spread of resistance genes via horizontal gene transfer, as demonstrated in environmental resistome studies (Martinez, 2009). Environmental reservoirs of antimicrobial resistance link human activity to microbial evolution, with implications that extend beyond environmental systems to animal and human health (Berendonk et al., 2015). Resistance traits maintained in environmental microbial communities can be mobilized into clinical settings, blurring traditional boundaries between environmental and medical microbiology.

Urban environments are additional arenas of microbial transformation. Modified temperature regimes, novel substrates, and high connectivity create distinct microbial assemblages that differ from those in surrounding natural systems (McDonnell & Hahs, 2015). These urban microbiomes reflect the convergence of human-associated and environmentally adapted microorganisms, underscoring the role of built environments as emerging ecological niches (Gilbert & Stephens, 2018).

## 6.10 TECHNOLOGICAL ADVANCES IN MICROBIAL ECOLOGY

Rapid advances in molecular and analytical technologies have transformed microbial ecology, enabling researchers to resolve community structure, metabolic potential, and functional activity with unprecedented resolution (Prosser, 2015). Shotgun metagenomics now enables reconstruction of microbial genomes directly from environmental samples, revealing taxonomic diversity, biosynthetic pathways, and ecological traits without cultivation, as demonstrated in early and contemporary metagenomic studies (Handelsman et al., 2007; Tyson et al., 2004).

Complementary approaches such as metatranscriptomics and metaproteomics provide insight into active metabolic processes by quantifying expressed genes and proteins under in situ conditions, while metabolomics links these molecular signatures to biochemical networks operating across spatial and temporal gradients (VerBerkmoes et al., 2009). SIP further assigns function to specific taxa by tracking the assimilation of isotopically labeled substrates, thereby linking metabolic activity to organismal identity within complex communities (Radajewski et al., 2000).

Single-cell genomics has illuminated the vast "microbial dark matter," enabling genome sequencing of rare or uncultured lineages that would otherwise go undetected (Rinke et al., 2013). Increasingly, these high-resolution datasets are integrated using microbial network modeling, machine learning, and trait-based ecosystem simulations to predict community responses to environmental change and identify keystone taxa that govern biogeochemical fluxes (Widder et al., 2016). Together, these technologies are reshaping the field by bridging genotypes, phenotypes, and ecosystem processes, ultimately enabling more predictive and mechanistic frameworks for understanding and managing microbial ecosystems.

## 6.11 FUTURE PERSPECTIVES

As environmental change accelerates, understanding how microbial communities respond and reorganize under shifting conditions is a central challenge for ecosystem science. Microbial processes operate at spatial and temporal scales that often differ markedly from those of plants and animals, yet they exert disproportionate influence on biogeochemical cycles, ecosystem resilience, and climate feedbacks. Future research must therefore prioritize frameworks that explicitly link microbial traits and interactions to emergent ecosystem properties. One promising direction is integrating trait-based ecology with microbial genomics and physiology. Identifying traits that reliably predict microbial responses to environmental gradients may help bridge the persistent gap between molecular data and ecosystem-scale processes. Similarly, long-term observational studies and experimental manipulations are essential for disentangling transient responses from sustained functional shifts, particularly in systems subject to chronic disturbance or rapid climate change. Microbiome engineering and synthetic ecology approaches offer additional opportunities, particularly in applied contexts such as agriculture, bioremediation, and carbon management. However, deliberate manipulation of microbial communities raises important ecological and ethical considerations, including unintended consequences, ecological stability, and evolutionary feedbacks. These challenges highlight the need for cautious, evidence-based approaches grounded in a deep understanding of microbial ecology rather than purely technological optimism. At a broader level, incorporating microbial processes into conceptual and predictive models of ecosystems remains a major frontier. This effort requires not only improved data but also shifts in perspective, recognizing microbes as active regulators of ecosystem dynamics rather than background components. Achieving this integration will depend on interdisciplinary collaboration across microbiology, biogeochemistry, ecology, and Earth-system science.

## 6.12 CONCLUSION

Microorganisms form the foundational biological infrastructure of ecosystems, regulating nutrient cycles, shaping community interactions, and mediating exchanges between the biosphere and the atmosphere. From extremophiles that initiate ecosystems under severe constraints to symbiotic and pathogenic microbes that influence host fitness and population dynamics, microbial life operates across scales and contexts that collectively define ecosystem function. Despite remarkable advances in observational and analytical tools, significant challenges remain in linking microbial diversity and metabolic potential to measurable ecosystem outcomes. Functional redundancy, environmental heterogeneity, and dynamic interactions complicate efforts to establish clear causal relationships. Yet these same complexities underpin the resilience and adaptability of microbial systems, enabling rapid responses to environmental change. Recognizing microbes as central agents in ecosystem processes is essential for advancing ecological theory and addressing pressing global challenges related to climate change, biodiversity loss, and environmental sustainability. A more complete integration of microbial ecology into ecosystem science will not only improve predictive understanding but also reshape how ecosystems are conceptualized. In this sense, microbes are not merely components of ecosystems—they are architects of the living world.

## REFERENCES


Allison, S., Wallenstein, M. & Bradford, M. (2010). Soil-carbon response to warming dependent on microbial physiology. *Nature Geosci* 3, 336–340.

Anderson, R. M., May, R. M., & Anderson, B. (2004). *Infectious diseases of humans: Dynamics and control*. Oxford University Press, UK.

Azam, F., Fenchel, T., Field, J. G., Gray, J. S., Meyer-Reil, L. A., & Thingstad, F. (1983). The ecological role of water-column microbes in the sea. *Marine Ecology Progress Series*, 10, 257–263.

Azam, F., & Malfatti, F. (2007). Microbial structuring of marine ecosystems. *Nature Reviews Microbiology*, 5(10), 782–791. https://doi.org/10.1038/nrmicro1747

Bais, H. P., Weir, T. L., Perry, L. G., Gilroy, S., & Vivanco, J. M. (2006). The role of root exudates in rhizosphere interactions with plants and other organisms. *Annual Review of Plant Biology*, 57, 233–266.

Baker-Austin, C., & Dopson, M. (2007). Life in acid: pH homeostasis in acidophiles. *Trends in Microbiology*, 15(4), 165–171. https://doi.org/10.1016/j.tim.2007.02.005

Bardgett, R. D., & van der Putten, W. H. (2014). Belowground biodiversity and ecosystem functioning. *Nature*, 515(7528), 505–511. https://doi.org/10.1038/nature13855

Bauer, J. E., Cai, W.-J., Raymond, P. A., Bianchi, T. S., Hopkinson, C. S., & Regnier, P. A. G. (2013). The changing carbon cycle of the coastal ocean. *Nature*, 504, 61–70.

Berendonk, T. U., et al. (2015). Tackling antibiotic resistance: The environmental framework. *Nature Reviews Microbiology*, 13(5), 310–317.

Berendsen, R. L., Pieterse, C. M. J., & Bakker, P. A. H. M. (2012). The rhizosphere microbiome and plant health. *Trends in Plant Science*, 17(8), 478–486. https://doi.org/10.1016/j.tplants.2012.04.001

Berg, I. A., Kockelkorn, D., Ramos-Vera, W. H., Say, R. F., Zarzycki, J., Hügler, M., ... Fuchs, G. (2010). Autotrophic carbon fixation in archaea. *Nature Reviews Microbiology*, 8(6), 447–460. https://doi.org/10.1038/nrmicro2365

Boetius, A., Ravenschlag, K., Schubert, C. J., Rickert, D., Widdel, F., Gieseke, A., et al. (2000). A marine microbial consortium apparently mediating anaerobic oxidation of methane. *Nature*, 407, 623–626.

Boyd, P. W., & Trull, T. W. (2007). Understanding the export of biogenic particles in oceanic waters. *Progress in Oceanography*, 72, 276–312. https://doi.org/10.1016/j.pocean.2006.10.007

Boyd, P.W., Lennartz, S.T., Glover, D.M. & Doney, S.C. (2015). Biological ramifications of climate-change-mediated oceanic multi-stressors. *Nature Climate Change*, 5, 71–79.

Bronstein, J. L. (2009). The evolution of facilitation and mutualism. *Journal of Ecology*, 97, 1160–1170.

Brum, J. R., & Sullivan, M. B. (2015). Rising to the challenge: Accelerated pace of discovery transforms marine virology. *Nature Reviews Microbiology*, 13(3), 147–159. https://doi.org/10.1038/nrmicro3404

Brune, A. (2014). Symbiotic digestion of lignocellulose in termite guts. *Nature Reviews Microbiology*, 12, 168–180.

Cavanaugh, C. M., Gardiner, S. L., Jones, M. L., Jannasch, H. W., & Waterbury, J. B. (1981). Prokaryotic cells in the hydrothermal vent tube worm *Riftia pachyptila*. *Science*, 213, 340–342.

Charlson, R. J., Lovelock, J. E., Andreae, M. O., & Warren, S. G. (1987). Oceanic phytoplankton, atmospheric sulphur, cloud albedo and climate. *Nature*, 326(6114), 655–661. https://doi.org/10.1038/326655a0

Cockell, C. S., & Raven, J. A. (2004). Zones of photosynthetic potential on Mars and the early Earth. *Icarus*, 169, 300–310.

Cockell, C. S., et al. (2016). Habitability: A review. *Astrobiology*, 16, 89–117.

Cole, J. J., Prairie, Y. T., Caraco, N. F., McDowell, W. H., Tranvik, L. J., Striegl, R. G., et al. (2007). Plumbing the global carbon cycle: Integrating inland waters into the terrestrial carbon budget. *Ecosystems*, 10, 172–185. https://doi.org/10.1007/s10021-006-9013-8

Conrad, R. (2009). The global methane cycle: Microbial and environmental perspectives. *Microbiology and Molecular Biology Reviews*, 73(3), 457–479.

Crowther TW, Todd-Brown KE, Rowe CW, Wieder WR, Carey JC, Machmuller MB, Snoek BL, Fang S, Zhou G, Allison SD, Blair JM, Bridgham SD, Burton AJ, Carrillo Y, Reich PB, Clark JS, Classen AT, Dijkstra FA, Elberling B, Emmett BA, Estiarte M, Frey SD, Guo J, Harte J, Jiang L, Johnson BR, Kröel-Dulay G, Larsen KS, Laudon H, Lavallee JM, Luo Y, Lupascu M, Ma LN, Marhan S, Michelsen A, Mohan J, Niu S, Pendall E, Peñuelas J, Pfeifer-Meister L, Poll C, Reinsch S, Reynolds LL, Schmidt IK, Sistla S, Sokol NW, Templer PH, Treseder KK, Welker JM, Bradford MA. Quantifying global soil carbon losses in response to warming. *Nature*, 540(7631):104–108.

D'Amico, S., Collins, T., Marx, J. C., Feller, G., & Gerday, C. (2006). Psychrophilic microorganisms: Challenges for life. *EMBO Reports*, 7(4), 385–389.

Davidson, E. A., & Janssens, I. A. (2006). Temperature sensitivity of soil carbon decomposition and feedbacks to climate change. *Nature*, 440(7081), 165–173.

Diaz, R. J., & Rosenberg, R. (2008). Spreading dead zones and consequences for marine ecosystems. *Science*, 321, 926–929.

Douglas, A. E. (2010). *The symbiotic habit*. Princeton University Press, New Jersey, USA.

Dubilier, N., Bergin, C., & Lott, C. (2008). Symbiotic diversity in marine animals. *Nature Reviews Microbiology*, 6, 725–740.

Ducklow, H. W., Steinberg, D. K., & Buesseler, K. O. (2001). Upper ocean carbon export and the biological pump. *Oceanography*, 14, 50–58.

Edwards, K. J., Becker, K., & Colwell, F. (2014). The deep, dark energy biosphere: Intraterrestrial life on Earth. *Annual Review of Earth and Planetary Sciences*, 42, 551–568.

Falkowski, P. G. (2012). Ocean science: The power of plankton. *Nature*, 483, S17–S20.

Falkowski, P. G., Barber, R. T., & Smetacek, V. (1998). Biogeochemical controls and feedbacks on ocean primary production. *Science*, 281(5374), 200–206.

Falkowski, P. G., Fenchel, T., & Delong, E. F. (2008). The microbial engines that drive Earth's biogeochemical cycles. *Science*, 320(5879), 1034–1039.

Feller, G., & Gerday, C. (2003). Psychrophilic enzymes: Hot topics in cold adaptation. *Nature Reviews Microbiology*, 1, 200–208.

Field, C. B., Behrenfeld, M. J., Randerson, J. T., & Falkowski, P. G. (1998). Primary production of the biosphere: Integrating terrestrial and oceanic components. *Science*, 281, 237–240.

Fierer, N. (2017). Embracing the unknown: Disentangling the complexities of the soil microbiome. *Nature Reviews Microbiology*, 15, 579–590.

Fisher, M. C., et al. (2012). Emerging fungal threats to animal, plant and ecosystem health. *Nature*, 484, 186–194.

Flint, H. J., Bayer, E. A., Rincon, M. T., Lamed, R., & White, B. A. (2008). Polysaccharide utilization by gut bacteria. *Nature Reviews Microbiology*, 6, 121–131.

Flynn, K. J., Mitra, A., Anestis, K., Anschütz, A. A., Calbet, A., Ferreira, G. D., … Zubkov, M. V. (2019). Mixotrophic protists and a new paradigm for marine ecology: Where does plankton research go now? *Journal of Plankton Research*, 41(4), 375–391.

Friedrich, C. G., Rother, D., Bardischewsky, F., Quentmeier, A., & Fischer, J. (2001). Oxidation of reduced inorganic sulfur compounds by bacteria: Emergence of a common mechanism?. *Applied and Environmental Microbiology*, 67(7), 2873–2882.

Friesen, M. L., et al. (2011). Microbially mediated plant functional traits. *New Phytologist*, 190, 716–727.

Gadd, G. M. (2010). Metals, minerals and microbes: Geomicrobiology and bioremediation. *Microbiology*, 156, 609–643.

Galloway, J.N., Dentener, F.J., Capone, D.G. et al. (2004). Nitrogen Cycles: Past, Present, and Future. *Biogeochemistry*, 70, 153–226.

Galloway, J. N., et al. (2008). Transformation of the nitrogen cycle: Recent trends and questions. *Science*, 320, 889–892.

Geisen, S., Mitchell, E. A. D., Wilkinson, D. M., Adl, S., Bonkowski, M., Brown, M. W., … Lara, E. (2018). Soil protists: A fertile frontier in soil biology research. *FEMS Microbiology Reviews*, 42(3), 293–323.

Gilbert, J. A., & Stephens, B. (2018). Microbiology of the built environment. *Nature Reviews Microbiology*, 16, 661–670.

Goodfellow, M., & Williams, S. T. (1983). Ecology of actinomycetes. *Annual Review of Microbiology*, 37(1), 189–216.

Handelsman,J.,etal.(2007).Metagenomics:Application of genomics to uncultured microorganisms. *Microbiology and Molecular Biology Reviews*, 71, 669–685.

Herndl, G. J., Reinthaler, T., Teira, E., van Aken, H., Veth, C., Pernthaler, A., & Pernthaler, J. (2005). Contribution of Archaea to total prokaryotic production in the deep Atlantic Ocean. *Applied and Environmental Microbiology*, 71, 2303–2309.

Hibbing, M. E., Fuqua, C., Parsek, M. R., & Peterson, S. B. (2010). Bacterial competition: Surviving and thriving in the microbial jungle. *Nature Reviews Microbiology*, 8(1), 15–25.

Hobbie, J. E., & Hobbie, E. A. (2006). $^{15}N$ in symbiotic fungi and plants estimates nitrogen and carbon flux rates in arbuscular mycorrhizal symbioses. *Ecology*, 87(4), 816–822.

Horikoshi, K. (2006). *Alkaliphiles: Genetic properties and applications*. Springer, Tokyo.

Hupfer, M. and Lewandowski, J. (2008), Oxygen Controls the Phosphorus Release from Lake Sediments – a Long-Lasting Paradigm in Limnology. *International Review of Hydrobiology*, 93: 415–432.

Imfeld, G., & Vuilleumier, S. (2012). Measuring the effects of pesticides on bacterial communities. *Environmental Science & Technology*, 46, 105–114.

Jannasch, H. W., & Mottl, M. J. (1985). Geomicrobiology of deep-sea hydrothermal vents. *Science*, 229, 717–725.

Johnson, D. B., & Hallberg, K. B. (2003). The microbiology of acidic mine waters. *Research in Microbiology*, 154, 466–473.

Jørgensen, B. B., & Boetius, A. (2007). Feast and famine—microbial life in the deep-sea bed. *Nature Reviews Microbiology*, 5, 770–781.

Karl, D. M., Wirsen, C. O., & Jannasch, H. W. (1980). Deep-sea primary production at the Galapagos hydrothermal vents. *Science*, 207, 1345–1347.

Knittel, K., & Boetius, A. (2009). Anaerobic oxidation of methane: Progress with an unknown process. *Annual Review of Microbiology*, 63(1), 311–334. https://doi.org/10.1146/annurev.micro.61.080706.093130

Kowalchuk, G. A., & Stephen, J. R. (2001). Ammonia-oxidizing bacteria: A model for molecular microbial ecology. *Annual Review of Microbiology*, 55(1), 485–529. https://doi.org/10.1146/annurev.micro.55.1.485

Lam, P., & Kuypers, M. M. M. (2011). Microbial nitrogen cycling processes in oxygen minimum zones. *Annual Review of Marine Science*, 3, 317–345.

Lehmann, J., & Kleber, M. (2015). The contentious nature of soil organic matter. *Nature*, 528(7580), 60–68. https://doi.org/10.1038/nature16069

Liu J., Weinbauer MG., Maier C., Minhan Dai M., Gattuso J.P. (2010), Effect of ocean acidification on microbial diversity and on microbe-driven biogeochemistry and ecosystem functioning. *Aquatic Microbial Ecology*, 61:291–305.

Locey, K. J., & Lennon, J. T. (2016). Scaling laws predict global microbial diversity. *Proceedings of the National Academy of Sciences*, 113(21), 5970–5975. https://doi.org/10.1073/pnas.1521291113

Lori, M., et al. (2017). Organic farming enhances soil microbial abundance and activity. *Proceedings of the Royal Society B*, 284, 20171549.

Lugtenberg, B., & Kamilova, F. (2009). Plant-growth-promoting rhizobacteria. *Annual Review of Microbiology*, 63, 541–556.

Madigan, M. T., Bender, K. S., Buckley, D. H., Sattley, W. M., & Stahl, D. A. (2021). *Brock biology of microorganisms* (16th ed.). Pearson.

Makarova, K. S., Aravind, L., Wolf, Y. I., Tatusov, R. L., Minton, K. W., Koonin, E. V., & Daly, M. J. (2001). Genome of the extremely radiation-resistant bacterium *Deinococcus radiodurans* viewed from the perspective of comparative genomics. *Microbiology and Molecular Biology Reviews*, 65(1), 44–79.

Martinez, J. L. (2009). Environmental pollution by antibiotics and antibiotic resistance determinants. *Environmental Pollution*, 157(11), 2893–2902. https://doi.org/10.1016/j.envpol.2009.05.051

McDonnell, M. J., & Hahs, A. K. (2015). Adaptation and resilience of cities. *Nature*, 518, 229–237.

Moran, N. A., McCutcheon, J. P., & Nakabachi, A. (2008). Genomics and evolution of heritable bacterial symbionts. *Annual Review of Genetics*, 42(1), 165–190.

Morris, J. J., Lenski, R. E., & Zinser, E. R. (2012). The Black Queen Hypothesis: Evolution of dependencies through adaptive gene loss. mBio, 3(2), e00036–e00012.

Muyzer, G., & Stams, A. J. M. (2008). The ecology and biotechnology of sulphate-reducing bacteria. *Nature Reviews Microbiology*, 6(6), 441–454.

Nadell, C. D., Xavier, J. B., & Foster, K. R. (2009). The sociobiology of biofilms. *FEMS Microbiology Reviews*, 33(1), 206–224. https://doi.org/10.1111/j.1574-6976.2008.00150.x

Nealson, K. H., & Ghiorse, W. C. (2001). Geomicrobiology: Exploring the interface between the biosphere and the geosphere. *American Journal of Science*, 301, 599–623.

Offre, P., Spang, A., & Schleper, C. (2013). Archaea in biogeochemical cycles. *Annual Review of Microbiology*, 67(1), 437–457. https://doi.org/10.1146/annurev-micro-092412-155614

Oldroyd, G. E. D., & Dixon, R. (2014). Biotechnological solutions to the nitrogen problem. *Current Opinion in Biotechnology*, 26, 19–24.

Oren, A. (2011). Thermodynamic limits to microbial life at high salt concentrations. *Environmental Microbiology*, 13, 1908–1923.

Paerl, H. W., & Otten, T. G. (2013). Harmful cyanobacterial blooms: Causes, consequences, and controls. *Microbial Ecology*, 65(4), 995–1010. https://doi.org/10.1007/s00248-012-0159-y

Partensky, F., Hess, W. R., & Vaulot, D. (1999). Prochlorococcus, a marine photosynthetic prokaryote of global significance. *Microbiology and Molecular Biology Reviews*, 63(1), 106–127.

Philippot, L., Raaijmakers, J. M., Lemanceau, P., & van der Putten, W. H. (2013). Going back to the roots: The microbial ecology of the rhizosphere. *Nature Reviews Microbiology*, 11(11), 789–799.

Prosser, J. I. (2015). Dispersing misconceptions and identifying opportunities for the use of "omics" in soil microbial ecology. *Nature Reviews Microbiology*, 13(7), 439–446.

Radajewski, S., et al. (2000). Stable-isotope probing as a tool in microbial ecology. *Nature*, 403, 646–649.

Raich, J.W. and Schlesinger, W.H. (1992). The global carbon-dioxide flux in soil respiration and its relationship to vegetation and climate. *Tellus B: Chemical and Physical Meteorology*, 44(2), pp. 81–99.

Ravishankara A.R., Daniel J.S., Portmann R.W. (2009). Nitrous oxide (N2O): the dominant ozone-depleting substance emitted in the 21st century. *Science*. 326(5949):123–125. doi: 10.1126/science.1176985.

Raymond, J., Siefert, J. L., Staples, C. R., & Blankenship, R. E. (2004). The natural history of nitrogen fixation. *Molecular Biology and Evolution*, 21(3), 541–554.

Reeburgh, W. S. (2007). Oceanic methane biogeochemistry. *Chemical Reviews*, 107, 486–513.

Richardson, A. E., & Simpson, R. J. (2011). Soil microorganisms mediating phosphorus availability. *Plant Physiology*, 156(3), 989–996.

Rillig, M. C., & Mummey, D. L. (2006). Mycorrhizas and soil structure. *Soil Biology & Biochemistry*, 38(8), 2061–2070.

Rinke, C., et al. (2013). Insights into the phylogeny and coding potential of microbial dark matter. *Nature*, 499, 431–437.

Rodríguez, H., & Fraga, R. (1999). Phosphate solubilizing bacteria and their role in plant growth promotion. *Biotechnology Advances*, 17(4–5), 319–339.

Rodriguez R.J., White J.F. Jr, Arnold A.E., (2009), Redman R.S. Fungal endophytes: diversity and functional roles. *New Phytol*. 182(2):314–330. doi: 10.1111/j.1469-8137.2009.02773.x

Rothschild, L. J., & Mancinelli, R. L. (2001). Life in extreme environments. *Nature*, 409, 1092–1101.

Schimel, J. P., & Schaeffer, S. M. (2012). Microbial control over carbon cycling in soil. *Soil Biology & Biochemistry*, 48, 229–234.

Schopf, J. W. (2006). Fossil evidence of Archaean life. *Philosophical Transactions of the Royal Society B: Biological Sciences*, 361(1470), 869–885.

Schuur, E. A. G., McGuire, A. D., Schädel, C., Grosse, G., Harden, J. W., Hayes, D. J., ... Turetsky, M. R. (2015). Climate change and the permafrost carbon feedback. *Nature*, 520(7546), 171–179.

Segers, R. (1998). Methane production and methane consumption: A review of processes underlying wetland methane fluxes. *Biogeochemistry*, 41, 23–51.

Seitzinger, S. P., Harrison, J. A., Böhlke, J. K., Bouwman, A. F., Lowrance, R., Peterson, B., et al. (2006). Denitrification across landscapes and waterscapes: A synthesis. *Ecological Applications*, 16, 2064–2090.

Simard, S. W., Beiler, K. J., Bingham, M. A., Deslippe, J. R., Philip, L. J., & Teste, F. P. (2012). Mycorrhizal networks: Mechanisms, ecology and modelling. *New Phytologist*, 195(3), 542–555.

Simon, M., Grossart, H.-P., Schweitzer, B., & Ploug, H. (2002). Microbial ecology of organic aggregates in aquatic ecosystems. *Aquatic Microbial Ecology*, 28, 175–211.

Smith, S. E., & Read, D. J. (2008). *Mycorrhizal symbiosis* (3rd ed.). Academic Press, London.

Sommer, U., Adrian, R., De Senerpont Domis, L., Elser, J. J., Gaedke, U., Ibelings, B., et al. (2012). Beyond the Plankton Ecology Group (PEG) model: Mechanisms driving plankton succession. *Annual Review of Ecology, Evolution, and Systematics*, 43, 429–448.

Stahl, D. A., & de la Torre, J. R. (2012). Physiology and diversity of ammonia-oxidizing archaea. *Annual Review of Microbiology*, 66(1), 83–101.

Stetter, K. O. (1999). Extremophiles and their adaptation to hot environments. *FEBS Letters*, 452(1–2), 22–25.

Strous, M., Fuerst, J. A., Kramer, E. H. M., Logemann, S., Muyzer, G., van de Pas-Schoonen, K. T., ... Jetten, M. S. M. (1999). Missing lithotroph identified as new planctomycete. *Nature*, 400(6743), 446–449.

Suttle, C. A. (2007). Marine viruses—Major players in the global ecosystem. *Nature Reviews Microbiology*, 5(10), 801–812.

Thauer, R. K., Kaster, A. K., Seedorf, H., Buckel, W., & Hedderich, R. (2008). Methanogenic archaea: Ecologically relevant differences in energy conservation. *Nature Reviews Microbiology*, 6(8), 579–591. https://doi.org/10.1038/nrmicro1931

Tranvik, L. J., et al. (2009). Lakes and reservoirs as regulators of carbon cycling and climate. *Limnology and Oceanography*, 54, 2298–2314.

Treseder, K. K., & Lennon, J. T. (2015). Fungal traits that drive ecosystem dynamics on land. *Microbiology and Molecular Biology Reviews*, 79(2), 243–262.

Tyson, G. W., et al. (2004). Community structure and metabolism through metagenomics. *Nature*, 428, 37–43.

Vacheron J., Desbrosses G., Bouffaud M.L., (2013) Touraine B., Moënne-Loccoz Y., Muller D., Legendre L., Wisniewski-Dyé F., Prigent-Combaret C. Plant growth-promoting rhizobacteria and root system functioning. *Front Plant Sci*. Sep 17;4:356. doi: 10.3389/fpls.2013.00356.

Van der Heijden, M. G. A., Martin, F. M., Selosse, M. A., & Sanders, I. R. (2015). Mycorrhizal ecology and evolution: The past, the present, and the future. *New Phytologist*, 205(4), 1406–1423.

VerBerkmoes N.C., Denef V.J., Hettich R.L., Banfield J.F. (2009). Systems biology: Functional analysis of natural microbial consortia using community proteomics. *Nat Rev Microbiol*. 7(3):196–205. doi: 10.1038/nrmicro2080.

Weitz, J. S., & Wilhelm, S. W. (2012). Ocean viruses and their effects on microbial communities and biogeochemical cycles. *F1000 Biology Reports*, 4, 17.

Wetzel, R. G. (2001). *Limnology: Lake and river ecosystems* (3rd ed.). Academic Press, San Diego.

Whitman, W. B., Coleman, D. C., & Wiebe, W. J. (1998). Prokaryotes: The unseen majority. *Proceedings of the National Academy of Sciences*, 95(12), 6578–6583.

Widder, S., et al. (2016). Challenges in microbial ecology: Building predictive understanding. *ISME Journal*, 10, 2557–2568.

Wieder, W., Cleveland, C., Smith, W. et al. (2015). Future productivity and carbon storage limited by terrestrial nutrient availability. *Nature Geosci*, 8, 441–444.

Wilhelm, S. W., & Suttle, C. A. (1999). Viruses and nutrient cycles in the sea. *BioScience*, 49, 781–788.

Wright, J. J., Konwar, K. M., & Hallam, S. J. (2012). Microbial ecology of expanding oxygen minimum zones. *Nature Reviews Microbiology*, 10, 381–394. https://doi.org/10.1038/nrmicro2778

Zumft, W. G. (1997). Cell biology and molecular basis of denitrification. *Microbiology and Molecular Biology Reviews*, 61(4), 533–616.

CHAPTER 7

# Soil Microbiome

## NATURAL, AGRICULTURAL, AND CONTAMINATED SOILS

*Acacio Aparecido Navarrete, Giselle Gomes Monteiro Fracetto, Paulo Ademar Avelar Ferreira, Talles Renan Souza Nicoletti, Kelly Hidalgo Martinez, Gabrielle Marques Inácio, and Valéria Maia de Oliveira*

## 7.1 INTRODUCTION

The soil microbiome plays a central role in sustaining terrestrial ecosystems, yet our understanding of its global distribution, drivers, and functions remains incomplete. Rapid advances in soil microbiology – particularly in molecular, ecological, and network-based approaches – have revealed that natural soils harbor extraordinarily diverse microbial communities shaped by edaphic factors, vegetation, climate, and disturbance regimes. Bacteria, archaea, fungi, protists, and viruses interact within highly heterogeneous soil environments, driving almost all soil functions, including carbon turnover, nutrient cycling, and the regulation of greenhouse gases such as methane. Comparative studies across biomes show pronounced spatial heterogeneity but also patterns of convergence across successional stages, while extreme environments host uniquely adapted microbial assemblages. At the same time, global change factors and land-use intensification are altering microbial composition, network complexity, and functional resilience, underscoring the need for baseline assessments and conservation strategies that integrate plant–soil–microbe interactions.

Human activities further reshape soil microbiomes through agriculture, contamination, and mechanical disturbance. Intensive farming practices modify soil structure, nutrient dynamics, and aeration, often reducing microbial diversity and stability, whereas sustainable management and organic inputs can promote complex microbial networks and multifunctionality. In contaminated soils, hydrocarbons and heavy metals (HMs) select for specialized microbial taxa capable of metabolizing or immobilizing pollutants through diverse biochemical pathways, making these communities key indicators of soil health and essential agents in bioremediation. Modern omics and cultivation-based methods now enable detailed characterization of microbial structure, function, and adaptive mechanisms, facilitating the development of targeted remediation approaches such as biostimulation and bioaugmentation. Together, these insights highlight the ecological and biogeochemical significance of soil microbiomes in natural, agricultural, and polluted environments, while pointing toward integrated strategies for their conservation and restoration.



DOI: 10.1201/9781003637950-7

## 7.2 NATURAL SOIL MICROBIOME

The direction of soil microbiology has been developing very rapidly in recent years, but many studies remain highly specialized, targeting specific taxonomic or functional microbial groups and concentrating on regional or local scales. As a result, a comprehensive global view of the soil microbiome is still taking shape (Guerra et al., 2021; Bahram et al., 2018; Baldrian et al., 2023). Building upon this growing yet fragmented body of knowledge, recent research highlights that soils contain highly diverse microbial communities shaped by environmental factors and varying widely across biomes, with extreme ecosystems hosting particularly distinct groups. Modern molecular and network-based tools have deepened the understanding of their interactions and roles in key soil processes. Disturbances, climate change, and land-use intensification alter microbial composition, functions, and resilience, and comparisons with natural soils show notable losses under human impact. Effective conservation and restoration of native soil microbiomes therefore require baseline assessment, monitoring, and integration of plant–microbe relationships to sustain soil biodiversity and ecosystem functioning.

### 7.2.1 Diversity and Structure of Microbial Communities in Natural Soils

Soils constitute a highly heterogeneous habitat composed of a complex mixture of minerals, organic matter, and interconnected water- and air-filled pore spaces (Orgiazzi et al., 2016) (Figure 7.1). This environment supports a vast abundance and diversity of organisms that drive essential soil processes and functions. Among them, microorganisms inhabiting the bulk soil and the rhizosphere – the zone surrounding plant roots – are the key contributors (Philippot et al., 2013). These microbes represent some of the most abundant and diverse soil biota, accounting for an estimated 3–4% of Earth's total biomass, or up to 20 Gt of carbon (C) (Bar-On et al., 2018). Through interactions with their abiotic environmental factors, other soil organisms, and plants, they play fundamental roles in major biological processes. In fact, 80–90% of soil functions are carried out by the soil microbiome, including bacteria, fungi, archaea, viruses, and protists (Nannipieri et al., 2003). For instance, a seminal metagenomic and small-subunit rRNA study conducted across prairie, desert, and rainforest soils revealed exceptionally high estimated richness of bacterial, archaeal, and fungal operational taxonomic units (OTUs) and showed that soil viruses were taxonomically diverse and markedly different from viral communities known from other environments (Fierer et al., 2007).

Natural soil microbiome is dominated generally by bacteria in terms of abundance and diversity, with common phyla including Proteobacteria, Actinobacteria, Acidobacteria, Chloroflexi, and Firmicutes (Ko et al., 2017). *Archaea*, though lower in abundance, can play pivotal roles (e.g., nitrogen oxidation activities and methanogenesis) and are increasingly recognized in soil microbiome studies (Cao et al., 2025). Fungi and micro-eukaryotes provide additional trophic levels and functional niches, e.g., saprotrophic fungi, mycorrhizal fungi, protists. The viral component, although less studied, is gaining attention as regulators of microbial populations and gene flow (Fierer et al., 2007).

Numerous studies indicate that edaphic factors strongly shape microbial community structure and diversity. For instance, in an alpine grassland on the Tibetan Plateau, soil pH was found to be the key environmental variable for predicting network-level features of the microbial co-occurrence network: higher pH enhanced associations among microorganisms, and *Archaea* held a central network role relative to bacteria (Chen et al., 2021). Along an aridity gradient in temperate grasslands of Inner Mongolia, aridity (moisture stress) weakened diversity correlations between bacteria and fungi/*Archaea*, indicating that moisture and aridity are major drivers of microbial community patterns (Liu et al., 2023).

Organic matter content, soil texture (clay vs sand), and soil moisture also play important roles in shaping the soil microbiome. One study of the Taojia River Basin found that clay content, pH, and moisture were the key environmental variables explaining microbial diversity under different land-use types (He et al., 2023). Additionally, vegetation type (plant species composition, richness, functional groups) impacts microbial communities: for example, in a long-term biodiversity experiment in Germany, plant functional group (legumes vs grasses vs herbs) identity affected bacterial richness (highest in legume plots) and archaeal and arbuscular mycorrhizal fungi composition, more so than plant species richness per se (Dassen et al., 2017).

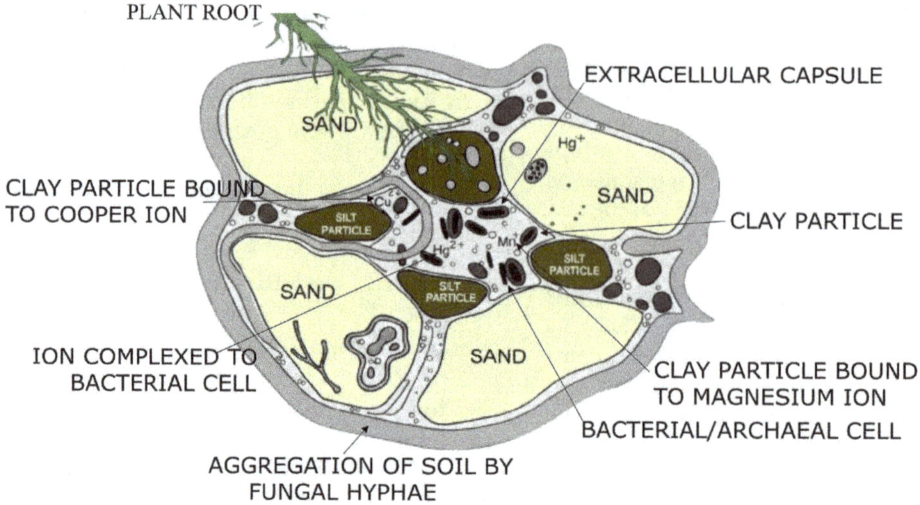


Figure 7.1 The soil habitat and structure. Within the soil structure, minerals, organic compounds, and microorganisms are among the major solid components of soils and these components are constantly interacting with one another to act as a unified system.

Comparative studies across biomes help reveal broad-scale patterns and exceptions in soil microbial community structure. For example, the early metagenomic study mentioned above compared prairie, desert, and rainforest soils and found minimal taxonomic overlap among these sites for bacteria, archaea, and fungi, suggesting high spatial heterogeneity at the global scale (Fierer et al., 2007). Another recent review of microbial succession across successional stages (primary, secondary forest, grassland) compiled 5,184 soil samples and found that both bacterial and fungal β-diversity decreased with successional age, indicating convergence in community composition across very different biomes (Yu et al., 2025). Although fewer in number, studies in extreme biomes (e.g., tundra, deserts) report unique assemblages and adaptations, e.g., cold-region wetlands – a high-throughput sequencing study found distinct shifts in Proteobacteria, Actinobacteriota, Acidobacteriota along a retrogressive succession in cold wetlands (Ding and Yu, 2025).

### 7.2.2 Ecological and Biogeochemical Functions

Microbial communities are key drivers of soil biogeochemical cycles, supporting processes such as carbon sequestration and turnover, nitrogen fixation, nitrification and denitrification, phosphorus solubilization, and sulfur oxidation and reduction. For instance, a review of microbial biomass stoichiometry at global scale showed microbes maintain homeostasis relative to soil elemental concentrations (C:N:P:S), thus underscoring their role in nutrient cycling (Chen et al., 2021).

Global biogeochemical cycles are mainly driven by soil microorganisms that feed on base compounds of carbon, such as methane ($CH_4$) or carbon dioxide ($CO_2$) (Raghoebarsin et al., 2006). $CH_4$ is the most abundant hydrocarbon in the atmosphere (Caldwell et al., 2008). Due to its absorption characteristics, $CH_4$ manifests positive radiative forcing, being a greenhouse gas that contributes to the regulation of temperature on the surface of the planet. The amount of $CH_4$ emitted from an ecosystem is the result of the balance between the production of $CH_4$ (methanogenesis) and the consumption (oxidation) of this gas (methanotrophy). Therefore, the emission of $CH_4$ into the atmosphere is determined by the activity of methanogenic and methanotrophic microorganisms.

Amazon floodplain ecosystems include open water and intermittent flood forest and agricultural systems with different water types. They are a significant natural source of $CH_4$ in the tropics. When soils are flooded and become anoxic, $CH_4$ is produced by methanogenesis, while microbially mediated aerobic and anaerobic oxidation of $CH_4$ serves as the primary biological sink of this greenhouse gas. Bento et al. (2021) showed that soil physicochemical factors, Amazonian water types, land use, soil depth, and seasonality can structure the active methanogenic and methanotrophic microbial communities in soil. These microbial communities were also modulated by other factors such as the presence of nitrifiers like *Nitrososphaera* and *Nitrospira*. Additionally, the authors highlighted floodplains of the Amazon rivers with different hydrogeochemical characteristics and origin of organic matter acting simultaneously as strong sources and sinks of $CH_4$. A signature was revealed across methanotrophic microbial communities in soils from all the three areas of floodplain periodically flooded with black, white, and clear water from large Amazonian rivers, with a disproportionate role of NC10 phylum in $CH_4$ mitigation in natural and agricultural Amazonian floodplains.

In turn, decomposition of organic matter and subsequent mineralization of nutrients are a core microbial function in soils. Higher microbial diversity has been linked to enhanced decomposition and ecosystem functioning (Chen et al., 2021). Spatial heterogeneity, such as differences in aggregate size, influences decomposition rates by creating distinct microbial habitats and carbon turnover patterns in macro-aggregates compared with micro-aggregates (Frąc et al., 2020).

Additionally, microbial communities underpin many ecosystem services, including nutrient availability for plants, soil structure maintenance, disease suppression, and carbon storage. Iqbal et al. (2025) describe how microbial communities enhance soil structure and thereby support plant growth and resilience. In natural ecosystems, soil microbial-driven processes support ecosystem productivity, resistance to invasion, and nutrient-use efficiency.

### 7.2.3 Dynamics, Transformations, Conservation, and Restoration of Natural Soil Microbiomes

Natural soil microbiomes are becoming subject to anthropogenic pressures: climate warming, $CO_2$ increase, drought, nitrogen deposition, land-use conversion. For example, a study in Australian native grassland soil found that elevated $CO_2$ and warming influenced archaeal and fungal community diversity after five years (Hayden et al., 2012). Land-use intensification (cultivation, continuous cropping) has been shown to reduce microbial network complexity and alter community composition compared to perennial grasslands (Romdhane et al., 2022).

Comparing undisturbed "natural" soils with anthropised soils (agriculture, urban, degraded) reveals differences in microbial diversity, community structure, network connectivity, functional potential, and resilience. For instance, a study in Uruguay found significant losses of fungal diversity and necromass recycling traits in cropland soils compared to native grasslands (Giménez et al., 2025). Another study found land-use legacies strongly shaping microbial composition and nutrient-cycling functions in shifting cultivation fields of Thailand (Arunrat et al., 2025).

Finally, a forward-looking discussion is needed on how to conserve and restore native soil microbiomes in natural and degraded systems. While much of soil microbiome-restoration literature focuses on agriculture, the principles apply to natural soils: baseline characterization of native communities, monitoring microbial indicators, managing disturbance regimes, using native vegetation to re-establish microbial habitats, possibly microbial inoculation (though less tested in natural soils), and recognizing the spatial/temporal complexity of microbial recovery. Muchalak et al. (2025) indicated that during tree succession, cooperative interactions and functional redundancy increased, reinforcing the role of natural forest succession in restoring soil functionality, and plant–soil feedback during forest successional process in a tropical ecotonal tension zone. The persistence of anthropogenic legacies (e.g., elevated Ca and Mg in pastures) suggested that active restoration, such as organic amendments, may accelerate recovery. The study of Muchalak and collaborators (2025) underscores the importance of ecotones as biodiversity reservoirs and advocates for soil microbial indicators (network modularity, enzymatic profiles) in restoration monitoring.

#### *7.2.3.1 Agricultural Soil Microbiome*

Land-use change, primarily driven by agricultural expansion, has triggered substantial shifts in ecosystem structure and functioning (Gomes et al., 2021). Excessive fertilizer inputs, intensive mechanical soil disturbance, and the improper application of pesticides have contributed to soil degradation, environmental pollution, eutrophication of water bodies, and increased greenhouse gas emissions (Gebhardt et al., 2023). In this context, the adoption of sustainable agricultural practices becomes essential to ensure global food security while safeguarding biodiversity and maintaining key ecosystem services (Lal, 2023).

Soil microbiomes constitute a major driver of edaphic processes that underpin the sustainability of agricultural production (Jansson et al., 2023). They perform critical functions, including climate regulation, nutrient cycling, soil structure formation and stabilization, plant growth promotion, enhancement of plant tolerance to abiotic stresses, pest and disease suppression, and pollutant degradation (Hartmann and Six, 2023). Consequently, soil microbial communities represent some of the most sensitive indicators of land-use conversion and disturbance, being strongly shaped by soil physicochemical properties (Fierer and Jackson, 2006), climate (Yao et al., 2017), agricultural use intensity (Thomson et al., 2015), and vegetation composition (Guo et al., 2016).

Tillage systems impose disturbances of varying intensities, generating contrasting edaphic environments that directly affect microbiome structure and composition (Xue et al., 2022). In conservation systems such as no-till, soil aggregates promote the physical protection of organic carbon and create heterogeneous microhabitats with spatial variability in nutrient availability for microorganisms and microfauna. These conditions tend to support more complex microbial communities, with greater diversity and structured interactions, ultimately enhancing nutrient cycling efficiency (Wilpiszeski et al., 2019).

Conversely, conventional tillage continuously exposes deeper soil layers to wetting–drying fluctuations, promoting macro-aggregate disruption and pore-network collapse (Liang et al., 2025). This process intensifies erosion, accelerates organic matter decomposition, and consequently reduces soil organic carbon stocks (Xue et al., 2022). Such structural alterations directly modify the physical habitat of microorganisms, affecting fungi particularly, whose hyphal networks are fragmented by mechanical disturbance (Orrù et al., 2021). Moreover, the rapid nutrient release resulting from the mineralization of previously protected organic matter favors opportunistic and copiotrophic microorganisms (Degrune et al., 2017). This enhanced mineralization can reshape bacterial and fungal community composition as well as soil fauna dynamics. Hence, reductions in soil biodiversity and a relative increase in aerobic microorganisms are commonly observed, reflecting a more oxidized and biologically unstable environment (Schmidt et al., 2018; Xin et al., 2018).

Long-term fertilization strongly affects soil microbial communities, with outcomes varying across cropping systems (conventional and no-till). In a system with legumes that received phosphorus, nitrogen, and manure inputs for 36 years, bacterial richness and diversity increased, whereas in a grain-legume rotation system, both metrics declined, indicating that crop rotation systems may be less resistant to environmental fluctuations (Kong et al., 2023). Additionally, the authors demonstrated that fertilization effects on the soil microbiome and crop productivity were attributable not only to increased soil nutrient availability but also to microbial community modulation through the enrichment of specific fertilization-responsive taxa (Kong et al., 2023).

The type of fertilization (mineral vs organic) can induce shifts in predominant microbial life-history strategies, with consequences for multiple ecosystem functions (Zhang et al., 2019). Luo et al. (2023) showed that organic fertilization accelerated the resilience of copiotrophic bacterial communities to biodiversity loss compared with unfertilized and mineral-fertilized soils and strengthened the positive relationships between microbially mediated soil functions and the diversity and complexity of bacterial network structures, rather than fungal networks. This highlights organic management as a key strategy for conserving microbiome complexity and mitigating the negative impacts of biodiversity loss on multiple agroecosystem functions.

Beyond fertilization, another conventional farming practice with direct impacts on the soil microbiome is mechanical soil management, including plowing and harrowing. A single soil-compaction event directly affects microbiome resilience, as demonstrated by Longepierre et al. (2021) across four growing seasons in different management systems. Anaerobic prokaryotes increased in compacted soils due to reduced air permeability and gas diffusion. Furthermore, soil compaction enhanced the abundance of denitrification-related genes due to diminished soil aeration (Fracetto et al., 2017). On the other hand, aerobic prokaryotes and plant-associated fungi were negatively affected. The cumulative negative effects of compaction on the soil microbiome and its functions can directly impair crop productivity, underscoring the need for strategies that enhance the sustainability of cropping systems.

Land-use systems play a significant role on soil microbiome structure and composition. Over three consecutive years, Kraut-Cohen et al. (2020) monitored microbial composition and activity in soils under conventional tillage, minimum tillage, and no-till. Intensive tillage promoted substantial disturbances in biological, chemical, and physical soil properties. These alterations compromised ecosystem services by reducing overall microbial activity, destabilizing microbial community composition, and decreasing the abundance of heterotrophic bacteria involved in the decomposition of complex organic matter. Additionally, a relative increase in specific aerobic taxa was observed, indicating a more oxidative and biologically unstable environment.

The soil microbiome plays an essential role in maintaining the health of soils, plants, animals, and humans, forming one of the pillars of the One Health concept. Microbiome diversity directly influences agricultural productivity, ecological stability, and the quality of food consumed by humans and livestock. In animals, exposure to and ingestion of soil contribute to the formation and modulation of the gut microbiota, affecting digestion, immunity, and behavior. In humans, biodiverse soils are associated with reduced allergies and inflammatory diseases, strengthening the immune system and contributing to overall well-being. Conversely, soil degradation and declining microbial diversity increase the risk of pathogen emergence and dissemination. Therefore, preserving soil microbial diversity is fundamental to ensuring environmental and public health (Banerjee and van der Heijden, 2023).

### *7.2.3.2 The Microbiome of Soils Contaminated with Hydrocarbons and Heavy Metals*

#### 7.2.3.2.1 SOIL POLLUTION CAUSED BY HYDROCARBONS AND HEAVY METALS

Soils are matrices that directly suffer the effects of anthropogenic activity. Contaminated soils are defined as those containing higher concentration of a given substance than that observed in their natural conditions, without necessarily being harmful to human or to the environment (Rodríguez-Eugenio et al., 2018). Polluted soils, in contrast, contain substances at concentrations that exceed safe thresholds, resulting in damage to living organisms (Rodríguez-Eugenio et al., 2018).

Activities such as agriculture, the oil and gas industry, and mining are major sources of soil pollutants (Brady and Weil, 2008). These substances tend to accumulate in areas associated with industrial or agricultural operations, including gas stations, wastewater disposal sites, crop areas, factories, landfills, and mining regions (Brady and Weil, 2008; Rodríguez-Eugenio et al., 2018).

Pollution of water systems caused by spills and improper disposal of crude oil and its products often leads to soil contamination, as polluted water may be used for irrigation or may infiltrate directly (Koshlaf and Ball, 2017). Consequently, hydrocarbon pollution represents a serious and widespread problem caused by the increased activity of the oil industry (Goyal et al., 2014; Koshlaf and Ball, 2017). Remediation of such contamination imposes extremely high costs for agricultural producers (Das and Chandran, 2011; Koshlaf and Ball, 2017). Furthermore, polycyclic aromatic hydrocarbons (PAHs), the most recalcitrant and harmful portion of oil, offer a high risk to human and environmental health, due to their mutagenic and carcinogenic properties (Latif et al., 2010; Kim et al., 2013).

Industrialization has also intensified pollution caused by a group of metals and metalloids called HMs, defined as toxic elements with high atomic mass (densities greater than 4.5 $g/cm^3$) such as arsenic (As), cadmium (Cd), chromium (Cr), copper (Cu), mercury (Hg), lead (Pb), nickel (Ni), and zinc (Zn) (Alloway, 2013; Rodríguez-Eugenio et al., 2018). HMs contamination is a serious problem, as these pollutants are difficult to remove and tend to bioaccumulate throughout the food chain, bringing serious impacts to ecosystems (Ali and Khan, 2019; Shetty et al., 2025). Exposure to HMs can cause damage to body tissue and chronic health complications in the immune, respiratory, nervous, endocrine, and cardiovascular systems (Mohammed Abdul et al., 2015; Alvarez et al., 2021; Shetty et al., 2025). Both PAHs and HMs are among the most common and hazardous pollutants in soils, causing severe environmental and economic consequences (Haque et al., 2022).

#### 7.2.3.2.2 IMPACTS OF CONTAMINANTS ON THE SOIL MICROBIOME AND CHANGES IN DIVERSITY

The microbiome associated with contaminated soil plays a critical role in the biogeochemical cycling of pollutants and in mitigating their toxic effects (Wang et al., 2025). Upon exposure to contaminants, indigenous microbial communities undergo deep compositional and functional shifts, favoring resistant microorganisms with the metabolic capacity to tolerate or transform these compounds (Ying and Wei, 2019). These microorganisms synergistically interact within complex microbial communities, where their collective functional diversity determines the ecosystem's resilience and capacity for pollutant detoxification through processes such as bioaccumulation, bioleaching, detoxification, and biotransformation (Ying and Wei, 2019; Selim and Harun, 2025).

Structural and functional diversity of microbiomes is strong indicators of soil health. Structural diversity refers to the number and abundance of species (or other taxonomic rank of choice), whereas functional diversity reflects the metabolic repertoire, i.e., genes coding for substrate use and enzymatic activity, and related pathways (Avidano et al., 2005). Biological indicators (bioindicators) help monitor environmental conditions by associating community composition or function with specific disturbances (De Cáceres et al., 2010; Astudillo-García et al., 2019). Healthy soils maintain diverse microbial communities capable of nutrient cycling, while contaminated soils often exhibit reduced richness and dominance of specialized taxa. Consequently, taxonomic or functional shifts can serve as bioindicators of contamination intensity and type (Astudillo-García et al., 2019; Yang et al., 2023).

In soil areas contaminated with PHAs, bacterial taxa such as Alphaproteobacteria often dominate with abundant genera including *Lysobacter*, *Dietzia*, *Pseudomonas*, and *Extensimonas*, and among fungi, the phylum Ascomycota is the most prevalent (Wang et al., 2021). In a recent study, diesel oil and gasoline contamination stimulated the growth of *Rhodanobacter*, *Sphingomonas*, *Burkholderia*, *Sphingobium*, and *Mycobacterium*, as well as the fungal genus *Penicillium* (Borowik et al., 2025). In soils contaminated with biodiesel B20, a mixture composed of 20% biodiesel and 80% fossil diesel, the bacterial families Xanthobacteraceae, Burkholderiaceae, and Acetobacteraceae showed greater relative abundance, while among the Archaea, the phyla Euryarchaeota, Halobacterota, and Nanoarchaeota predominated (Hidalgo-Martinez et al., 2024). Studies of soils polluted by naphthalene, dioctyl phthalate, diesel, and crude oil consistently reported the enrichment of hydrocarbon-degrading bacteria such as *Pseudomonas*, *Arthrobacter* and *Paenibacillus* (Panov et al., 2013).

In HMs-contaminated soils, pollutants such as Pb, Cd, and Zn have been shown to decrease bacterial metabolic activity (Fajardo et al., 2019). Communities shifted from Pseudomonadota-dominant (in control soils) to Bacillota-dominant in polluted sites (Fajardo et al., 2019). Genera commonly found in HMs-affected soils include *Solirubrobacter*, *Geobacter*, *Edaphobacter*, *Pseudomonas*, *Gemmatimonas*, *Nitrosomonas*, *Xanthobacter*, *Sphingomonas*, *Pedobacter*, *Ktedonobacter*, *Bryobacter*, *Nitrospira*, and *Steroidobacter*, in addition to members of the archaeal order Nitrososphaerales (Hemmat-Jou et al., 2018; Yang et al., 2022). Long-term exposure to Pb and Zn reduces microbial richness and alters community composition, with the decrease of bacterial taxa *Marmoricola* and *Nocardioides*, and the fungal genus *Gibberella* that can be considered indicators of sensitivity to these metals (Pan et al., 2020).

##### 7.2.3.2.3 MECHANISMS OF MICROBIAL RESPONSE TO PAH AND HM EXPOSURE

Bacteria can act in PAH degradation through metabolism or cometabolism (Peng et al., 2008). Different pathways are involved in this process, including phototrophic, aerobic, anaerobic, and chemotrophic pathways (Hidalgo et al., 2020). In the aerobic pathway, peroxidases and oxygenase enzymes, such as toluene dioxygenase (TDO), naphthalene dioxygenase (NDO), and biphenyl dioxygenase (BPDO), play key roles in these transformations (Peng et al., 2008; Hidalgo et al., 2020; Kumari and Das, 2023). First, a dioxygenase adds hydroxyl groups to the aromatic ring forming cis-dihydrodiol; then a cis-dihydrodiol dehydrogenase converts this compound into a dihydroxylated intermediate (such as catechol), which will be further cleaved by dioxygenases (Hidalgo et al., 2020). Cleavage can occur between the OH groups (intradiol pathway, when they are in the ortho position) or outside them (extradiol pathway, when in the meta position), and this process is repeated for each ring of the molecule. The intermediates generated further enter the tricarboxylic acid cycle (TCA) cycle, ultimately resulting in carbon dioxide and water (Hidalgo et al., 2020; Kumari and Das, 2023). In anaerobic degradation, other compounds, such as nitrate or sulfate, function as final electron acceptors. The process begins with an activation reaction, and in the case of naphthalene this activation can occur by carboxylation or by methylation followed by the addition of fumarate (Hidalgo et al., 2020). Both routes lead to the intermediate 2-naphthoyl-CoA, and degradation proceeds with aryl-CoA reductases (Hidalgo et al., 2020). Subsequently, the aromatic ring is reduced by a process similar to beta-oxidation, being dearomatized by the enzyme 2-naphthoyl-CoA reductase (NCR), a flavoprotein with a flavin mononucleotide cofactor.

Biosurfactant production further enhances degradation by reducing the surface tension of hydrophobic substances, thus increasing pollutant solubility and bioavailability (Kumari and Das, 2023). Fungi are also capable of degrading PAHs through the cytochrome P-450 systems or by using extracellular ligninolytic enzymes such as laccase, lignin peroxidase and manganese peroxidase (Peng et al., 2008). In soils contaminated with hydrocarbons, the type of microbial biodegradation is primarily determined by the availability of oxygen, with aerobic pathways generally being faster and more efficient (Kebede et al., 2021). In oxygenated zones, the presence of aerobic bacteria and fungi is predominant, and consequently, aerobic hydrocarbon degradation pathways prevail (Kebede et al., 2021). In deeper soil layers, saturated with water or compacted, oxygen becomes scarcer, and anaerobic degradation pathways become dominant (Laczi et al., 2020).

In the case of HMs, microbial tolerance relies on mechanisms including cell wall accumulation, membrane transport, complexation, and redox reactions (Nanda et al., 2019). Bacteria can absorb HMs in two different ways: by chemiosmotic transport (ATP-independent) or by specific ATP-dependent transport, with the cell wall being an important component for interacting with metal ions (Nanda et al., 2019). Gram-positive bacteria accumulate greater quantities of metals due to negatively charged carboxyl groups in their cell walls (Nanda et al., 2019). Extracellular compounds like exopolysaccharides (EPS), siderophores, and glutathione can bind metal ions, preventing intracellular toxicity, while intracellular metallothioneins sequester metals into inactive forms (Yin et al., 2019). Fungal species can accumulate HMs either by binding them to cell wall components, such as chitin, chitosan, and polysaccharides, or through intracellular sequestration by polyphosphates and metal-binding proteins such as metallothioneins (Yin et al., 2019).

##### 7.2.3.2.4 METHODOLOGICAL APPROACHES TO ASSESSING THE SOIL MICROBIOME

Assessing microbiomes in contaminated soils requires integrating physical, chemical, and biological methods. Physicochemical parameters such as pH, organic matter concentration, soil redox state, and pollutant concentration strongly influence the community composition and the state of microbial activity (Zhang et al., 2016; Chen et al., 2023).

Modern molecular and omics approaches are crucial for determining ecological parameters, such as structural and functional diversity, that reflect the true complexity of soil microbiomes. Techniques such as 16S rRNA gene amplicon sequencing, shotgun metagenomic and metatranscriptomics, allow for the recovery of genomes, identification of taxonomic composition and prevalent metabolic pathways, and of tolerance and resistance mechanisms (Quince et al., 2017; Ruppert et al., 2019; Francioli et al., 2021). Complementarily, classical enrichment and cultivation methodologies are essential for analyzing the morphological, biochemical, and metabolic characteristics of microorganisms, as well as preserving isolates with potential in biotechnological approaches (Clagnan et al., 2024).

##### 7.2.3.2.5 MICROBIOME-CENTERED BIOREMEDIATION STRATEGIES

Microorganisms isolated from contaminated environments are valuable for bioremediation, as they have a versatile and robust metabolic capacity for consuming these substances and mitigating damage to the ecosystem (Zhang et al., 2019; Gupta et al., 2020). These microbes can also produce biosurfactants (Dvořák et al., 2017) with broad biotechnological applications, including pollutant cleanup, new drug development, and cosmetics (Soberón-Chávez and Maier, 2011).

Two main strategies are used in microbial-based bioremediation: biostimulation and/or bioaugmentation. Biostimulation enhances the activity of native microbial communities by adjusting pH and adding nutrients, while bioaugmentation introduces specialized microbial strains or microbial consortia, capable of efficiently degrading specific pollutants (Roy et al., 2018; Haleyur et al., 2019). Thus, by integrating microbiome data with omics-based and cultivation methodologies, it becomes possible to design targeted and sustainable remediation approaches (Figure 7.2).

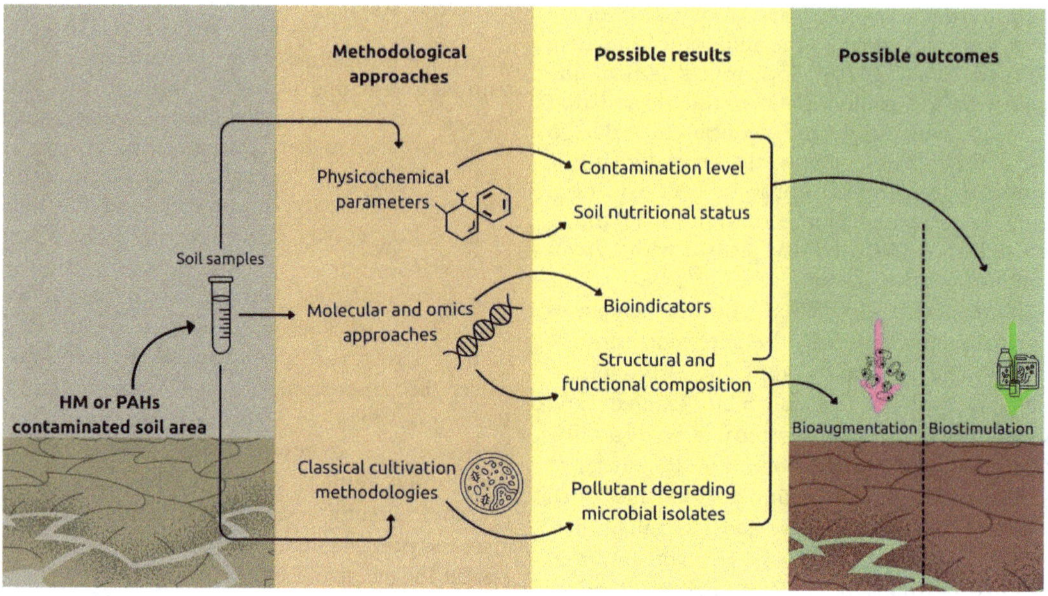


Figure 7.2 Strategies for bioremediation of contaminated soils using microbiome information. Different integrated methodologies can provide a range of important outcomes for the application of microbial metabolism-based techniques to mitigate the harmful effects of pollutants.

## 7.3 FINAL REMARKS

Across natural, agricultural, and contaminated environments, soil microbiomes emerge as highly diverse, functionally indispensable, yet increasingly vulnerable components of terrestrial ecosystems. Environmental filters, land-use change, and pollution reshape their structure, functions, and resilience, while specialized taxa in contaminated soils reveal powerful metabolic capacities for biodegradation and detoxification. Advances in molecular, omics, and cultivation approaches now enable deeper insight into these complex communities, supporting conservation, sustainable management, and bioremediation. Strengthening integrative strategies that protect and restore soil microbial diversity is essential for sustaining ecosystem functioning, agricultural productivity, and environmental health.

## REFERENCES

Ali, H., & Khan, E. (2019). Trophic transfer, bioaccumulation, and biomagnification of non-essential hazardous heavy metals and metalloids in food chains/webs—Concepts and implications for wildlife and human health. *Human and Ecological Risk Assessment: An International Journal*, 25(6), 1353–1376.

Alloway, B. J. (Ed.). (2013). *Heavy metals in soils: Trace metals and metalloids in soils and their bioavailability* (3rd ed.). Dordrecht, Springer Netherlands.

Alvarez, C. C., Bravo Gómez, M. E., & Hernández Zavala, A. (2021). Hexavalent chromium: Regulation and health effects. *Journal of Trace Elements in Medicine and Biology*, 65, 126729.~~Elsevier.~~

Arunrat, N., Mhuantong, W., & Sereenonchai, S. (2025). Land-use legacies shape soil microbial communities and nutrient cycling functions in rotational shifting cultivation fields of Northern Thailand. *Microbial Ecology*, 88, 102.

Astudillo-García, C., Hermans, S. M., Stevenson, B., Buckley, H. L., & Lear, G. (2019). Microbial assemblages and bioindicators as proxies for ecosystem health status: Potential and limitations. *Applied Microbiology and Biotechnology*, 103(16), 6407–6421.

Avidano, L., Gamalero, E., Cossa, G. P., & Carraro, E. (2005). Characterization of soil health in an Italian polluted site by using microorganisms as bioindicators. *Applied Soil Ecology*, 30(1), 21–33.

Bahram, M., Hildebrand, F., Forslund, S. K., Anderson, J. L., Soudzilovskaia, N. A., Bodegom, P. M., … et al. (2018). Structure and function of the global topsoil microbiome. *Nature*, 560, 233–237.

Baldrian, P., López-Mondéjar, R., & Kohout, P. (2023). Forest microbiome and global change. *Nature Reviews Microbiology*, 21, 487–501.

Banerjee, S., & van der Heijden, M. G. (2023). Soil microbiomes and One Health. *Nature Reviews Microbiology*, 21(1), 6–20.

Bar-On, Y. M., Phillips, R., & Milo, R. (2018). The biomass distribution on Earth. *Proceedings of the National Academy of Sciences*, 115, 6506–6511.

Bento, M. S., Barros, D. J., Araújo, M. G. S., et al. (2021). Active methane processing microbes and the disproportionate role of NC10 phylum in methane mitigation in Amazonian floodplains. *Biogeochemistry*, 156, 293–317.

Borowik, A., Wyszkowska, J., Zaborowska, M., & Kucharski, J. (2025). Regulation of the microbiome in soil contaminated with diesel oil and gasoline. *International Journal of Molecular Sciences*, 26(13), 6491.

Brady, N. C., & Weil, R. R. (2008). *The nature and properties of soils* Upper Saddle River, New Jersey, (Vol. 13, pp. 662–710). Prentice Hall.

Caldwell, S. L., Laidler, J. R., Brewer, E. A., Eberly, J. E., Sandborgh, S. C. (2008). Anaerobic oxidation of methane: Mechanisms, bioenergetics, and the ecology of associated microorganisms. *Environmental Science & Technology*, 42(18), 6791–6799.

Cao, X., Cui, Q., Li, D., Liu, Y., Liu, K., & Li, Z. (2025). Characteristics of soil microbial community structure in different land-use types of the Huanghe Alluvial Plain. *Microorganisms*, 13, 273.

Chen, G., Yuan, M., Ma, B., & Ren, Y. (2023). Responses of petroleum contamination at different sites to soil physicochemical properties and indigenous microbial communities. *Water, Air, & Soil Pollution*, 234(8), 494.

Chen, H., Ma, K., Huang, Y., Yao, Z., & Chu, C. (2021). Stable soil microbial functional structure responding to biodiversity loss based on metagenomic evidences. *Frontiers in Microbiology*, 12, 716764.

Clagnan, E., Costanzo, M., Visca, A., Di Gregorio, L., Tabacchioni, S., Colantoni, E., ... & Bevivino, A. (2024). Culturomics- and metagenomics-based insights into soil microbiome preservation and application for sustainable agriculture. *Frontiers in Microbiology*, 15, 1473666.

Das, N., & Chandran, P. (2011). Microbial degradation of petroleum hydrocarbon contaminants: An overview. *Biotechnology Research International*, 2011, 941810.

Dassen, S., Cortois, R., Martens, H., de Hollander, M., Kowalchuk, G. A., van der Putten, W. H., & De Deyn, G. B. (2017). Differential responses of soil bacteria, fungi, archaea and protists to plant species richness and plant functional group identity. *Molecular Ecology*, 26(15), 4085–4098.

De Cáceres, M., Legendre, P., & Moretti, M. (2010). Improving indicator species analysis by combining groups of sites. *Oikos*, 119(10), 1674–1684.

Degrune, F., Theodorakopoulos, N., Colinet, G., Hiel, M. P., Bodson, B., Taminiau, B., ... & Hartmann, M. (2017). Temporal dynamics of soil microbial communities below the seedbed under two contrasting tillage regimes. *Frontiers in Microbiology*, 8, 1127.

Ding, J., & Yu, S. (2025). Mechanisms of soil microbial community adaptation in cold-region wetlands under retrogressive succession. *Life*, 15, 817.

Dvořák, P., Nikel, P. I., Damborský, J., & de Lorenzo, V. (2017). Bioremediation 3.0: Engineering pollutant-removing bacteria in the times of systems biology. *Biotechnology Advances*, 35(7), 845–866.

Fajardo, C., Costa, G., Nande, M., Botías, P., García-Cantalejo, J., & Martín, M. (2019). Pb, Cd, and Zn soil contamination: Functional and structural impacts on the microbiome. *Applied Soil Ecology*, 135, 56–64.

Fierer, N., Breitbart, M., Nulton, J., Salamon, P., Lozupone, C., Jones, R., ... & Jackson, R. B. (2007). Metagenomic and small-subunit rRNA analyses reveal the genetic diversity of bacteria, archaea, fungi, and viruses in soil. *Applied and Environmental Microbiology*, 73(21), 7059–7066.

Fierer, N., & Jackson, R. B. (2006). The diversity and biogeography of soil bacterial communities. *Proceedings of the National Academy of Sciences*, 103(3), 626–631.

Frąc, M., Lipiec, J., Usowicz, B., Oszust, K., & Brzezińska, M. (2020). Structural and functional microbial diversity of sandy soil under cropland and grassland. *PeerJ*, 8, e9501.

Fracetto, F. J., Fracetto, G. G., Bertini, S. C., Cerri, C. C., Feigl, B. J., & Neto, M. S. (2017). Effect of agricultural management on $N_2O$ emissions in Brazilian sugarcane. *Soil Biology and Biochemistry*, 109, 205–213.

Francioli, D., Lentendu, G., Lewin, S., & Kolb, S. (2021). DNA metabarcoding for the characterization of terrestrial microbiota—Pitfalls and solutions. *Microorganisms*, 9(2), 361.

Gebhardt, S., Van Dijk, J., Wassen, M. J., & Bakker, M. (2023). Agricultural intensity interacts with landscape arrangement in driving ecosystem services. *Agriculture, Ecosystems & Environment*, 357, 108692.

Giménez, M., Berenstecher, P., Ligrone, A., et al. (2025). Soil microbiome analysis of Uruguayan grasslands and croplands reveals losses of microbial diversity and necromass recycling traits. *Environmental Microbiome*, 20, 96.

Gomes, E., Inácio, M., Bogdzevič, K., Kalinauskas, M., Karnauskaitė, D., & Pereira, P. (2021). Future land-use changes and impacts on terrestrial ecosystem services: A review. *Science of the Total Environment*, 781, 146716.

Goyal, A., Sharma, V., Upadhyay, N., Gill, S., & Sihag, M. (2014). Flax and flaxseed oil: An ancient medicine and modern functional food. *Journal of Food Science and Technology*, 51, 1633–1653.

Guerra, C. A., Delgado-Baquerizo, M., Duarte, E., Marigliano, O., Görgen, C., Maestre, F. T., & Eisenhauer, N. (2021). Global projections of the soil microbiome in the Anthropocene. *Global Ecology and Biogeography*, 30, 987–999.

Guo, X., Chen, H. Y., Meng, M., Biswas, S. R., Ye, L., & Zhang, J. (2016). Effects of land-use change on soil microbial communities in a managed subtropical forest. *Forest Ecology and Management*, 373, 93–99.

Gupta, K., Biswas, R., & Sarkar, A. (2020). Advancement of omics: Prospects for bioremediation of contaminated soils. In *Microbial Bioremediation & Biodegradation*, Shah M. (Singapore: Springer), (pp. 113–142).

Haleyur, N., Shahsavari, E., Jain, S. S., Koshlaf, E., Ravindran, V. B., Morrison, P. D., ... & Ball, A. S. (2019). Influence of bioaugmentation and biostimulation on PAH degradation in aged contaminated soils: Bacterial community dynamics. *Journal of Environmental Management*, 238, 49–58.

Haque, S., Srivastava, N., Pal, D. B., Alkhanani, M. F., Almalki, A. H., Areeshi, M. Y., ... & Gupta, V. K. (2022). Functional microbiome strategies for bioremediation of petroleum-hydrocarbon and heavy metal contaminated soils: A review. *Science of the Total Environment*, 833, 155222.

Hartmann, M., & Six, J. (2023). Soil structure and microbiome functions in agroecosystems. *Nature Reviews Earth & Environment*, 4(1), 4–18.

Hayden, H. L., Mele, P. M., Bougoure, D. S., Allan, C. Y., Norng, S., Piceno, Y. M., ... & Hovenden, M. J. (2012). Changes in microbial community structure of bacteria, archaea and fungi in response to elevated $CO_2$ and warming in Australian grassland soil. *Environmental Microbiology*, 14(12), 3081–3096.

He, Z., Yuan, C., Chen, P., Rong, Z., Peng, T., Farooq, T. H., Wang, G., Yan, W., & Wang, J. (2023). Soil microbial community composition and diversity under different land uses in Taojia River Basin. *Forests*, 14, 1004.

Hemmat-Jou, M. H., Safari-Sinegani, A. A., Mirzaie-Asl, A., & Tahmourespour, A. A. (2018). Microbial communities in heavy metal–contaminated soils: A metagenomic analysis. *Ecotoxicology*, 27(9), 1281–1291.

Hidalgo, K. J., Sierra-Garcia, I. N., Dellagnezze, B. M., & de Oliveira, V. M. (2020). Metagenomic insights into biodegradation of polycyclic aromatic hydrocarbons in the oil supply chain. *Frontiers in Microbiology*, 11, 561506.

Hidalgo-Martinez, K., Giachini, A. J., Schneider, M., Soriano, A., Baessa, M. P., Martins, L. F., & de Oliveira, V. M. (2024). Shifts in soil microbiome dynamics in fuel-affected areas under bioremediation. *Environmental Science and Pollution Research*, 31(23), 33663–33684).

Iqbal, S., Begum, F., Nguchu, B. A., ... & et al. (2025). The invisible architects: Microbial communities and their transformative role in soil health and global climate changes. *Environmental Microbiome*, 20, 36.

Jansson, J. K., McClure, R., & Egbert, R. G. (2023). Soil microbiome engineering for sustainability in a changing environment. *Nature Biotechnology*, 41(12), 1716–1728.

Kebede, G., Tafese, T., Abda, E. M., Kamaraj, M., & Assefa, F. (2021). Factors influencing the bacterial bioremediation of hydrocarbon contaminants in the soil: Mechanisms and impacts. *Journal of Chemistry*, 2021, 9823362.

Kim, K. H., Jahan, S. A., Kabir, E., & Brown, R. J. (2013). A review of airborne polycyclic aromatic hydrocarbons (PAHs) and their human health effects. *Environment International*, 60, 71–80.

Ko, D., Yoo, G., Yun, ST. et al. (2017). Bacterial and fungal community composition across the soil depth profiles in a fallow field. *j ecology environ* 41: 34. https://doi.org/10.1186/s41610-017-0053-0

Kong, W., Qiu, L., Ishii, S., Jia, X., Su, F., Song, Y., ... & Wei, X. (2023). Contrasting response of soil microbiomes to long-term fertilization in various highland cropping systems. *ISME Communications*, 3(1), 81.

Koshlaf, E., & Ball, A. S. (2017). Soil bioremediation approaches for petroleum hydrocarbon polluted environments. *AIMS Microbiology*, 3(1), 25–42.

Kraut-Cohen, J., Zolti, A., Shaltiel-Harpaz, L., Argaman, E., Rabinovich, R., Green, S. J., & Minz, D. (2020). Effects of tillage practices on soil microbiome and agricultural parameters. *Science of the Total Environment*, 705, 135791.

Kumari, S., & Das, S. (2023). Bacterial enzymatic degradation of recalcitrant organic pollutants: Catabolic pathways and genetic regulations. *Environmental Science and Pollution Research*, 30(33), 79676–79705.

Laczi, K., Erdeine Kis, A., Szilágyi, Á., Bounedjoum, N., Bodor, A., Vincze, G. E., ... & Perei, K. (2020). New frontiers of anaerobic hydrocarbon biodegradation in the multi-omics era. *Frontiers in Microbiology*, 11, 590049.

Lal, R. (2023). Farming systems to return land for nature: It's all about soil health and re-carbonization of the terrestrial biosphere. *Farming System*, 1(1), 100002.

Latif, I. K., Karim, A. J., Zuki, A. B. Z., Zamri-Saad, M., Niu, J. P., & Noordin, M. M. (2010). Pulmonary modulation of benzo[a]pyrene-induced hemato- and hepatotoxicity in broilers. *Poultry Science*, 89(7), 1379–1388.

Liang, X., Rehman, S. U., Zhiqi, W., Raza, M. A., Haider, I., Khalid, M. H. B., ... & Zhongming, M. (2025). Impacts of conservation tillage on agricultural land development: A review. *Journal of Soil Science and Plant Nutrition*, 25(1), 428–449.

Liu, N., Hu, H., Ma, W., Deng, Y., Dimitrov, D., Wang, Q., ... & Wang, Z. (2023). Relationships between soil microbial diversities across an aridity gradient in temperate grasslands. *Microbial Ecology*, 85(3), 1013–1027.

Longepierre, M., Widmer, F., Keller, T., Weisskopf, P., Colombi, T., Six, J., & Hartmann, M. (2021). Limited resilience of the soil microbiome to mechanical compaction within four growing seasons of agricultural management. *ISME Communications*, 1, 44.

Luo, J., Banerjee, S., Ma, Q., Liao, G., Hu, B., Zhao, H., & Li, T. (2023). Organic fertilization drives shifts in microbiome complexity and keystone taxa increase the resistance of microbial-mediated functions to biodiversity loss. *Biology and Fertility of Soils*, 59(4), 441–458.

Mohammed Abdul, K. S., Jayasinghe, S. S., Chandana, E. P. S., Jayasumana, C., & De Silva, P. M. C. S. (2015). Arsenic and human health effects: A review. *Environmental Toxicology and Pharmacology*, 40(3), 828–846.

Muchalak, F., Freitas, A. S., Zagatto, L. F. G., Rocha, G. S., Martins, G. L., Silva-Zagatto, S. S., ... & Navarrete, A. A. (2025). Plant–soil microbiome feedback under post-pasture tropical ecotone forest regeneration. *Applied Soil Ecology*, 214, 106381.

Nanda, M., Kumar, V., & Sharma, D. K. (2019). Multimetal tolerance mechanisms in bacteria: The resistance strategies acquired by bacteria that can be exploited to clean up heavy metal contaminants from water. *Aquatic Toxicology*, 212, 1–10.

Nannipieri, P., Ascher, J., Ceccherini, M. T., Landi, L., Pietramellara, G., & Renella, G. (2003). Microbial diversity and soil functions. *European Journal of Soil Science*, 54, 655–670.

Orgiazzi, A., Bardgett, R. D., Barrios, E., Behan-Pelletier, V., Briones, M. J. I., Chotte, J.-L., ... et al. (2016). *Global soil biodiversity atlas*. Publications Office of the European Union, 2016, https://data.europa.eu/doi/10.2788/2613European ~~Commission~~.

Orrù, L., Canfora, L., Trinchera, A., Migliore, M., Pennelli, B., Marcucci, A., ... & Pinzari, F. (2021). How tillage and crop rotation change the distribution pattern of fungi. *Frontiers in Microbiology*, 12, 634325.

Pan, X., Zhang, S., Zhong, Q., Gong, G., Wang, G., Guo, X., & Xu, X. (2020). Effects of soil chemical properties and fractions of Pb, Cd, and Zn on bacterial and fungal communities. *Science of the Total Environment*, 715, 136904.

Panov, A. V., Esikova, T. Z., Sokolov, S. L., Kosheleva, I. A., & Boronin, A. M. (2013). Influence of soil pollution on the composition of a microbial community. *Microbiology*, 82(2), 241–248.

Peng, R. H., Xiong, A. S., Xue, Y., Fu, X. Y., Gao, F., Zhao, W., ... & Yao, Q. H. (2008). Microbial biodegradation of polyaromatic hydrocarbons. *FEMS Microbiology Reviews*, 32(6), 927–955.

Philippot, L., Raaijmakers, J. M., Lemanceau, P., & van der Putten, W. H. (2013). Going back to the roots: The microbial ecology of the rhizosphere. *Nature Reviews Microbiology*, 11, 789–799.

Quince, C., Walker, A. W., Simpson, J. T., Loman, N. J., & Segata, N. (2017). Shotgun metagenomics, from sampling to analysis. *Nature Biotechnology*, 35(9), 833–844.

Raghoebarsing, A. A., Pol, A., Van De Pas-Schoonen, K. T., ... & et al. (2006). A microbial consortium couples anaerobic methane oxidation to denitrification. *Nature*, 440(7086), 918–921.

Rodríguez-Eugenio, N., McLaughlin, M., & Pennock, D. (2018). *Soil pollution: A hidden reality*. FAO, Rome, 142.

Romdhane, S., Spor, A., Banerjee, S., ... & et al. (2022). Land-use intensification differentially affects bacterial, fungal and protist communities and decreases microbiome network complexity. *Environmental Microbiome*, 17, 1.

Roy, A., Dutta, A., Pal, S., Gupta, A., Sarkar, J., Chatterjee, A., ... & Kazy, S. K. (2018). Biostimulation and bioaugmentation of native microbial community accelerated bioremediation of oil refinery sludge. *Bioresource Technology*, 253, 22–32.

Ruppert, K. M., Kline, R. J., & Rahman, M. S. (2019). Past, present, and future perspectives of environmental DNA (eDNA) metabarcoding: A systematic review. *Global Ecology and Conservation*, 17, e00547.

Schmidt, R., Gravuer, K., Bossange, A. V., Mitchell, J., & Scow, K. (2018). Long-term use of cover crops and no-till shift soil microbial community life strategies in agricultural soil. *Plos One*, 13(2), e0192953.

Selim, S., & Harun-Ur-Rashid, M. (2025). Harnessing microbial diversity for effective remediation of heavy metal-contaminated soils. *Applied Soil Ecology*, 215, 106473.

Shetty, B. R., Jagadeesha, P. B., & Salmataj, S. A. (2025). Heavy metal contamination and its impact on the food chain: Exposure, bioaccumulation, and risk assessment. *CyTA–Journal of Food*, 23(1), 2438726.

Soberón-Chávez, G., & Maier, R. M. (2011). Biosurfactants: A general overview. In *Biosurfactants: From genes to applications*, Gloria Soberón-Chávez, Editor. Springer, Heidelberg, (pp. 1–11).

Thomson, B. C., Tisserant, E., Plassart, P., Uroz, S., Griffiths, R. I., Hannula, S. E., ... & Lemanceau, P. (2015). Soil conditions and land use intensification effects on soil microbial communities across a range of European field sites. *Soil Biology and Biochemistry*, 88, 403–413.

Wang, M., Dowling, D. N., & Germaine, K. J. (2025). Bioremediation of diesel-contaminated soil: Hydrocarbon degradation and soil toxicity reduction by constructed bacterial consortia. *Applied Sciences*, 15(18), 10143.

Wang, M., Garrido-Sanz, D., Sansegundo-Lobato, P., Redondo-Nieto, M., Conlon, R., Martin, M., ... & Germaine, K. J. (2021). Soil microbiome structure and function in ecopiles used to remediate petroleum-contaminated soil. *Frontiers in Environmental Science*, 9, 624070.

Wilpiszeski, R. L., Aufrecht, J. A., Retterer, S. T., Sullivan, M. B., Graham, D. E., Pierce, E. M., ... & Elias, D. A. (2019). Soil aggregate microbial communities: Towards understanding microbiome interactions at biologically relevant scales. *Applied and Environmental Microbiology*, 85(14), e00324–19.

Xin, X. L., Yang, W. L., Zhu, Q. G., Zhang, X. F., Zhu, A. N., & Zhang, J. B. (2018). Abundance and depth stratification of soil arthropods as influenced by tillage regimes. *Soil Use and Management*, 34(2), 286–296.

Xue, R., Wang, C., Zhao, L., Sun, B., & Wang, B. (2022). Agricultural intensification weakens the soil health index and stability of microbial networks. *Agriculture, Ecosystems & Environment*, 339, 108118.

Yang, Z. N., Liu, Z. S., Wang, K. H., Liang, Z. L., Abdugheni, R., Huang, Y., ... & Liu, S. J. (2022). Soil microbiomes divergently respond to heavy metals and polycyclic aromatic hydrocarbons in contaminated industrial sites. *Environmental Science and Ecotechnology*, 10, 100169.

Yang, P. F., Spanier, N., Aldredge, P., Shahid, N., Coleman, A., Lyons, J., & Langley, J. A. (2023). Will free-living microbial community composition drive biogeochemical responses to global change? *Biogeochemistry*, 162(3), 285–307.

Yao, M., Rui, J., Niu, H., Heděnec, P., Li, J., He, Z., ... & Li, X. (2017). The differentiation of soil bacterial communities along a precipitation and temperature gradient. *Catena*, 152, 47–56.

Yin, K., Wang, Q., Lv, M., & Chen, L. (2019). Microorganism remediation strategies towards heavy metals. *Chemical Engineering Journal*, 360, 1553–1563.

Ying, T. E. N. G., & Wei, C. H. E. N. (2019). Soil microbiomes: A promising strategy for contaminated soil remediation. *Pedosphere*, 29(3), 283–297.

Yu, Z., Zeng, X. M., Cheng, X., Zhang, Q., & Zhang, K. (2025). Patterns and environmental drivers of soil microbial succession. *Global Change Biology*, 31(9), e70475.

Zhang, B., Liang, A., Wei, Z., & Ding, X. (2019). No-tillage leads to higher resistance but lower resilience of soil multifunctionality to dry–wet disturbances. *Soil and Tillage Research*, 195, 104376.

Zhang, C., Nie, S., Liang, J., Zeng, G., Wu, H., Hua, S., ... & Xiang, H. (2016). Effects of heavy metals and soil physicochemical properties on wetland soil microbial biomass and bacterial community structure. *Science of the Total Environment*, 557, 785–790.

Zhang, B., Wang, S., Diao, M., Fu, J., Xie, M., Shi, J., ... & Borthwick, A. G. (2019). Microbial community responses to vanadium distributions in mining geological environments and bioremediation assessment. *Journal of Geophysical Research: Biogeosciences*, 124(3), 601–615.

CHAPTER 8

# Unveiling Lytic Polysaccharide Monooxygenases from Environmental Microbiome

## NATURE'S BIOCATALYST FOR BIOPROCESS ADVANCEMENT

*Chetana Akhand, Riddhi Singh, Sejal Bhairam, Rashi Bamrotwar, and Nishant A. Dafale*

## 8.1 INTRODUCTION

Microbial communities are pivotal to sustaining Earth's biological and geochemical microbiome. They drive essential ecosystem processes such as nutrient cycling, respiration, organic matter decomposition, and the regulation of elemental fluxes, particularly carbon, nitrogen, and phosphorus, through terrestrial and aquatic systems (Panthee et al., 2022). For instance, soil microbiomes play integral roles in sequestering carbon, mitigating greenhouse gas emissions, preventing erosion, detoxifying pollutants, suppressing phytopathogens, and supporting plant nutrition (Ossowicki et al., 2021). These ecological functions are often mediated by enzymes that degrade or modify complex carbohydrates. Collectively termed carbohydrate-active enzymes (CAZymes), these microbial biocatalysts are categorized into six major families based on their catalytic functions and substrate interactions (www.cazy.org/): Glycoside Hydrolases (GHs): Enzymes that hydrolyze glycosidic bonds in carbohydrates; Glycosyltransferases (GTs): Involved in the biosynthesis of complex carbohydrates via glycosidic bond formation; Polysaccharide Lyases (PLs): Cleave polysaccharides via a β-elimination mechanism; Carbohydrate Esterases (CEs): Remove ester-linked substitutions from carbohydrates; Carbohydrate-Binding Modules (CBMs): Non-catalytic domains that aid in substrate targeting; and Auxiliary activities (AAs): Oxidative enzymes that work in synergy with hydrolases, notably including Lytic Polysaccharide Monooxygenases (LPMOs) (Wardman et al., 2022).

Among these, GHs form the largest and most extensively studied class, especially in the context of lignocellulosic biomass (LCB) conversion. Conventional enzymatic cocktails used for biomass saccharification primarily comprise cellulases (endo-β-1,4-glucanase, cellobiohydrolase, β-glucosidase) and hemicellulases (xylanases, β-xylosidase, mannanases, and esterases), all belonging to the GH family (Haq et al., 2021). However, the tightly packed matrix of cellulose, hemicellulose, and lignin within LCB creates a significant barrier to enzymatic hydrolysis, limiting the accessibility and efficiency of traditional GH-based systems (Srivastava et al., 2023b). This challenge has led to a paradigm shift in biomass processing; the incorporation of LPMOs, a class of copper-dependent oxidative enzymes that complement hydrolytic CAZymes by breaking glycosidic bonds via oxidative mechanisms (Srivastava et al., 2024b).

LPMOs, classified under the AA family, are grouped into subfamilies AA9–AA11 and

DOI: 10.1201/9781003637950-8

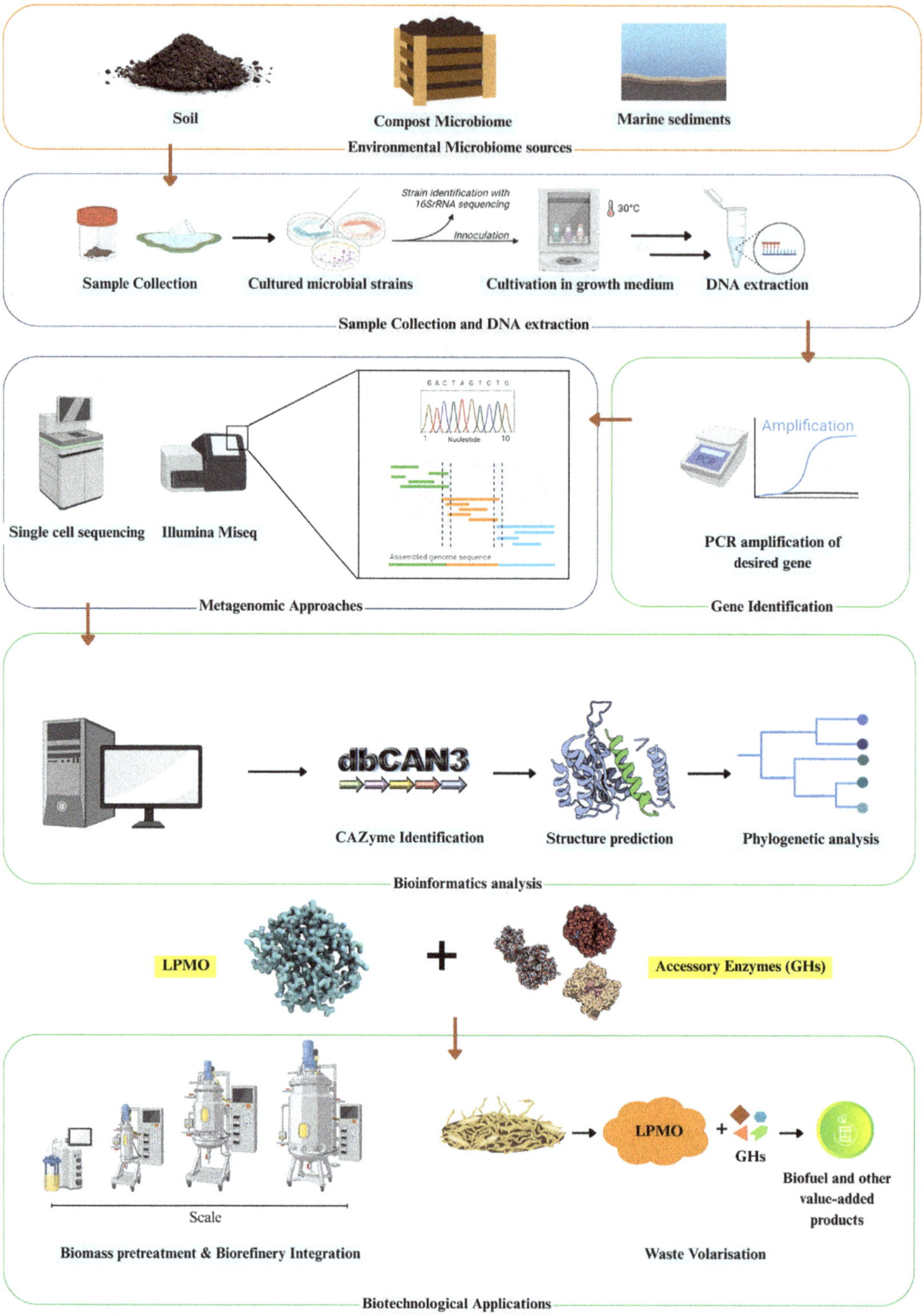


Figure 8.1 Workflow illustrating the identification of LPMOs potent microbes from environmental microbiomes and their application for biofuel and value-added product generation.

AA13–AA17 in the CAZy database, based on structural and taxonomic criteria (Srivastava et al., 2023a). Their integration into enzyme cocktails has significantly enhanced saccharification yields while lowering enzyme loading and process costs. Several studies underscore the synergistic impact of LPMOs. Müller et al. (2015) demonstrated a 60% improvement in saccharification of steam-exploded birchwood xylan when LPMOs were combined with cellulases. Notably, the study also suggested that lignin in LCB may act as an electron donor for LPMO activity, transforming what was previously considered a barrier into a functional component. Similarly, Harris et al. (2010) and Hu et al. (2014) observed enhanced biomass degradation when LPMOs were applied to pretreated substrates such as corn stover, poplar, and pine, with performance strongly influenced by pretreatment strategies and substrate composition. In another example, Angeltveit et al. (2024) evaluated fungal AA9s (TtAA9E from *Thermothielavioides terrestris* and TaAA9A from *Thermoascus aurantiacus*) in conjunction with a commercial cellulase blend (Celluclast 1.5 L) and β-glucosidase (NZ-BG). Their study revealed that LPMO supplementation enhanced cellulose conversion efficiency, particularly at higher dry matter content and extended reaction times.

Despite these advances, several challenges hinder the full-scale industrial implementation of LPMOs, notably the controlled supply of oxygen or hydrogen peroxide as electron donors (Chylenski et al., 2019). Ongoing research focused on understanding LPMO mechanisms, substrate specificity, redox partners, and environmental diversity is crucial to overcoming these barriers.

LPMOs have emerged as key oxidative enzymes in LCB breakdown, often sourced from environmental microbiomes rich in uncultured or underexplored microbial taxa. Diverse ecosystems such as soil, compost, and marine sediments provide valuable reservoirs for biocatalyst mining. Figure 8.1 presents a schematic overview of the pipeline used for LPMO discovery and application, from sample collection and microbial cultivation to metagenomic sequencing, CAZyme annotation, structural modeling, and downstream integration with accessory enzymes (GHs) for biomass pretreatment and valorization. This integrative approach offers a strategic framework for identifying robust LPMOs for biotechnological use in sustainable bioprocesses. The present chapter explores the world of LPMOs, from the classification, catalytic mechanisms, and ecological origins to their evolving role in biomass valorization. Given their natural abundance in environmental microbiomes, LPMOs represent promising tools for next-generation bioprocessing and sustainable biotechnological innovation.

### 8.1.1 Lytic Polysaccharide Monooxygenases (LPMOs): Structure, Function, and Mechanism

Understanding the structure, function, and catalytic mechanism of LPMOs is crucial for evaluating the substrate specificity and synergistic interactions with hydrolytic enzymes in lignocellulolytic cocktails (Srivastava et al., 2024a). Despite their transformative role in biomass deconstruction, comprehensive knowledge of LPMOs remains limited. This limitation is largely due to the relatively recent emergence of LPMO research, as well as the ongoing development of advanced analytical tools capable of capturing their dynamic oxidative activity with high precision. LPMOs belong to the AA CAZyme family class within the CAZy database, a family that was formally recognized only after 2010, following the discovery of their distinct copper-dependent oxidative cleavage mechanism (Vaaje-Kolstad et al., 2010). Prior to this, LPMOs were misclassified as GHs, due to their cellulose-degrading activity, but lacked a clear mechanistic identity. It was only through structural and biochemical studies that their oxidoreductive mode of action was elucidated, prompting the establishment of a separate CAZy class for these enzymes.

#### *8.1.1.1 Environmental and Phylogenetic Diversity*

LPMOs were originally misclassified as GHs, specifically GH61, due to their polysaccharide-degrading abilities. The pivotal work by Vaaje-Kolstad et al. (2010) on a bacterial metalloenzyme, CBM33, elucidated its role in enhancing chitin degradation via oxidative cleavage supported by $O_2$ and a reducing agent. This discovery laid the foundation for the establishment of a new class within CAZymes, AA enzymes, unifying oxidatively cleaving enzymes under the LPMO umbrella. As a result, GH61 and CBM33 were reclassified as AA9 and AA10, respectively (Srivastava et al., 2020a). Today, the CAZy database recognizes eight AA families of LPMOs (AA9–AA11, AA13–AA17), each with distinct structural and functional attributes. Among these, AA9 and AA10 are the most extensively studied. Both families exhibit a characteristic β-sandwich fold and harbor conserved residues at the substrate-binding interface (Yu et al., 2023). Fungal AA9s primarily target cellulose and hemicellulose and are highly effective against recalcitrant polysaccharides. Incorporation of AA9s into commercial cellulase blends, such as Cellic CTec3 and Accellerase TRIO, has enhanced saccharification efficiency by 30–40% (Long et al., 2022). Structural studies reveal that AA9s exhibit substrate oxidation at either the C1 or C4 position, typically involving a $Cu^{2+}$ ion coordinated with either tyrosine (C4 oxidation) or phenylalanine (C1 oxidation) (Yu et al., 2023). Promising examples of AA9 application include enzymes from *Ceriporiopsis subvermispora*, which increased sugar yield by up to 83% when added to Celluclast-treated wheat straw (Bissaro et al., 2018). Two other AA9s, CsLPMO9A and CsLPMO9B, demonstrated significant synergistic effects with cellulases, boosting the release of glucose, cellobiose, and xylose by as much as 83%, 52%, and 81%, respectively (Long et al., 2021). TausLPMO9B from *Thielavia australiensis* displayed stability in slightly acidic pH and efficacy on pretreated lignocellulosic substrates, further proving its industrial relevance (Calderaro et al., 2020). Interestingly, certain AA9s have shown potential even under anaerobic conditions. In a 40-day anaerobic digestion study involving four substrates (Avicel, birchwood, spruce wood, and lignin-rich residue), NcLPMO9C exhibited activity in the absence of oxygen, suggesting a need for deeper investigation into LPMO behavior under anoxic environments (Costa et al., 2019).

Bacterial AA10s exhibit broader substrate specificity, including activity on both cellulose and chitin. Their catalytic centers, often supported by phenylalanine and histidine residues, target primarily the C1 position but can also act on both C1 and C4. AA10s frequently occur alongside CBMs, which tether LPMOs to substrates and enhance catalytic efficiency (Zhang et al., 2023). For instance, BaLPMO10 from *Bacillus amyloliquefaciens* increased sugar yield by 40.5% when paired with cellulase to degrade ethylenediamine-pretreated *Caragana korshinskii* (Guo et al., 2023). Recent studies have further expanded the functional scope of AA10s. CgLPMO10A and CgLPMO10B from *Cellulomonas gelida* displayed novel oxidative and substrate specificities. CgLPMO10A oxidized C1 in both cellulose and chitin, while CgLPMO10B demonstrated both C1 and C4 oxidation and possessed the lowest redox potential among known AA10s, suggesting advanced oxidase capability (Turunen et al., 2025). Similarly, three AA10s from *Cellulomonas flavigena* and one from *Cellulomonas fimi* exhibited thermostability (40–60°C) and refolding capability, valuable for industrial use. Synergy was observed among *C. flavigena* enzymes but not with the *C. fimi* variant, emphasizing strain-specific compatibility (Li et al., 2021). Chitin-specific AA10s remain relatively underexplored. Enriched soil microbiota was mined to identify M2822, a novel chitin-active LPMO with a unique phylogenetic profile. Its inclusion with commercial chitinase increased chitin oligosaccharide yield by 83.6% (Zhang et al., 2023).

AA11 family members, identified in 2014, are fungal LPMOs primarily active on chitin. AfAA11A from Aspergillus fumigatus deviates structurally from typical LPMOs, lacking a conventional Cu-binding site but maintaining catalytic activity on par with bacterial AA10s. Here, Tyr133 and Ala69 coordinate the metal center instead of the usual tyrosine–phenylalanine pair (Støpamo et al., 2021). AA13 enzymes, isolated from fungi, are unique among LPMOs for targeting starch via C1 oxidation of $\alpha(1\rightarrow4)$ glycosidic linkages. Their structure includes a shallow surface groove suited to the helical conformation of starch, and many are associated with CBM20 modules. AA13s have the potential to synergize with amylases to produce value-added starch derivatives (Vu et al., 2019).

More recently, the discovery of AA14 (xylan-active), AA15 (insect gut-derived), AA16, and AA17 has expanded the known diversity of LPMOs. AA14 enzymes from *Pycnoporus coccineus* exhibit activity on cellulose–xylan complexes (Tuveng et al., 2023). AA15s, found in insect digestive and respiratory tracts, are believed to play roles in chitin remodeling and may have supported lignocellulose digestion in ancient insect diets (Vandhana et al., 2022). AA16s and AA17s, though still under exploration, represent promising additions to the LPMO family, with AaAA16 showing C1-oxidation of cellulose (Filiatrault-Chastel et al., 2019). With advances in metagenomics and bioinformatics, continued discovery and characterization of novel LPMOs will undoubtedly broaden our understanding of their ecological roles and industrial potential.

#### 8.1.1.2 *Copper-Centered Catalysis, Regioselectivity, and Role of Electron Donors in LPMO Activity*

LPMOs possess a conserved mono-copper active site anchored by the "histidine brace," comprising two histidines, one at the N-terminus and another internal, forming a surface-exposed T-shaped geometry (Srivastava et al., 2020b). In AA9 LPMOs, His1 and His86 coordinate the copper ion, while AA10s typically involve His1 and His109, supported by His32 (Kumar et al., 2024). The copper site mediates oxidative cleavage via two primary mechanisms: molecular oxygen ($O_2$)-driven and hydrogen peroxide ($H_2O_2$)-driven catalysis. In the $O_2$ pathway, reduction of Cu(II) to Cu(I) enables $O_2$ binding to form a Cu-superoxo species. Without substrate, this intermediate forms $H_2O_2$ and regenerates Cu(II); in the presence of polysaccharides, it generates a Cu–OOH species that attacks the glycosidic bond. Alternatively, the $H_2O_2$ mechanism bypasses the superoxo step, producing a Cu–OH species that initiates direct hydroxylation, although this poses a higher risk of oxidative inactivation (Wang et al., 2020).

LPMOs exhibit distinct regioselectivity: Type 1 enzymes oxidize the C1 position (e.g., PaLPMO9E), Type 2 target C4 (e.g., MtLPMO9J), and Type 3 perform dual C1/C4 oxidation (e.g., PaLPMO9H) (Yu et al., 2023). This specificity, dictated by loop architecture around the copper center, determines the nature of oxidation products, aldonic acids from C1 and ketoaldoses from C4, and significantly impacts downstream saccharification and product processing. Initiation of LPMO catalysis requires an external electron donor to reduce Cu(II) to Cu(I). Small molecules like ascorbic acid, gallic acid, and cysteine are commonly used but may accelerate enzyme inactivation via excess $H_2O_2$ formation. Biological donors such as cellobiose dehydrogenase (CDH) offer controlled electron release and minimize enzyme damage, preserving activity over extended periods (Adnan et al., 2023). Structural studies have shown that uncontrolled oxidation, particularly with ascorbate, causes damage around the copper center, leading to enzyme inactivation (Christensen et al., 2023). Additionally, lignin has been proposed as a natural electron source within biomass, playing a dual role by both impeding and facilitating enzymatic hydrolysis (Harris et al., 2010). Together, these mechanistic insights into copper coordination, oxidative chemistry, and electron donor dynamics are central to optimizing LPMO functionality in industrial biorefineries.

### 8.1.2 Mining of LPMOs from Environmental Microbiomes

Environmental microbiomes represent a rich and underexplored reservoir for discovering functionally diverse LPMOs with potential industrial relevance. Microorganisms adapted to complex and variable ecosystems such as compost, forest litter, marine sediments, and extreme habitats (e.g., thermophilic or saline environments) often produce lignocellulose-degrading enzymes with enhanced stability, catalytic efficiency, and substrate versatility. These niches offer a strategic advantage in mining LPMOs that are not only active under harsh industrial conditions, such as high temperatures, variable pH, or high salt concentrations, but also capable of synergizing with existing enzyme cocktails for efficient biomass deconstruction. Functional metagenomics, combined with high-throughput screening and omics-guided approaches, is increasingly employed to identify novel LPMO genes and elucidate their catalytic properties. This section highlights key studies where LPMOs have been successfully mined from environmental samples and validated for their lignocellulolytic potential, emphasizing their structural diversity, stability traits, and substrate preferences.

#### 8.1.2.1 *Thermostable and Halotolerant LPMOs from Extreme Microbiomes*

Environmental microbiomes, especially those subjected to extreme conditions, are promising reservoirs for robust, industrially relevant LPMOs. Such enzymes often display enhanced thermostability, broad pH tolerance, and salt-resistance, key attributes for effective biomass deconstruction under industrial settings. A thermostable, C1-specific bacterial LPMO (MgLPMO10) was identified from a high-temperature metagenomic dataset enriched with thermophiles. MgLPMO10 remained catalytically active up to 80°C and performed comparably to the widely used fungal TaLPMO9A when co-applied with the commercial cellulase cocktail Celluclast. It exhibited robust performance under varied pH and salt conditions and produced exclusively C1-oxidized oligosaccharides, confirming its specificity (Tuveng et al., 2020). Thermophile-rich compost samples were collected during the thermophilic stage from three composting facilities in Portugal: Terra Fértil (TF), Lipor (LP), and Gintegral (GN). Metagenomic analysis revealed widespread distribution of lignocellulolytic enzymes, including LPMOs, across the different compost types (Srivastava et al., 2021a). Notably, LP and GN samples showed higher abundance of CAZyme-encoding genes compared to TF, correlating with the organic composition of the input materials (Santos-Pereira et al., 2023). In another study, compost systems utilizing different raw materials, sludge with hay (SD/HY), pig manure with cane marrow (PM/CM), and dairy manure with cane marrow (DM/CM), were compared for LPMO gene abundance. The DM/CM compost showed the highest levels of AA10 family genes, attributed to the cellulose-rich plant residues in manure and favorable enrichment conditions, such as those found in ruminant digestive systems (Qin et al., 2023). A native AA9 LPMO was directly purified from the thermophilic fungus *T. aurantiacus*, which is well known for producing heat-stable enzymes. This LPMO retained oxidative activity after prolonged incubation at 60–70°C and synergistically enhanced sugar release when used with hydrolytic enzymes and ascorbate (Fritsche et al., 2020). The C1-specific MtLPMO9J from *Myceliophthora thermophila* was cloned and expressed in *Pichia pastoris* to evaluate its thermophilic potential. It displayed broad compatibility with various electron donors, including ascorbic acid, gallic acid, and lignin-derived compounds, and enhanced cellulose degradation up to ~70°C (Kadowaki et al., 2018). A comparative analysis of three *M. thermophila* AA9s, MtLPMO9B, MtLPMO9E, and MtLPMO9F, revealed substrate-specific activity and varying regioselectivity. MtLPMO9F was the most versatile, showing activity on both cellulose and xyloglucan, while MtLPMO9B and MtLPMO9F generated both C1- and C4-oxidized products. MtLPMO9E was restricted to C1 oxidation (Frommhagen et al., 2016).

Two halotolerant AA9 enzymes, PsLPMOA and PsLPMOB, were identified from the mangrove-derived fungus *Pestalotiopsis* sp. Nci6. These enzymes showed increased activity in saline conditions, likely an adaptation to their native high-salt environment. Both LPMOs efficiently released C1-oxidized oligosaccharides from cellulose, making them attractive for biomass conversion in saline or marine settings (Patel et al., 2016). An AA9 LPMO from the thermophilic fungus *Malbranchea cinnamomea* was structurally characterized to understand the basis of its regioselectivity and thermostability. The study highlighted functional features of dual C1/C4 oxidizing enzymes and contributed to the broader mapping of structure–function relationships in heat-tolerant LPMOs (Mazurkewich et al., 2021).

#### *8.1.2.2 Forest Microbiomes as Reservoirs of LPMO Diversity*

Forests, with their abundant LCB and complex microbial ecosystems, serve as rich ecological niches for mining novel LPMOs. Several studies have demonstrated the presence and functional diversity of LPMOs in forest-associated fungi and microbial consortia. Truong et al. (2024) mined multiple putative LPMOs from soil associated with white-rot fungi in Vietnam's Cuc Phuong Primary Forest. A total of 31 full-length proteins were identified through HMM-based annotation. Two genes, when heterologously expressed in E. coli, showed chitin-active LPMO activity in enzyme assays, suggesting the presence of functional AA10-type LPMOs. Hage et al. (2021) conducted genomic and transcriptomic analysis across 50 fungal species within the Polyporales order, revealing a significant expansion of AA9-encoding genes in early-diverging white-rot lineages. Co-expression of AA9s with CAZymes, CBMs, and oxidoreductases under lignocellulose-inducing conditions confirmed their involvement in wood-litter decomposition. These findings indicate that forest litter environments are hotspots for AA9 diversity.

In another multi-omics investigation, Hori et al. (2018) analyzed the decayed wood of *Pinus contorta* and detected abundant expression of AA9 LPMOs at both transcriptomic and proteomic levels. These enzymes co-occurred with other cellulolytic and oxidative enzymes, indicating a coordinated enzymatic system for effective lignocellulose degradation. Predominant taxa included Basidiomycota fungi and bacterial phyla such as Acidobacteria, Actinobacteria, and Proteobacteria. Liu et al. (2017) isolated two AA9-type LPMOs (HiLPMO9H and HiLPMO9I) from *Heterobasidion irregulare*, a phytopathogenic white-rot fungus. Both enzymes exhibited C1-specific oxidative cleavage of cellulose and synergistically enhanced CBHI-mediated hydrolysis. Gene expression studies showed significant upregulation in the presence of spruce wood, confirming their functional relevance during natural wood decay. Couturier et al. (2018) characterized two AA14 LPMOs (PcAA14A and PcAA14B) from forest-dwelling white- and brown-rot fungi. These enzymes targeted xylan, an abundant hemicellulose in woody biomass, and enhanced glucose yield by 25–40% when supplemented in enzyme cocktails. This revealed the potential of AA14s in synergizing the hydrolytic deconstruction of complex plant cell walls.

#### *8.1.2.3 Marine Sediments: Unveiling Salt-Tolerant and Redox-Flexible LPMOs*

Marine ecosystems, particularly deep-sea and cold-seawater sediments, represent untapped niches for discovering robust LPMOs adapted to extreme conditions such as high pressure, low temperature, and saline environments. Hoang et al. (2024) discovered MaLPMO9K from *Talaromyces sedimenticola*, a deep-sea fungal strain isolated from the Mariana Trench. The enzyme displayed oxidative activity on cellulose substrates even in the absence of external electron donors, suggesting a built-in redox flexibility. Its broad substrate spectrum and peroxidase-like activity highlight the biochemical adaptability of deep-sea LPMOs. Jiang et al. (2022) elucidated a chitin oxidation system involving AA10 LPMO (PpLPMO10A) in *Pseudoalteromonas prydzensis* ACAM 620, a bacterium from Antarctic marine waters. Functional and structural analyses revealed $H_2O_2$-dependent peroxygenase activity and broad substrate recognition. PpLPMO10A increased chitinase accessibility, demonstrating an ecologically significant mechanism for marine carbon and nitrogen cycling. Fowler et al. (2019) characterized TtAA10A, an AA10 LPMO from *Teredinibacter turnerae*, a shipworm symbiont. The enzyme contained a C-terminal CBM10 domain targeting crystalline cellulose and a unique sodium-binding pocket that likely contributes to salt stability. These adaptations underscore the enzyme's potential in salt-tolerant biocatalysis. Zhou and Suginta (2021) provided mechanistic insights into a four-domain AA10 LPMO from *Vibrio campbellii*, featuring a catalytic domain linked to a CBM73 chitin-binding module. Structural analysis revealed domain–domain interactions that facilitate substrate recognition and activation, offering a modular blueprint for LPMO design in marine contexts.

### 8.1.3 Functional and Evolutionary Diversity of LPMOs

LPMOs have undergone significant functional diversification across kingdoms, with evolutionary traces found in early fungal lineages and a broad distribution across bacteria. The earliest evolutionary presence of fungal LPMOs has been documented in basal lineages including Chytridiomycota, *Rhizophlyctis rosea* (Chytridiomycetes), and *Gonapodya prolifera* (Monoblepharidomycetes) as well as in higher fungi such as Ascomycota and Basidiomycota (Lange et al., 2019). AA11 LPMOs have also been identified in basal fungi, indicating early functional emergence (Støpamo et al., 2021).

Among the fungal LPMOs, the AA9 family is the most diverse and extensively studied. Despite their broad catalytic activity on cellulose, xyloglucan, glucomannan, and xylan, the functional diversity of AA9 LPMOs at the species level remains inadequately explored (Qin, Zou, et al., 2023). Sequence variation within this family presents a significant barrier to establishing consistent functional characterizations (Mathieu et al., 2023). AA9 LPMOs are especially prevalent among basidiomycetes, with 8 of the 38 characterized enzymes affiliated with Basidiomycota (Grace Barrios-Gutiérrez et al., 2023). AA9s are usually active on cellulose, though some also utilize cello-oligosaccharides. However, AA9s, only exhibiting activity on C-1 bonds, degrading cello-oligosaccharides, have not been seen so far (Zhang, 2020). Additionally, several AA9 LPMOs exhibit activity against hemicellulosic substrates, particularly xyloglucans, xylans, and glucomannans (Zhang, 2020).

The AA11 family, classified in 2014, was the next to be defined. Initially identified as chitin-active, AA11 LPMOs are functionally under-characterized, although polymeric chitin degradation has been demonstrated for select members (Hall et al., 2025). The catalytic core typically includes a histidine brace with conserved tyrosine and alanine residues. These enzymes also utilize $H_2O_2$ efficiently as a co-substrate and remain active in peroxygenase mode under controlled peroxide concentrations (Nohr, 2023).

The AA13 family, also exclusive to fungi, was described shortly after and contains more than 85 ascomycetous sequences. It is known for its activity on starch. Interestingly, AA13 enzymes show unique methylation at the N-terminal histidine residue, distinguishing their structure from LPMOs identified in different fungal species (Lo Leggio et al., 2015). In 2018, the AA14 family was reported, comprising PcAA14A and PcAA14B from *P. coccineus*. These LPMOs are active on heteroxylans embedded within lignocellulose and require synergistic enzymes such as GH30 xylanases to deconstruct wood fibers effectively. They do not act on soluble xylan alone (Vandhana et al., 2022). AA14 genes have been detected in white-rot fungi (*Pleurotus ostreatus*, *Phanerochaete chrysosporium*, *Trametes versicolor*), brown-rot fungi (*Serpula lacrymans*, *Coniophora puteana*, *Postia placenta*), and certain ascomycetes (Xylariaceae, Hypocreaceae). The absence of CBMs in AA14s likely contributed to their delayed discovery, as they escaped the "module-walk" strategy that identified AA11 and AA13 (Couturier et al., 2018). The AA16 family was more recently introduced, with *AaAA16* from *Aspergillus aculeatus* X273 characterized as a cellulose-active LPMO (Filiatrault-Chastel et al., 2019). However, further investigations have indicated that AA16s may exhibit only low catalytic activity on cellulose and instead act as $H_2O_2$ suppliers for co-expressed AA9 LPMOs (Tuveng et al., 2023).

Unlike fungal LPMOs, bacterial LPMOs are primarily from the AA10 family and have a broader phylogenetic representation and unique structural features. The first reported AA10 protein, CHB1, was identified in *Streptomyces olivaceoviridis* (Sagarika et al., 2022). AA10s include cellulose-active enzymes that are exhibiting exclusive C1 oxidation or dual C1/C4 activity; notably, no strict C4-specific AA10 LPMOs have been identified (Zhou & Zhu, 2020). Additional bacterial AA10s, such as *CelS2* from *Streptomyces coelicolor*, *BlAA10A* from *Bacillus licheniformis*, and E8 from *Thermobifida fusca*, work synergistically with hydrolytic enzymes for efficient biomass breakdown (Book et al., 2014). Taxonomic analyses show that AA10s are distributed across numerous bacterial phyla. CBM domains associated with these enzymes include cellulose-binding modules (CBM2, CBM3) and chitin-binding modules (CBM5, CBM12). Actinobacteria dominate cellulose-active CBM-AA10s (94%), while chitin-specific CBM-containing AA10s are mostly found in Firmicutes and Proteobacteria (Book et al., 2014).

The multiplicity of LPMO genes in fungal genomes has been hypothesized as an evolutionary adaptation to the chemical heterogeneity of LCB (Tuveng et al., 2023). Nonetheless, the biological significance of this gene expansion and how it relates to enzyme function remains a major area of investigation.

### 8.1.4 Biotechnological Applications of Environmental LPMOs

Over the last few decades, LPMOs have significantly influenced biomass processing, offering novel enzymatic routes to overcome the recalcitrance of lignocellulose. LPMOs mark a critical advancement in industrial biotechnology. Their unique mechanism of action, which oxidatively cleaves glycosidic bonds in crystalline polysaccharides, has redefined how we approach the saccharification of plant biomass. The strategic industrial application of LPMOs began with their integration into commercial enzyme formulations designed for biomass hydrolysis (Rani Singhania et al., 2021). Novozymes, a leader in enzyme biotechnology, progressively enhanced its cellulase cocktails to improve efficiency on recalcitrant substrates. The widely used Celluclast®, composed primarily of classical cellulases, was soon surpassed by Cellic® CTec2, which incorporated a significantly higher concentration of β-glucosidases, thus resolving the cellobiose accumulation bottleneck. The breakthrough came with Cellic® CTec3, an advanced formulation containing higher concentrations of AA9 family LPMOs, enabling deeper and more efficient access to cellulose embedded in complex biomass matrices (Rani Singhania et al., 2021). This marked a shift from purely hydrolytic enzymology to one that also leverages oxidative cleavage is a synergistic mechanism that unlocks tightly packed polysaccharide chains and enhances overall hydrolysis.

LPMOs, originally characterized for their action on cellulose, particularly in its crystalline form, are copper-dependent monooxygenases that oxidatively cleave C-1 and/or C-4 glycosidic bonds, using $O_2$ or $H_2O_2$ as co-substrates. The oxidative activity distorts the crystalline lattice, increasing accessibility to classical cellulases and leading to accelerated saccharification. More recently, AA9s are also seen acting on xylan, thereby expanding their utility beyond cellulose. Ipsen et al. (2021) demonstrated the activity of AA9 LPMOs on xylan-rich biomass, highlighting the relevance of LPMOs in the degradation of hemicellulosic components tightly interlinked with cellulose. These findings broaden the biocatalytic potential of LPMOs in deconstructing heteropolysaccharide-rich plant cell walls (Ipsen et al., 2021).

A critical application of LPMOs lies in their role as pretreatment agents in enzymatic biomass processing. Unlike traditional chemical pretreatments, which can be energy-intensive and generate inhibitory byproducts, LPMO-driven oxidation specifically targets crystalline regions of cellulose, disrupting the hydrogen-bonding network without significant damage to fiber integrity. This results in a more accessible substrate for hydrolytic enzymes, greatly enhancing sugar release. Studies have shown that pretreatment of spruce with AA9 LPMOs from *M. thermophila* or *Scytalidium thermophilum* resulted in a 20% increase in sugar release (Moreau et al., 2019). Similarly, the synergistic action of LPMOs with hydrolases has been reported for several lignocellulosic feedstocks, including wheat straw, poplar, sugarcane bagasse, lodgepole pine, and maize stover (Sagarika et al., 2022). In each case, the oxidative modification of cellulose by LPMOs significantly boosted saccharification efficiency when compared to the use of cellulases alone. Furthermore, Sharma et al. (2019) highlighted that LPMOs, when used as part of a synergistic enzymatic blend, accelerate the liberation of fermentable sugars by oxidatively attacking recalcitrant regions that resist hydrolytic cleavage (Sharma et al., 2019). This synergy is essential in biorefinery applications, where maximizing sugar yield from heterogeneous biomass is crucial for cost-effective biofuel production (Jhariya et al., 2021; Srivastava et al., 2021b).

Beyond conventional biofuel applications, LPMOs have demonstrated potential in nanocellulose production, offering a sustainable pretreatment strategy for generating nanoscale cellulose fibers. The fungal enzyme *Pa*LPMO9E, for instance, has been effectively used to oxidize cellulose at the C1 position, facilitating mechanical fibrillation while maintaining fiber morphology and crystalline integrity (Moreau et al., 2019). This treatment not only enhances the liberation of nanofibrils but also preserves the mechanical properties and functional characteristics of nanofibrillated cellulose. The application of LPMOs in this context reflects their broader industrial potential beyond saccharification, in sectors ranging from bioplastics to biomedical materials.

As new substrates, electron donors, and redox partners continue to be elucidated, the full potential of LPMOs in sustainable biomass utilization is only beginning to be realized. Their adoption in modern biorefineries underscores not only their functional importance but also their economic and environmental value in the global transition to bio-based industries.

## 8.2 CONCLUSION

LPMOs have emerged as pivotal enzymes in overcoming the recalcitrance of LCB, offering a paradigm shift in biocatalytic stubble transformation through oxidative cleavage. The current chapter highlights their structural features, catalytic versatility, and synergistic role with GHs. Emphasis on metagenomic mining from extreme, forest, and marine microbiomes underscores the untapped diversity of robust, industrially promising LPMOs. Furthermore, insights into electron donor specificity, substrate binding, and regioselectivity deepen the insights into the mechanistic complexity of LPMOs. Despite rapid progress, knowledge gaps remain in substrate recognition, inactivation pathways, and in vivo redox partners. Future integration of omics, protein engineering, and structural biology will be key to harnessing LPMOs for scalable biorefinery applications. The present chapter positions environmental microbiomes as rich reservoirs for discovering next-generation LPMOs, accelerating innovation in biomass valorization and sustainable biofuel production.

## REFERENCES


Adnan, M., Ma, X., Xie, Y., Waheed, A., & Liu, G. (2023). Heterologously expressed cellobiose dehydrogenase acts as efficient electron-donor of lytic polysaccharide monooxygenase for cellulose degradation in trichoderma reesei. *International Journal of Molecular Sciences*, 24(24), 17202. https://doi.org/10.3390/ijms242417202

Angeltveit, C. F., Várnai, A., Eijsink, V. G. H., & Horn, S. J. (2024). Enhancing enzymatic saccharification yields of cellulose at high solid loadings by combining different LPMO activities. *Biotechnology for Biofuels and Bioproducts*, 17(1), 39. https://doi.org/10.1186/s13068-024-02485-6

Bissaro, B., Várnai, A., Røhr, Å. K., & Eijsink, V. G. H. (2018). Oxidoreductases and reactive oxygen species in conversion of lignocellulosic biomass. *Microbiology and Molecular Biology Reviews*, 82(4), e00029–18. https://doi.org/10.1128/MMBR.00029-18

Book, A. J., Yennamalli, R. M., Takasuka, T. E., Currie, C. R., Phillips, G. N., & Fox, B. G. (2014). Evolution of substrate specificity in bacterial AA10 lytic polysaccharide monooxygenases. *Biotechnology for Biofuels*, 7(1), 109. https://doi.org/10.1186/1754-6834-7-109

Calderaro, F., Keser, M., Akeroyd, M., Bevers, L. E., Eijsink, V. G. H., Várnai, A., & Van Den Berg, M. A. (2020). Characterization of an AA9 LPMO from Thielavia australiensis, TausLPMO9B, under industrially relevant lignocellulose saccharification conditions. *Biotechnology for Biofuels*, 13(1), 195. https://doi.org/10.1186/s13068-020-01836-3

Christensen, I. A., Eijsink, V. G. H., Stepnov, A. A., Courtade, G., & Aachmann, F. L. (2023). Following the fate of lytic polysaccharide monooxygenases under oxidative conditions by NMR spectroscopy. *Biochemistry*, 62(12), 1976–1993. https://doi.org/10.1021/acs.biochem.3c00089

Chylenski, P., Bissaro, B., Sørlie, M., Røhr, Å. K., Várnai, A., Horn, S. J., & Eijsink, V. G. H. (2019). Lytic polysaccharide monooxygenases in enzymatic processing of lignocellulosic biomass. *ACS Catalysis*, 9(6), 4970–4991. https://doi.org/10.1021/acscatal.9b00246

Costa, T. H. F., Eijsink, V. G. H., & Horn, S. J. (2019). The use of lytic polysaccharide monooxygenases in anaerobic digestion of lignocellulosic materials. *Biotechnology for Biofuels*, 12(1), 270. https://doi.org/10.1186/s13068-019-1611-8

Couturier, M., Ladevèze, S., Sulzenbacher, G., Ciano, L., Fanuel, M., Moreau, C., Villares, A., Cathala, B., Chaspoul, F., Frandsen, K. E., Labourel, A., Herpoël-Gimbert, I., Grisel, S., Haon, M., Lenfant, N., Rogniaux, H., Ropartz, D., Davies, G. J., Rosso, M.-N., Walton, P.H., Henrissat B., Berrin, J.-G. (2018). Lytic xylan oxidases from wood-decay fungi unlock biomass degradation. *Nature Chemical Biology*, 14(3), 306–310. https://doi.org/10.1038/nchembio.2558

Filiatrault-Chastel, C., Navarro, D., Haon, M., Grisel, S., Herpoël-Gimbert, I., Chevret, D., Fanuel, M., Henrissat, B., Heiss-Blanquet, S., Margeot, A., & Berrin, J.-G. (2019). AA16, a new lytic polysaccharide monooxygenase family identified in fungal secretomes. *Biotechnology for Biofuels*, 12(1), 55. https://doi.org/10.1186/s13068-019-1394-y

Fowler, C. A., Sabbadin, F., Ciano, L., Hemsworth, G. R., Elias, L., Bruce, N., McQueen-Mason, S., Davies, G. J., & Walton, P. H. (2019). Discovery, activity and characterisation of an AA10 lytic polysaccharide oxygenase from the shipworm symbiont Teredinibacter turnerae. *Biotechnology for Biofuels*, 12(1), 232. https://doi.org/10.1186/s13068-019-1573-x

Fritsche, S., Hopson, C., Gorman, J., Gabriel, R., & Singer, S. W. (2020). Purification and characterization of a native lytic polysaccharide monooxygenase from Thermoascus aurantiacus. *Biotechnology Letters*, 42(10), 1897–1905. https://doi.org/10.1007/s10529-020-02942-w

Frommhagen, M., Koetsier, M. J., Westphal, A. H., Visser, J., Hinz, S. W. A., Vincken, J.-P., Van Berkel, W. J. H., Kabel, M. A., & Gruppen, H. (2016). Lytic polysaccharide monooxygenases from Myceliophthora thermophila C1 differ in substrate preference and reducing agent specificity. *Biotechnology for Biofuels*, 9(1), 186. https://doi.org/10.1186/s13068-016-0594-y

Grace Barrios-Gutiérrez, S., Inés Vélez-Mercado, M., Rodrigues Ortega, J., Da Silva Lima, A., Luiza Da Rocha Fortes Saraiva, A., Leila Berto, G., & Segato, F. (2023). Oxidative machinery of basidiomycetes as potential enhancers in lignocellulosic biorefineries: A lytic polysaccharide monooxygenases approach. *Bioresource Technology*, 386, 129481. https://doi.org/10.1016/j.biortech.2023.129481

Guo, X., An, Y., Lu, F., Liu, F., & Wang, B. (2023). Efficient secretory production of lytic polysaccharide monooxygenase BaLPMO10 and its application in plant biomass conversion. *International Journal of Molecular Sciences*, 24(11), 9710. https://doi.org/10.3390/ijms24119710

Hage, H., Miyauchi, S., Virágh, M., Drula, E., Min, B., Chaduli, D., Navarro, D., Favel, A., Norest, M., Lesage-Meessen, L., Bálint, B., Merényi, Z., De Eugenio, L., Morin, E., Martínez, A. T., Baldrian, P., Štursová, M., Martínez, M. J., Novotny, C., ... Rosso, M. (2021). Gene family expansions and transcriptome signatures uncover fungal adaptations to wood decay. *Environmental Microbiology*, 23(10), 5716–5732. https://doi.org/10.1111/1462-2920.15423

Hall, K. R., Elisa Rønnekleiv, S., Gautieri, A., Lilleås, H., Skaali, R., Rieder, L., Nikoline Englund, A., Landsem, E., Emrich-Mills, T. Z., Ayuso-Fernández, I., Kjendseth Røhr, Å., Sørlie, M., & Eijsink, V. G. H. (2025). Structure–function analysis of an understudied type of LPMO with unique redox properties and substrate specificity. *ACS Catalysis*, 15(12), 10601–10617. https://doi.org/10.1021/acscatal.5c03003

Haq, I., Qaisar, K., Nawaz, A., Akram, F., Mukhtar, H., Zohu, X., Xu, Y., Mumtaz, M., Rashid, U., Ghani, W., & Choong, T. (2021). Advances in valorization of lignocellulosic biomass towards energy generation. *Catalysts*, 11(3), 309. https://doi.org/10.3390/catal11030309

Harris, P.V., Welner, D., McFarland, K. C., Re, E., Navarro Poulsen, J.-C., Brown, K., Salbo, R., Ding, H., Vlasenko, E., Merino, S., Xu, F., Cherry, J., Larsen, S., & Lo Leggio, L. (2010). Stimulation of lignocellulosic biomass hydrolysis by proteins of glycoside hydrolase family 61: Structure and function of a large, enigmatic family. *Biochemistry*, 49(15), 3305–3316. https://doi.org/10.1021/bi100009p

Hoang, H., Liu, W., Zhan, W., Zou, S., Xu, L., Zhan, Y., Cheng, H., Chen, Z., Zhou, H., & Wang, Y. (2024). Reductant-independent oxidative cleavage of cellulose by a novel marine fungal lytic polysaccharide monooxygenase. *International Journal of Biological Macromolecules*, 276, 133929. https://doi.org/10.1016/j.ijbiomac.2024.133929

Hori, C., Gaskell, J., Cullen, D., Sabat, G., Stewart, P. E., Lail, K., Peng, Y., Barry, K., Grigoriev, I. V., Kohler, A., Fauchery, L., Martin, F., Zeiner, C. A., & Bhatnagar, J. M. (2018). Multi-omic analyses of extensively decayed pinus contorta reveal expression of a diverse array of lignocellulose-degrading enzymes. *Applied and Environmental Microbiology*, 84(20), e01133–e011318. https://doi.org/10.1128/AEM.01133-18

Hu, J., Arantes, V., Pribowo, A., Gourlay, K., & Saddler, J. N. (2014). Substrate factors that influence the synergistic interaction of AA9 and cellulases during the enzymatic hydrolysis of biomass. *Energy and Environmental Science*, 7(7), 2308–2315. https://doi.org/10.1039/C4EE00891J

Ipsen, J. Ø., Hallas-Møller, M., Brander, S., Lo Leggio, L., & Johansen, K. S. (2021). Lytic polysaccharide monooxygenases and other histidine-brace copper proteins: Structure, oxygen activation and biotechnological applications. *Biochemical Society Transactions*, 49(1), 531–540. https://doi.org/10.1042/BST20201031

Jhariya, U., Dafale, N. A., Srivastava, S., Bhende, R. S., Kapley, A., & Purohit, H. J. (2021). Understanding ethanol tolerance mechanism in saccharomyces cerevisiae to enhance the bioethanol production: Current and future prospects. *BioEnergy Research*, 14(2), 670–688. https://doi.org/10.1007/s12155-020-10228-2

Jiang, W.-X., Li, P.-Y., Chen, X.-L., Zhang, Y.-S., Wang, J.-P., Wang, Y.-J., Sheng, Q., Sun, Z.-Z., Qin, Q.-L., Ren, X.-B., Wang, P., Song, X.-Y., Chen, Y., & Zhang, Y.-Z. (2022). A pathway for chitin oxidation in marine bacteria. *Nature Communications*, 13(1), 5899. https://doi.org/10.1038/s41467-022-33566-5

Kadowaki, M. A. S., Várnai, A., Jameson, J.-K., T. Leite, A. E., Costa-Filho, A. J., Kumagai, P. S., Prade, R. A., Polikarpov, I., & Eijsink, V. G. H. (2018). Functional characterization of a lytic polysaccharide monooxygenase from the thermophilic fungus Myceliophthora thermophila. *Plos One*, 13(8), e0202148. https://doi.org/10.1371/journal.pone.0202148

Kumar, A., Singh, A., Sharma, V. K., Goel, A., & Kumar, A. (2024). The upsurge of lytic polysaccharide monooxygenases in biomass deconstruction: Characteristic functions and sustainable applications. *The FEBS Journal*, 291(23), 5081–5101. https://doi.org/10.1111/febs.17063

Lange, L., Pilgaard, B., Herbst, F.-A., Busk, P. K., Gleason, F., & Pedersen, A. G. (2019). Origin of fungal biomass degrading enzymes: Evolution, diversity and function of enzymes of early lineage fungi. *Fungal Biology Reviews*, 33(1), 82–97. https://doi.org/10.1016/j.fbr.2018.09.001

Li, J., Solhi, L., Goddard-Borger, E. D., Mathieu, Y., Wakarchuk, W. W., Withers, S. G., & Brumer, H. (2021). Four cellulose-active lytic polysaccharide monooxygenases from Cellulomonas species. *Biotechnology for Biofuels*, 14(1), 29. https://doi.org/10.1186/s13068-020-01860-3

Liu, B., Olson, Å., Wu, M., Broberg, A., & Sandgren, M. (2017). Biochemical studies of two lytic polysaccharide monooxygenases from the white-rot fungus Heterobasidion irregulare and their roles in lignocellulose degradation. *Plos One*, 12(12), e0189479. https://doi.org/10.1371/journal.pone.0189479

Lo Leggio, L., Simmons, T. J., Poulsen, J.-C. N., Frandsen, K. E. H., Hemsworth, G. R., Stringer, M. A., Von Freiesleben, P., Tovborg, M., Johansen, K. S., De Maria, L., Harris, P. V., Soong, C.-L., Dupree, P., Tryfona, T., Lenfant, N., Henrissat, B., Davies, G. J., & Walton, P. H. (2015). Structure and boosting activity of a starch-degrading lytic polysaccharide monooxygenase. *Nature Communications*, 6(1), 5961. https://doi.org/10.1038/ncomms6961

Long, L., Hu, Y., Sun, F., Gao, W., Hao, Z., & Yin, H. (2022). Advances in lytic polysaccharide monooxygenases with the cellulose-degrading auxiliary activity family 9 to facilitate cellulose degradation for biorefinery. *International Journal of Biological Macromolecules*, 219, 68–83. https://doi.org/10.1016/j.ijbiomac.2022.07.240

Long, L., Sun, L., Ding, D., Chen, K., Lin, Q., & Ding, S. (2021). Two C1-oxidizing lytic polysaccharide monooxygenases from Ceriporiopsis subvermispora enhance the saccharification of wheat straw by a commercial cellulase cocktail. *Process Biochemistry*, 110, 243–250. https://doi.org/10.1016/j.procbio.2021.08.013

Mathieu, Y., Raji, O., Bellemare, A., Di Falco, M., Nguyen, T. T. M., Viborg, A. H., Tsang, A., Master, E., & Brumer, H. (2023). Functional characterization of fungal lytic polysaccharide monooxygenases for cellulose surface oxidation. *Biotechnology for Biofuels and Bioproducts*, 16(1), 132. https://doi.org/10.1186/s13068-023-02383-3

Mazurkewich, S., Seveso, A., Hüttner, S., Brändén, G., & Larsbrink, J. (2021). Structure of a C1/C4-oxidizing AA9 lytic polysaccharide monooxygenase from the thermophilic fungus *Malbranchea cinnamomea*. *Acta Crystallographica Section D Structural Biology*, 77(8), 1019–1026. https://doi.org/10.1107/S2059798321006628

Moreau, C., Tapin-Lingua, S., Grisel, S., Gimbert, I., Le Gall, S., Meyer, V., Petit-Conil, M., Berrin, J.-G., Cathala, B., & Villares, A. (2019). Lytic polysaccharide monooxygenases (LPMOs) facilitate cellulose nanofibrils production. *Biotechnology for Biofuels*, 12(1), 156. https://doi.org/10.1186/s13068-019-1501-0

Müller, G., Várnai, A., Johansen, K. S., Eijsink, V. G. H., & Horn, S. J. (2015). Harnessing the potential of LPMO-containing cellulase cocktails poses new demands on processing conditions. *Biotechnology for Biofuels*, 8(1), 187. https://doi.org/10.1186/s13068-015-0376-y

Nohr, H. (2023). *Lytic Polysaccharide Monooxygenases (LPMOs): A Comprehensive Overview of Emerging Families and Their Implications*. Norwegian University of Life Sciences, As, Norway.

Ossowicki, A., Raaijmakers, J. M., & Garbeva, P. (2021). Disentangling soil microbiome functions by perturbation. *Environmental Microbiology Reports*, 13(5), 582–590. https://doi.org/10.1111/1758-2229.12989

Panthee, B., Gyawali, S., Panth Ossowickiee, P., & Techato, K. (2022). Environmental and human microbiome for health. *Life*, 12(3), 456. https://doi.org/10.3390/life12030456

Patel, I., Kracher, D., Ma, S., Garajova, S., Haon, M., Faulds, C. B., Berrin, J.-G., Ludwig, R., & Record, E. (2016). Salt-responsive lytic polysaccharide monooxygenases from the mangrove fungus Pestalotiopsis sp. NCi6. *Biotechnology for Biofuels*, 9(1), 108. https://doi.org/10.1186/s13068-016-0520-3

Qin, X., Chen, X., & Li, Q. (2023). The evolution and profile of AA10 lytic polysaccharide monooxygenase coupled with cellulose decomposition in different composting microenvironments. *Research Square*, https://doi.org/10.21203/rs.3.rs-3478593/v1

Qin, X., Zou, J., Yang, K., Li, J., Wang, X., Tu, T., Zhang, J., Su, X., Yao, B., Huang, H., & Luo, H. (2023). Discovering functional diversity of lytic polysaccharide monooxygenases from the thermophilic fungus *Myceliophthora thermophila* and their application in lignocellulosic biomass degradation. *ACS Sustainable Chemistry & Engineering*, 11(29), 10929–10937. https://doi.org/10.1021/acssuschemeng.3c03014

Rani Singhania, R., Dixit, P., Kumar Patel, A., Shekher Giri, B., Kuo, C.-H., Chen, C.-W., & Di Dong, C. (2021). Role and significance of lytic polysaccharide monooxygenases (LPMOs) in lignocellulose deconstruction. *Bioresource Technology*, 335, 125261. https://doi.org/10.1016/j.biortech.2021.125261

Sagarika, M. S., Parameswaran, C., Senapati, A., Barala, J., Mitra, D., Prabhukarthikeyan, S. R., Kumar, A., Nayak, A. K., & Panneerselvam, P. (2022). Lytic polysaccharide monooxygenases (LPMOs) producing microbes: A novel approach for rapid recycling of agricultural wastes. *Science of the Total Environment*, 806, 150451. https://doi.org/10.1016/j.scitotenv.2021.150451

Santos-Pereira, C., Sousa, J., Costa, Â. M. A., Santos, A. O., Rito, T., Soares, P., Franco-Duarte, R., Silvério, S. C., & Rodrigues, L. R. (2023). Functional and sequence-based metagenomics to uncover carbohydrate-degrading enzymes from composting samples. *Applied Microbiology and Biotechnology*, 107(17), 5379–5401. https://doi.org/10.1007/s00253-023-12627-9

Sharma, H. K., Xu, C., & Qin, W. (2019). Biological pretreatment of lignocellulosic biomass for biofuels and bioproducts: An overview. *Waste and Biomass Valorization*, 10(2), 235–251. https://doi.org/10.1007/s12649-017-0059-y

Srivastava, S., Bombaywala, S., Jakhesara, S. J., Patil, N. V., Joshi, C. G., Purohit, H. J., & Dafale, N. A. (2023b). Potential of camel rumen derived Bacillus subtilis and Bacillus velezensis strains for application in plant biomass hydrolysis. *Molecular Genetics and Genomics*, 298(2), 361–374. https://doi.org/10.1007/s00438-022-01987-y

Srivastava, S., & Dafale, N. A. (2024a). Genomic dissection of Niallia sp. For potential application in lignocellulose hydrolysis and bioremediation. *Archives of Microbiology*, 206(1), 2. https://doi.org/10.1007/s00203-023-03728-0

Srivastava, S., & Dafale, N. A. (2024b). Tailored microbial consortium producing hydrolytic enzyme cocktail for maximum saccharification of wheat straw. *Bioresource Technology*, 399, 130560. https://doi.org/10.1016/j.biortech.2024.130560

Srivastava, S., Dafale, N. A., Jakhesara, S. J., Joshi, C. G., Patil, N. V., & Purohit, H. J. (2021a). Unraveling the camel rumen microbiome through metaculturomics approach for agriculture waste hydrolytic potential. *Archives of Microbiology*, 203(1), 107–123. https://doi.org/10.1007/s00203-020-02010-x

Srivastava, S., Dafale, N., & Purohit, H. (2020a). Insight into the lytic polysaccharide monooxygenases for plant biomass valorization. https://doi.org/10.22541/au.158379523.32546673

Srivastava, S., Dafale, N. A., & Purohit, H. J. (2020b). Functional genomics assessment of lytic polysaccharide mono-oxygenase with glycoside hydrolases in Paenibacillus dendritiformis CRN18. *International Journal of Biological Macromolecules*, 164, 3729–3738. https://doi.org/10.1016/j.ijbiomac.2020.08.147

Srivastava, S., Dafale, N. A., Tulsani, N., Jakhesara, S. J., Patil, N. V., Joshi, C. G., & Purohit, H. J. (2021b). Evaluation of designed consortium SNH-1 for efficient hydrolysis of agriculture waste to benefit bioethanol production. *Journal of Cleaner Production*, 288, 125601. https://doi.org/10.1016/j.jclepro.2020.125601

Srivastava, S., Jhariya, U., Purohit, H. J., & Dafale, N. A. (2023a). Synergistic action of lytic polysaccharide monooxygenase with glycoside hydrolase for lignocellulosic waste valorization: A review. *Biomass Conversion and Biorefinery*, 13(10), 8727–8745. https://doi.org/10.1007/s13399-021-01736-y

Støpamo, F. G., Røhr, Å. K., Mekasha, S., Petrović, D. M., Várnai, A., & Eijsink, V. G. H. (2021). Characterization of a lytic polysaccharide monooxygenase from Aspergillus fumigatus shows functional variation among family AA11 fungal LPMOs. *Journal of Biological Chemistry*, 297(6), 101421. https://doi.org/10.1016/j.jbc.2021.101421

Truong, N.-H., Le, T.-T.-H., Nguyen, H.-D., Nguyen, H.-T., Dao, T.-K., Tran, T.-M.-N., Tran, H.-L., Nguyen, D.-T., Nguyen, T.-Q., Phan, T.-H.-T., Do, T.-H., Phan, N.-H., Ngo, T.-C.-N., & Vu, V.-V. (2024). Sequence and structure analyses of lytic polysaccharide monooxygenases mined from metagenomic DNA of humus samples around white-rot fungi in Cuc Phuong tropical forest, Vietnam. *PeerJ*, 12, e17553. https://doi.org/10.7717/peerj.17553

Turunen, R., Tuveng, T. R., Forsberg, Z., Schiml, V. C., Eijsink, V. G. H., & Arntzen, M. Ø. (2025). Functional characterization of two AA10 lytic polysaccharide monooxygenases from *Cellulomonas gelida*. *Protein Science*, 34(3), e70060. https://doi.org/10.1002/pro.70060

Tuveng, T. R., Jensen, M. S., Fredriksen, L., Vaaje-Kolstad, G., Eijsink, V. G. H., & Forsberg, Z. (2020). A thermostable bacterial lytic polysaccharide monooxygenase with high operational stability in a wide temperature range. *Biotechnology for Biofuels*, 13(1), 194. https://doi.org/10.1186/s13068-020-01834-5

Tuveng, T. R., Østby, H., Tamburrini, K. C., Bissaro, B., Hegnar, O. A., Stepnov, A. A., Várnai, A., Berrin, J., & Eijsink, V. G. H. (2023). Revisiting the AA14 family of lytic polysaccharide monooxygenases and their catalytic activity. *FEBS Letters*, 597(16), 2086–2102. https://doi.org/10.1002/1873-3468.14694

Vaaje-Kolstad, G., Westereng, B., Horn, S. J., Liu, Z., Zhai, H., Sørlie, M., & Eijsink, V. G. H. (2010). An oxidative enzyme boosting the enzymatic conversion of recalcitrant polysaccharides. *Science*, 330(6001), 219–222. https://doi.org/10.1126/science.1192231

Vandhana, T. M., Reyre, J., Sushmaa, D., Berrin, J., Bissaro, B., & Madhuprakash, J. (2022). On the expansion of biological functions of lytic polysaccharide monooxygenases. *New Phytologist*, 233(6), 2380–2396. https://doi.org/10.1111/nph.17921

Vu, V. V., Hangasky, J. A., Detomasi, T. C., Henry, S. J. W., Ngo, S. T., Span, E. A., & Marletta, M. A. (2019). Substrate selectivity in starch polysaccharide monooxygenases. *Journal of Biological Chemistry*, 294(32), 12157–12166. https://doi.org/10.1074/jbc.RA119.009509

Wang, B., Wang, Z., Davies, G. J., Walton, P. H., & Rovira, C. (2020). Activation of $O_2$ and $H_2$ $O_2$ by Lytic Polysaccharide Monooxygenases. *ACS Catalysis*, 10(21), 12760–12769. https://doi.org/10.1021/acscatal.0c02914

Wardman, J. F., Bains, R. K., Rahfeld, P., & Withers, S. G. (2022). Carbohydrate-active enzymes (CAZymes) in the gut microbiome. *Nature Reviews Microbiology*, 20(9), 542–556. https://doi.org/10.1038/s41579-022-00712-1

Yu, X., Zhao, Y., Yu, J., & Wang, L. (2023). Recent advances in the efficient degradation of lignocellulosic metabolic networks by lytic polysaccharide monooxygenase. *Acta Biochimica et Biophysica Sinica*, 55(4), 529–539. https://doi.org/10.3724/abbs.2023059

Zhang, H., Zhou, Z., Lou, T., Xiang, R., Zhang, D., Wang, D., & Wang, S. (2023). Advance in heterologous expression of biomass-degrading auxiliary activity 10 family of lytic polysaccharide monooxygenases. *Fermentation*, 9(9), 795. https://doi.org/10.3390/fermentation9090795

Zhang, R. (2020). Functional characterization of cellulose-degrading AA9 lytic polysaccharide monooxygenases and their potential exploitation. *Applied Microbiology and Biotechnology*, 104(8), 3229–3243. https://doi.org/10.1007/s00253-020-10467-5

Zhou, Y., Suginta, W. (2021). Chitin binding study of a lytic polysaccharide monooxygenase from the marine bacterium Vibrio campbellii. 7th International Conference on Biochemistry and Molecular Biology. 2021.

Zhou, X., & Zhu, H. (2020). Current understanding of substrate specificity and regioselectivity of LPMOs. *Bioresources and Bioprocessing*, 7(1), 11. https://doi.org/10.1186/s40643-020-0300-6

CHAPTER 9

# Genomic Diversity and Novel Findings of Chitinase-Producing Bacteria in Yok Don National Park

*Dinh Minh Tran*

## 9.1 INTRODUCTION

Microbial genomics has become an essential pillar of modern sustainable agriculture, providing high-resolution insight into the metabolic capacities, ecological functions, and plant-associated traits of soil and rhizosphere microorganisms (Berg et al. 2020). Advances in whole-genome sequencing have accelerated the discovery of microbial genes responsible for nutrient cycling, phytohormone production, stress tolerance, and disease suppression (Fitzpatrick et al. 2018; van Dijk et al. 2018). Comparative genomics has revealed key determinants of nitrogen fixation, including *nif* and *fix* gene clusters in *Rhizobium* and *Azotobacter* (Dos Santos et al. 2012), and has clarified microbial contributions to phosphorus, potassium, and micronutrient solubilization through genes encoding phytases, organic-acid biosynthesis, and silicate-solubilizing enzymes (Rodríguez and Fraga 1999). Similarly, pathways governing microbial synthesis of auxins, cytokinins, gibberellins, and 1-aminocyclopropane-1-carboxylate (ACC) deaminase underpin improvements in root development, plant vigor, and abiotic stress tolerance (Glick 2012). Genome-enabled mining of microbial secondary metabolite biosynthetic gene clusters (BGCs) has further advanced the identification of novel antimicrobial compounds from genera such as *Bacillus* and *Streptomyces* (Rutledge and Challis 2015). Complementarily, microbial elicitors, including flagellin, lipopolysaccharides, and chitin fragments, are increasingly recognized as important activators of induced systemic resistance in plants (Pieterse et al. 2014). Genomic studies also shed light on microbial stress-adaptation strategies, such as genes for osmoprotectant biosynthesis, exopolysaccharide formation, sodium antiporters, heat-shock proteins, and metal resistance (Berendsen et al. 2012; Etesami and Beattie 2018; Rajkumar et al. 2012; Vurukonda et al. 2016). Therefore, microbial genomics plays a crucial role in modern agriculture by enabling the identification, characterization, and utilization of plant-beneficial microbes, holding immense potential to enhance nutrient efficiency, stress resilience, disease resistance, and overall crop productivity in the face of climate change and increasing global food demand.

Chitin, the second most abundant biopolymer in nature after cellulose, is a major structural component of arthropod exoskeletons, crustacean shells, fungal cell walls, and nematode eggshells (Dhillon et al. 2013; Synowiecki and Al-Khateeb 2003). Chitinases (EC 3.2.1.14) are enzymes that hydrolyze chitin into soluble oligosaccharides. These enzymes are widely distributed in organisms and play essential roles in nutrient cycling, development, defense, and ecological interactions. Most characterized chitinases were reported to be glycoside hydrolase (GH) families 18 and 19 (GH18 and



DOI: 10.1201/9781003637950-9

GH19) based on conserved catalytic domains and structural features (Drula et al. 2022). Bacterial chitinases suppressed soilborne pathogens, including a wide range of phytopathogenic fungi and nematodes (Pentekhina et al. 2020; Tran et al. 2018, 2022a, b, 2023a; Veliz et al. 2017). In addition to their chitinolytic activity, chitin-degrading bacteria also synthesize plant growth-promoting metabolites and diverse antifungal compounds (Do et al. 2023; Dukare et al. 2020; Gu et al. 2017; Tran et al. 2023a, b). Owing to this combination of properties, chitin-degrading bacteria have gained significant attention as environmentally friendly bio-agents for nutrient recycling, soil health, ecological balance, and sustainable agriculture, functioning effectively as plant growth-promoting inoculants and biological control agents (Le and Yang 2019).

The Central Highlands possesses great agricultural potential and is the main region for producing valuable export crops of Vietnam, including coffee and black pepper. However, long-term dependence on chemical fertilizers has resulted in declining soil health, nutrient imbalance, and environmental degradation. Harnessing indigenous beneficial microorganisms offers a promising approach to restoring soil fertility while reducing chemical inputs. To this end, we undertook a long-term program to isolate, characterize, and sequence chitinase-producing bacteria from Yok Don National Park, the largest dry-dipterocarp forest ecosystem in Vietnam. This unique environment supports microbial assemblages shaped by seasonal drought, nutrient limitation, and high ecological complexity, providing ideal conditions for the evolution of specialized metabolic traits. Our studies identified numerous bacteria with strong chitinolytic capacity, plant-growth-promoting activity, antifungal potential, and, in several cases, previously undescribed taxonomic features (Tran et al. 2023a, 2024b, 2025b). This chapter summarizes the genomic characteristics of chitinolytic bacteria isolated from Yok Don National Park, including *Paenibacillus chitinolyticus* YSY-3.1, *Paenibacillus tyrfis* YSS-72.2.G2, *Paenibacillus elgii* YSY-1.2, *Paenibacillus illinoisensis* YWY-3.1, *Brevibacillus brevis* YSY-3.4, *Brevibacillus formosus* YSY-2.2, *Chitiniphilus shinanonensis* YWS-5.2, and a possibly novel species *Paenibacillus* sp. YSY-4.3. These genomes revealed novel enzyme architectures, diverse metabolic strategies, and great potential for applications in agriculture, ecology, and industry.

## 9.2 THE GENOMIC CHARACTERISTICS OF *PAENIBACILLUS CHITINOLYTICUS* YSY-3.1

*P. chitinolyticus* YSY-3.1 exhibited high chitinase activity, strong biofilm formation, broad extracellular hydrolytic enzyme activities (proteases, cellulases, and amylases), and the ability to produce siderophores and phytohormones at a high level, including indole-3-acetic acid (IAA), gibberellic acid ($GA_3$), and zeatin. The *P. chitinolyticus* YSY-3.1 genome included 6.57 Mb with 52.2% GC content. DFAST annotation predicted 6,115 protein-coding genes and 79 tRNAs, while PATRIC annotation identified 6,631 protein-coding genes, 79 tRNAs, and 3 rRNAs. Of the 6,631 protein-coding genes predicted by PATRIC, 2,883 were hypothetical proteins and 3,743 were assigned specific functions. Average nucleotide identity (ANI) analysis showed 96.53% similarity to *P. chitinolyticus* KCCM 41400, confirming species-level classification. The genome contained 219 carbohydrate-active enzymes (CAZymes), including 89 GHs, 60 glycosyl transferases (GTs), 9 polysaccharide lyases (PLs), 31 carbohydrate esterases (CEs), 11 auxiliary activity enzymes (AAs), and 25 carbohydrate-binding modules (CBMs) (Table 9.1). Notably, the genome harbored the novel chitinolytic system composed of 22 enzymes, including 14 GH18 chitinases, one GH19 chitinase, 2 GH20 β-N-acetylglucosaminidases, and 5 AA10 lytic polysaccharide monooxygenases. Of the enzymes, the GH19 chitinase possessed a unique loop structure distinct from known GH19 enzymes; one chitinase contained two GH18 catalytic domains; and another featured three GH18 catalytic domains (Figure 9.1A and Table 9.2), contrasting with typical bacterial chitinases, which generally contain only one catalytic domain (Tran et al. 2023a). These findings underscored the remarkable genomic and functional diversity of *P. chitinolyticus* YSY-3.1, supporting great potential for biotechnological and agricultural applications.

## 9.3 THE GENOMIC CHARACTERISTICS OF *PAENIBACILLUS TYRFIS* YSS-72.2.G2

*P. tyrfis* YSS-72.2.G2 exhibited strong chitinase production, broad hydrolytic enzyme activities, diverse plant-growth-promoting traits and pronounced antagonism against phytopathogenic fungi (Tran

TABLE 9.1
*General genomic features of chitinase-producing bacteria.*

| Features | YSY-3.1 | YSS-72.2.G2 | YSY-1.2 | YWY-3.1 | YSY-3.4 | YSY-2.2 | YWS-5.2 | YSY-4.3 |
|---|---|---|---|---|---|---|---|---|
| Genome size (Mb) | 6.57 | 7.76 | 8.24 | 6.45 | 6.19 | 6.19 | 4.3 | 5.63 |
| GC content (%) | 52.2 | 53.4 | 52.8 | 47.3 | 47.3 | 47.3 | 66.3 | 49.3 |
| Protein-coding genes | 6,631 | 6,948 | 7,498 | 5,781 | 5,790 | 5,791 | 3,874 | 5,056 |
| tRNAs/rRNAs | 79/3 | 82/5 | 87/1 | 77/1 | 110/1 | 115/1 | 60/3 | 48/1 |
| CAZymes | 219 | 234 | 258 | 338 | 121 | 101 | 122 | 359 |
| GHs | 89 | 108 | 122 | 195 | 29 | 29 | 48 | 187 |
| GTs | 60 | 58 | 58 | 44 | 38 | 37 | 39 | 46 |
| PLs | 9 | 9 | 11 | 13 | | | 2 | 13 |
| CEs | 31 | 44 | 44 | 37 | 25 | 22 | 11 | 37 |
| AAs | 11 | 7 | 8 | | 11 | 11 | 5 | 5 |
| CBMs | 25 | 8 | 15 | 46 | 18 | 2 | 17 | 71 |

**Note**: YSY-3.1, *Paenibacillus chitinolyticus* YSY-3.1; YSS-72.2.G2, *Paenibacillus tyrfis* YSS-72.2.G2; YSY-1.2, *Paenibacillus elgii* YSY-1.2; YWY-3.1, *Paenibacillus illinoisensis* YWY-3.1; YSY-3.4, *Brevibacillus brevis* YSY-3.4; YSY-2.2, *Brevibacillus formosus* YSY-2.2; YWS-5.2, *Chitiniphilus shinanonensis* YWS-5.2; YSY-4.3, *Paenibacillus* sp. YSY-4.3.

et al. 2023a). The genome spanned 7.76 Mb with 53.4% GC content and encoded 6,948 protein-coding genes (4,018 functional proteins and 2,930 hypothetical proteins), 5 rRNAs, 82 tRNAs, and 4 ncRNAs. The strain harbored 234 genes related to CAZymes, including 108 GHs, 58 GTs, 44 CEs, 9 PLs, 7 AAs, and 8 CBMs (Table 9.1). The chitinolytic system consisted of 9 GH18 chitinases, 1 GH19 chitinase, 3 GH20 β-N-acetylglucosaminidases, and 2 AA10 lytic polysaccharide monooxygenases (Table 9.2). Secondary metabolite mining identified 28 BGCs, including NRPS, PKS, lanthipeptide, and metallophore pathways. Fourteen clusters showed no similarity to known BGCs (Table 9.3), suggesting the capacity to produce novel antimicrobial metabolites. The genome also contained 15 genes associated with heavy-metal resistance, siderophore production, and IAA biosynthesis, supporting environmental adaptation and plant-beneficial functions. Overall, the genome of *P. tyrfis* YSS-72.2.G2 revealed a metabolically versatile bacterium with substantial potential for biocontrol, biofertilization, enzymes, and secondary metabolites exploration (Tran et al. 2024a).

## 9.4 THE GENOMIC CHARACTERISTICS OF *PAENIBACILLUS ELGII* YSY-1.2

*P. elgii* YSY-1.2 exhibited strong chitinase activity, notable plant-growth-promoting traits, and antifungal biocontrol potential (Tran et al. 2023a). The whole genome contained 8.24 Mb with 52.8% GC content and encoded 7,498 proteins, 87 tRNAs, and one rRNA. ANI analysis showed 95.65% identity to *P. elgii* AC13, confirming species-level classification. The genome possessed 61 genes associated with environmental adaptation and plant-growth promotion, including resistance to $Cd^{2+}$, $Zn^{2+}$, $As^{3+}$, and $Cr^{6+}$; pathways for phosphate, zinc, and potassium solubilization; nitrogen reduction; siderophore production and iron acquisition; ACC metabolism; and IAA biosynthesis pathway. A total of 258 CAZymes were identified from the genome, including 122 GHs, 58 GTs, 44 CEs, 11 PLs, 8 AAs, and 15 CBMs (Table 9.1). The chitin-degrading enzyme system contained 9 GH18 chitinases, 1 GH19 chitinase, 4 β-N-acetylhexosaminidases (2 GH3 and 2 GH20), and 2 AA10 lytic polysaccharide

TABLE 9.2

*Chitinolytic system of chitinase-producing bacteria.*

| No. | Bacteria | GH18 | GH19 | GH20 | GH3 | AA10 | Novelties |
|---|---|---|---|---|---|---|---|
| 1 | *Paenibacillus chitinolyticus* YSY-3.1 | 14 | 1 | 2 | | 5 | A chitinase possessing two GH18 catalytic domains and a chitinase possessing three GH18 catalytic domains |
| 2 | *Paenibacillus tyrfis* YSS-72.2.G2 | 9 | 1 | 3 | | 2 | |
| 3 | *Paenibacillus elgii* YSY-1.2 | 9 | 1 | 2 | 2 | 2 | |
| 4 | *Paenibacillus illinoisensis* YWY-3.1 | 6 | | 1 | | 1 | |
| 5 | *Brevibacillus brevis* YSY-3.4 | 6 | | | 1 | 1 | |
| 6 | *Brevibacillus formosus* YSY-2.2 | 6 | | | 1 | 1 | |
| 7 | *Chitiniphilus shinanonensis* YWS-5.2 | 22 | 3 | 1 | 1 | 1 | 9 new chitinases |
| 8 | *Paenibacillus* sp. YSY-4.3 | 8 | | 1 | 3 | 3 | GH18–GH5 fusion chitinase |

**Note**: GH18, family 18 chitinase; GH19, family 19 chitinase; GH20, family 20 β-N-acetylhexosaminidase; GH3, family 3 β-N-acetylhexosaminidase; AA10, AA10 lytic polysaccharide monooxygenase.

TABLE 9.3

*Secondary metabolite biosynthetic gene cluster diversity of chitinase-producing bacteria.*

| No. | Bacteria | Total BGCs | Number of novel BGCs | Notable pathways |
|---|---|---|---|---|
| 1 | *Paenibacillus chitinolyticus* YSY-3.1 | ND | | |
| 2 | *Paenibacillus tyrfis* YSS-72.2.G2 | 28 | 14 | NRPS, PKS, lanthipeptides |
| 3 | *Paenibacillus elgii* YSY-1.2 | 32 | 17 | Penisin, bacillopaline, octapeptin, fusaricidin |
| 4 | *Paenibacillus illinoisensis* YWY-3.1 | 7 | 3 | Lasso peptide, opine-like metallophore |
| 5 | *Brevibacillus brevis* YSY-3.4 | 13 | 5 | Gramicidin, tyrocidine, bacitracin, petrobactin |
| 6 | *Brevibacillus formosus* YSY-2.2 | 13 | 5 | NRPS, PKS, metallophores |
| 7 | *Chitiniphilus shinanonensis* YWS-5.2 | 9 | 2 | Hybrid NRPS/metallophore/T1PKS |
| 8 | *Paenibacillus* sp. YSY-4.3 | 15 | 10 | RiPPs, PKS, NRPS |

**Note**: ND, data not shown.

monooxygenases (Table 9.2). In addition, 32 BGCs were identified from the genome, including penisin, bacillopaline, octapeptin, lankacidin, puwainaphycin, paenilipoheptin, fusaricidin, cystothiazole, macrobrevin, bacillibactin, laterocidine, paenilarvin, and thermoactinoamide classes, and 17 of the BGCs showed no similarity to known pathways (Table 9.3), suggesting the potential for novel antimicrobial compound production. Overall, the genome of *P. elgii* YSY-1.2 revealed a promising resource for applications in agriculture, biotechnology, and enzyme discovery (Tran et al. 2024d).

## 9.5 THE GENOMIC CHARACTERISTICS OF *PAENIBACILLUS ILLINOISENSIS* YWY-3.1

*P. illinoisensis* YWY-3.1 was a strong chitinolytic, cellulolytic, and amylolytic bacterium isolated from lake water in Yok Don National Park (Tran et al. 2023a). The whole genome of *P. illinoisensis* YWY-3.1 consisted of 6.45 Mb with 47.3% GC content and encoded 5,781 proteins, 77 tRNAs, and 1 rRNA. Phylogenomic comparisons showed 95.49% ANI and 87.1% digital DNA–DNA hybridization (dDDH) similarity to *P. illinoisensis* NBRC 15959, confirming species identity. The genome encoded an extensive repertoire of 338 CAZymes and comprised 195 GHs, 44 GTs, 13 PLs, 37 CEs, and 46 CBMs (Table 9.1). The chitinase system included 6 GH18 chitinases, 1 AA10 lytic polysaccharide monooxygenase, and 1 GH20 β-N-acetylhexosaminidase (Table 9.2). The genome of strain YWY-3.1 possessed 49 genes related to environmental adaptation and plant-growth promotion, including resistance determinants for arsenic, zinc, cobalt, magnesium, and chromium; biosynthetic pathways for the phytohormones IAA, $GA_3$, and zeatin; siderophore production systems; nitrate/nitrite reduction; and mechanisms for potassium and phosphate solubilization. Additionally, seven secondary metabolite BGCs were identified, including lasso peptide, terpene, NRP-metallophore/NRPS, and opine-like-metallophore classes. Of the BGCs, three showed no similarity to known pathways (Table 9.3), suggesting the capacity to produce novel antimicrobial metabolites. These genomic features highlighted *P. illinoisensis* YWY-3.1 as a valuable source of functional genes, particularly those involved in polysaccharide degradation, plant-growth promotion, and novel secondary metabolite biosynthesis, underscoring potential for future biotechnological and agricultural applications (Tran et al. 2024c).

## 9.6 THE GENOMIC CHARACTERISTICS OF *BREVIBACILLUS BREVIS* YSY-3.4

*B. brevis* YSY-3.4 was a strong chitin-degrading, plant-growth-promoting, and antifungal bacterium isolated from the dipterocarp forest ecosystem of Yok Don National Park (Tran et al. 2023a). The genome size of strain YSY-3.4 was 6.19 Mb with 47.3% GC content, and encoded 5,790 proteins, 110 tRNAs, and 1 rRNA. Phylogenomic comparison revealed 98.14% ANI and 91.7% dDDH similarity to *B. brevis* NBRC 100599, confirming species identity. The genome contained 121 CAZyme-related genes, including 29 GHs, 38 GTs, 25 CEs, 11 AAs, and 18 CBMs (Table 9.1). The chitinolytic machinery comprised 6 GH18 chitinases, 1 AA10 lytic polysaccharide monooxygenase, and 1 GH3 β-N-acetylhexosaminidase (Table 9.2). Secondary metabolite analysis identified 13 BGCs, including pathways for gramicidin, tyrocidine, petrobactin siderophore, bacitracin, and zwittermicin clusters. Five clusters showed no similarity to known BGCs (Table 9.3), indicating potential for the synthesis of novel natural products. Together, the genomic features of *B. brevis* YSY-3.4 revealed a metabolically versatile bacterium enriched in chitinases, phytohormone biosynthesis genes, diverse CAZymes, and antimicrobial pathways, underscoring great potential for applications in biocontrol, biofertilizer development, enzyme discovery, and natural-product innovation (Tran et al. 2025a).

## 9.7 THE GENOMIC CHARACTERISTICS OF *BREVIBACILLUS FORMOSUS* YSY-2.2

*B. formosus* YSY-2.2 was a strong chitinolytic and plant-growth-promoting bacterium isolated from soils in Yok Don National Park (Tran et al. 2023a). The whole genome of *B. formosus* YSY-2.2 spanned 6.19 Mb with 47.3% GC content, encoding 5,791 proteins, 115 tRNAs, and 1 rRNA. Phylogenomic comparisons showed 95.07% ANI and 72.6% dDDH similarity to *B. formosus* NF2, consistent with strain-level differentiation. The genome harbored 101 genes associated with CAZymes, including 29 GHs, 37 GTs, 22 CEs, 11 AAs, and 2 CBMs (Table 9.1). The chitinolytic system comprised 6 GH18 chitinases, 1 AA10 lytic polysaccharide monooxygenase, and 1 GH3 β-N-acetylglucosaminidase (Table 9.2). Plant-growth-promotion potential was reflected in 12 phytohormone-related genes, including 10 involved in IAA biosynthesis and 2 associated with $GA_3$ synthesis. In addition, secondary metabolite mining identified 13 BGCs spanning NRPS, PKS, lanthipeptide, siderophore, and metallophore classes. Five of these clusters exhibited no similarity to known BGCs (Table 9.3), indicating the potential to produce previously uncharacterized antimicrobial metabolites. These genomic features revealed *B. formosus* YSY-2.2 as a metabolically versatile and biotechnologically valuable bacterium with potential for applications in biocontrol, biofertilization, enzyme engineering, and natural-product discovery (Nguyen et al. 2025).

## 9.8 THE GENOMIC CHARACTERISTICS OF *CHITINIPHILUS SHINANONENSIS* YWS-5.2

*C. shinanonensis* YWS-5.2 is a valuable chitinolytic and plant-growth-promoting bacterium isolated from chitin flakes placed in the water of Suoi Hong Lake in Yok Don National Park. The genome of *C. shinanonensis* YWS-5.2 included 4.30 Mb with 66.3% GC content, encoding 3,874 proteins, 60 tRNAs, and 3 rRNAs. ANI (98.45%) and dDDH similarity (86.5%) to the reference strain NBRC 104970 (=SAY3) confirm strain-level identity. The genome harbored 122 CAZymes, comprising 48 GHs, 39 GTs, 2 PLs, 11 CEs, 5 AAs, and 17 CBMs (Table 9.1). The system of chitinolytic enzymes contained 28 enzymes, comprising 22 GH18 chitinases, 3 GH19 chitinases, 1 GH3 β-N-acetylhexosaminidase, 1 GH20 β-N-acetylhexosaminidase, and 1 AA10 lytic polysaccharide monooxygenase. Interestingly, 9 of those, including 7 GH18 chitinases and 2 GH19 chitinases, were absent from the reported chitinase system of *C. shinanonensis* (Figure 9.1B and Table 9.2), indicating genomic expansion and novelty. The genome contained 29 genes related to plant-growth promotion, including pathways for cellulose and starch degradation, phosphate and potassium solubilization, IAA biosynthesis, siderophore production, and biofilm formation. Genome mining using antiSMASH revealed 9 BGCs, including NRPS, aryl polyene, terpene, and hybrid NRPS/metallophore/T1PKS clusters. Two clusters displayed no similarity to known pathways (Table 9.3), suggesting the presence of previously uncharacterized antimicrobial or signaling secondary metabolites. Collectively, *C. shinanonensis* YWS-5.2 emerged as a genetically enriched and ecologically potent bacterium with strong potential for use in biofertilization, biocontrol, and the discovery of new enzymes and secondary metabolites (Tran et al. 2025b).

## 9.9 THE GENOMIC CHARACTERISTICS OF A POSSIBLY NOVEL *PAENIBACILLUS* SPECIES

*Paenibacillus* sp. YSY-4.3 was isolated from soils collected from Yok Don National Park. Experimental characterization showed that strain YSY-4.3 exhibited high activities of chitinase and cellulase and produced siderophore, IAA, and $GA_3$ with high levels. The genome of strain YSY-4.3 contained 5.63 Mb with a GC content of 49.3% and encoded 5,056 protein-coding sequences, 48 tRNAs, and 1 rRNA. Phylogenetic analysis of the full-length 16S rRNA gene showed only 98.82% identity to *Paenibacillus woosongensis* B2_4, and phylogenomic metrics, including ANI (89.54%) and dDDH (67.4%), exhibited well below accepted species thresholds, indicating that YSY-4.3 represents a novel *Paenibacillus* species. The genome possessed 359 CAZyme-related genes, comprising 187 GHs, 46 GTs, 13 PLs, 37 CEs, 5 AAs, and 71 CBMs (Table 9.1). The chitinolytic system included 15 enzymes, consisting of 8 GH18 chitinases, 3 GH3 β-N-acetylhexosaminidases, 1 GH20 β-N-acetylhexosaminidase, and 3 AA10 lytic polysaccharide monooxygenases. Notably, one chitinase, PsChiQ, harbored a GH18 catalytic domain at the N-terminus and a GH5 cellulase catalytic domain at the C-terminus (Figure 9.1A and Table 9.2), representing a novel chitinase architecture with potential bifunctional activity. In addition to carbohydrate metabolism, the genome contained 51 genes associated with siderophore production, biosynthesis of IAA and $GA_3$, nitrogen reduction, ACC metabolism, and resistance to multiple heavy metals, highlighting substantial ecological adaptability. It also harbored 15 BGCs, spanning NRPS, PKS, RiPP-like, lanthipeptide, and metallophore classes, and 10 of these BGCs are predicted to be novel clusters (Table 9.3). Consequently, these genomic and functional traits demonstrated that strain YSY-4.3 is a taxonomically novel and metabolically versatile *Paenibacillus* species with promising potential for chitin degradation, secondary metabolite discovery, and applications in sustainable agriculture (Tran et al. 2024b).

## 9.10 CONCLUSIONS AND FURTHER PERSPECTIVES

Genomic analyses of the isolated bacteria from Yok Don National Park revealed a highly diverse and metabolically versatile microbial community with biotechnological potential. Across the genera *Paenibacillus*, *Brevibacillus*, and *Chitiniphilus*, the bacteria shared conserved genomic signatures for

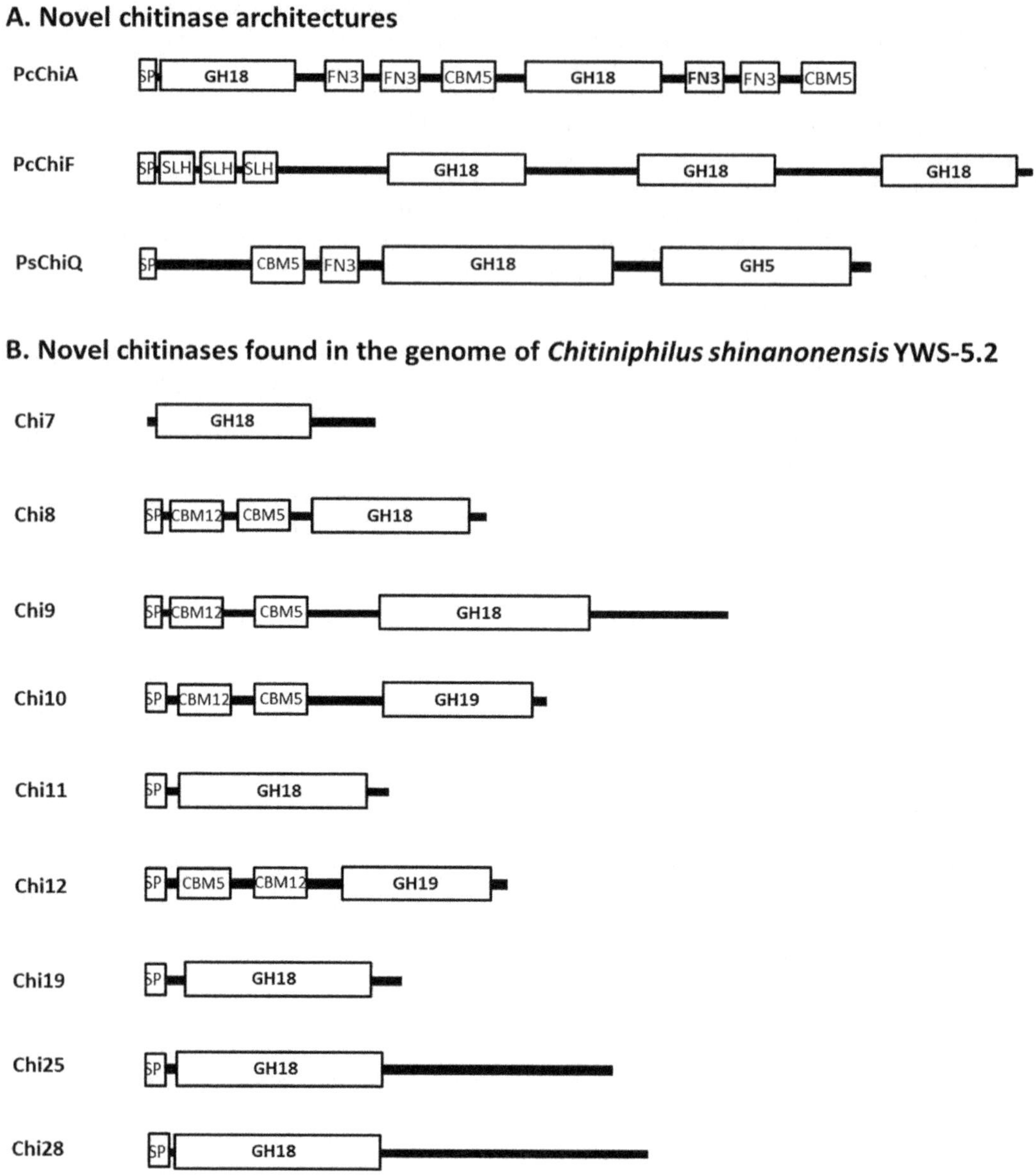


Figure 9.1 (A) Novel chitinases found in chitinase-producing bacteria and (B) new finding chitinases in *Chitiniphilus shinanonensis* YWS-5.2.

**Note**: SP, signal peptide sequence; FN3, fibronectin type III; SLH, surface layer homology; CBM5, carbohydrate-binding module family 5; CBM12, carbohydrate-binding module family 12; GH18, GH18 catalytic domain; GH5, GH5 catalytic domain; GH19, GH19 catalytic domain.

polysaccharide degradation, secondary metabolite biosynthesis, environmental resilience, and phytohormone production. The genomes encoded extensive CAZymes, including novel chitinolytic systems, featuring novel GH18 chitinases, the multiple catalytic domain enzymes, and rare bifunctional architectures such as the GH18GH5 fusion in *Paenibacillus* sp. YSY-4.3. These features highlighted the chitinolytic bacteria as powerful resources for sustainable agriculture and biotechnology. The high chitin-degrading capacity supported applications in chitin waste bioconversion, antifungal bioproduct development, and enzyme engineering. The nutrient-solubilizing, siderophore-producing, and phytohormone-generating capabilities highlighted promise as next-generation biofertilizers. The abundance of novel BGCs further underscored significant potential for natural-product discovery and synthetic biology innovation.

## ACKNOWLEDGMENTS

This work was supported by Vietnam National Foundation for Science and Technology Development (NAFOSTED) under grant numbers 106.04-2025.117 and 106.04-2019.337.

## REFERENCES


Berendsen RL, Pieterse CM, Bakker PA (2012) The rhizosphere microbiome and plant health. *Trends Plant Sci*. 17(8):478–486.

Berg G, Rybakova D, Fischer D, Cernava T, Vergès MC, Charles T, Schloter M (2020) Microbiome definition re-visited: Old concepts and new challenges. *Microbiome*. 8(1):103.

Dhillon GS, Kaur S, Brar SK, Verma M (2013) Green synthesis approach: Extraction of chitosan from fungus mycelia. *Crit Rev Biotechnol*. 33:379–403.

Do MC, Mai NTN, Tung HT, Hoang DK, Vu QL, Truong HP, Tran MD, Nguyen QV, Doan MD, Nguyen AD, Huynh TU, Nguyen TH, Do TO, Duong TN (2023) Phytohormones produced by *Pseudomonas otitidis* YWY-9.1 affected on the growth and development of Chrysanthemum plantlets. *Plant Cell Tiss Organ Cult*. 154:605–618.

Dos Santos PC, Fang Z, Mason SW, Setubal JC, Dixon R (2012) Distribution of nitrogen fixation genes in bacteria. *Microbiol Mol Biol Rev*. 76(3):342–358.

Drula E, Garron ML, Dogan S, Lombard V, Henrissat B, Terrapon N (2022) The carbohydrate-active enzyme database: Functions and literature. *Nucleic Acids Res*. 50:D571–D577.

Dukare A, Paul S, Arambam A (2020) Isolation and efficacy of native chitinolytic rhizobacteria for biocontrol activities against *Fusarium* wilt and plant growth promotion in pigeon pea (*Cajanus cajan* L.). *Egypt J Biol Pest Control*. 30:56.

Etesami H and Beattie GA (2018) Mining halophytes for plant growth–promoting halotolerant bacteria to enhance the salinity tolerance of non-halophytic crops. *Front Microbiol*. 9:148.

Fitzpatrick CR, Salas-González I, Conway JM, Dangl JL (2018) The plant microbiome: From ecology to reductionism and beyond. *Annu Rev Microbiol*. 72:81–100.

Glick BR (2012) Plant growth-promoting bacteria: Mechanisms and applications. *Scientifica*. 2012:963401.

Gu Q, Yang Y, Yuan Q, Shi G, Wu L, Lou Z, Huo R, Wu H, Borriss R, Gao X (2017) Bacillomycin D produced by *Bacillus amyloliquefaciens* is involved in the antagonistic interaction with the plant-pathogenic fungus *Fusarium graminearum*. *Appl Environ Microbiol*. 83:e01075–17.

Le B and Yang SH (2019) Microbial chitinases: Properties, current state and biotechnological applications. *World J Microbiol Biotechnol*. 35:144.

Mendes R, Garbeva P, Raaijmakers JM (2013) The rhizosphere microbiome: Significance of plant beneficial, plant pathogenic, and human pathogenic microorganisms. *FEMS Microbiol Rev*. 37(5):634–663.

Nguyen DS, Ngoc LNT, Vu BT, Pham TP, Pentekhina I, Tran DM (2025) Genome resource of chitinolytic *Brevibacillus formosus* YSY-2.2. *Ecol Genet Genom*. 35:100344.

Peiffer JA, Spor A, Koren O, Jin Z, Tringe SG, Dangl JL, Ley RE (2013) Diversity and heritability of the maize rhizosphere microbiome under field conditions. *PNAS*. 110(16):6548–6553.

Pentekhina I, Hattori T, Tran DM, Shima M, Watanabe T, Sugimoto H, Suzuki K (2020) Chitinase system of *Aeromonas salmonicida*, and characterization of enzymes involved in chitin degradation. *Biosci Biotechnol Biochem*. 84(9):1936–1947.

Pieterse CM, Zamioudis C, Berendsen RL, Weller DM, van Wees SC, Bakker PA (2014) Induced systemic resistance by beneficial microbes. *Annu Rev Phytopathol*. 52:347–375.

Rajkumar M, Ae N, Freitas H (2012) Endophytic bacteria and their potential to enhance heavy metal phytoextraction. *Chemosphere*. 86(5):609–616.

Rodríguez H and Fraga R (1999) Phosphate solubilizing bacteria and their role in plant growth promotion. *Biotechnol adv*. 17(4–5):319–339.

Rutledge PJ and Challis GL (2015) Discovery of microbial natural products by activation of silent biosynthetic gene clusters. *Nat Rev Microbiol*. 13(8):509–523.

Synowiecki J and Al-Khateeb NA (2003) Production, properties, and some new applications of chitin and its derivatives. *Crit Rev Food Sci Nutr*. 43:145–1471.

Tran DM, Do TO, Nguyen QV (2024a) Whole genome sequence data of *Paenibacillus tyrfis* YSS-72.2.G2, a chitinolytic bacterium newly isolated from a National Park of Vietnam. *Data Brief*. 53:110087.

Tran DM and Huynh TU (2025a) Genome resource of the chitinase-producer *Brevibacillus brevis* YSY-3.4 isolated from the dipterocarp forest ecosystem. *Microbiol Biotechnol Lett*. 53(1):121–124.

Tran DM, Huynh TU, Do TO, Ngoc LNT, Pentekhina I (2025b) Characterization and genome sequence of chitinase-producing *Chitiniphilus shinanonensis*. *Ecol Genet Genom*. 34:100320.

Tran DM, Huynh TU, Do TO, Nguyen AD (2024b) Isolation, plant growth-promoting properties, and whole-genome sequence of a novel *Paenibacillus* Species. *J Basic Microbiol*. 64(9):e2400119.

Tran DM, Huynh TU, Nguyen TH, Do TO, Nguyen QV, Nguyen AD (2022a) Molecular analysis of genes involved in chitin degradation from the chitinolytic bacterium *Bacillus velezensis*. *Antonie Van Leeuwenhoek*. 115:215–231.

Tran DM, Huynh TU, Nguyen TH, Do TO, Nguyen QV, Nguyen AD (2023a) Novel resources of chitinolytic bacteria isolated from Yok Don National Park, *Vietnam*. *J Appl Microbiol*. 134(8):lxad141.

Tran DM, Huynh TU, Nguyen TH, Do TO, Pentekhina I, Nguyen QV, Nguyen AD (2022b) Expression, purification, and basic properties of a novel domain structure possessing chitinase from *Escherichia coli* carrying the family 18 chitinase gene of *Bacillus velezensis* strain RB.IBE29. *Mol Biol Rep*. 49:4141–4148.

Tran DM, Nguyen TH, Huynh TU, Pentekhina I (2024c) Deciphering genome sequence of *Paenibacillus illinoisensis* strain YWY-3.1: A chitinase, cellulase, and amylase producer. *Ecol Genet Genom*. 33:100298.

Tran DM, Nguyen TH, Nguyen AD (2023b) Isolation, identification, and biological characterization of chitinase-producing actinomyces from Yok Don National Park. *Vietnam J Agri Sci*. 21(12):1517–1526.

Tran DM, Pham PT, Vu BT, Ngoc LNT (2024d) Data on annotation and analysis of genome sequence of *Paenibacillus elgii* YSY-1.2, a promising chitinase-producing, plant-growth-promoting, and biocontrol agent. *Data Brief*. 54:110285.

Tran DM, Sugimoto H, Nguyen DA, Watanabe T, Suzuki K (2018) Identification and characterization of chitinolytic bacteria isolated from a freshwater lake. *Biosci Biotechnol Biochem*. 82(2):343–355.

van Dijk EL, Jaszczyszyn Y, Naquin D, Thermes C (2018) The third revolution in sequencing technology. *Trends Genet*. 34(9):666–681.

Veliz EA, Martínez-Hidalgo P, Hirsch AM (2017) Chitinase-producing bacteria and their role in biocontrol. *AIMS Microbiol*. 3:689–705.

Vurukonda SS, Vardharajula S, Shrivastava M, Sk ZA (2016) Enhancement of drought stress tolerance in crops by plant growth–promoting rhizobacteria. *Microbiol Res*. 184:13–24.

CHAPTER 10

# Beyond the Rhizosphere

## UNDERSTANDING ENDOPHYTIC COMMUNITIES IN ENVIRONMENTAL MICROBIOMES

*Dhruvi K. Suvagiya, Sandeep Chovatiya, and Ramesh K. Kothari*

## 10.1 INTRODUCTION: THE EMERGING SIGNIFICANCE OF ENDOPHYTES

The intricate relationship between plants and microbes stands for one of nature's most dynamic ecological systems. For decades, the rhizosphere—the small zone of soil influenced by root secretions and microbial activity—has served as the primary area of investigation for plant-associated microbiota (Philippot *et al.*, 2013; Berendsen *et al.*, 2012). This region is a known hotspot for microbial diversity, nutrient transformation and biochemical exchange, all crucial for plant development and soil health (Mendes *et al.*, 2013). Nevertheless, a vast array of microorganisms, termed endophytes, colonize the internal tissues of plants, including stems, leaves, seeds and roots. These endophytes keep a delicate balance with their hosts, engaging in symbiotic, commensal or occasionally opportunistic associations that significantly influence plant physiology and ecosystem function (Hardoim *et al.*, 2015; Compant *et al.*, 2010).

### 10.1.1 Beyond the Rhizosphere: A Protected Niche

While the rhizosphere provides the external interface for microbial interactions, the endosphere (the internal plant environment) offers a selective and protected habitat (Hardoim *et al.*, 2008). Within this niche, microorganisms adapt to specific microenvironments distinct from bulk soil or rhizospheric conditions. Endophytic bacteria and fungi colonize these internal tissues without causing clear pathogenic symptoms, suggesting sophisticated mechanisms of molecular communication and host immune modulation (Compant *et al.*, 2010; Bulgarelli *et al.*, 2013). Understanding these interactions is fundamental to grasping how plants integrate microbial partners into their extended phenotype, affecting their stress tolerance, growth and disease resistance (Liu *et al.*, 2017; Fatima *et al.*, 2022).

### 10.1.2 Ecological and Environmental Relevance

Endophytic communities function as essential biological buffers within natural ecosystems, bolstering plants' capacity to adapt to fluctuating environmental conditions (Rodriguez *et al.*, 2009; Santoyo *et al.*, 2016). They contribute to the stability and sustainability of ecosystems by helping in nutrient mobilization, phytohormone production and the detoxification of contaminants (Singh *et al.*, 2020). Furthermore, endophytes are integral to biogeochemical cycling, connecting plant productivity, global element fluxes and soil microbial processes (Compant *et al.*, 2019). As environmental

DOI: 10.1201/9781003637950-10

microbiology advances towards a comprehensive systems-based understanding, recognizing the role of endophytes is imperative for completing the picture of plant–microbe–environment interactions (White *et al.*, 2019).

This chapter aims to illuminate endophytic communities within the context of environmental microbiomes, moving beyond the traditional focus on the rhizosphere to explore their extensive, untapped potential. By integrating applied and conceptual perspectives, this chapter provides a cohesive understanding of endophytic communities as crucial components of environmental microbiomes and significant contributors to sustainable ecosystem function.

## 10.2 THE RHIZOSPHERE–ENDOSPHERE CONTINUUM

The association between plants and microorganisms constitutes a gradient, not a relationship confined to a single niche, extending from the exterior soil environment to the interior of the plant. This gradient is conceptualized as three overlapping zones: the rhizosphere (soil next to the roots), the rhizoplane (the root surface) and the endosphere (internal plant tissues). Each zone is a distinct ecological niche with unique physical, chemical and biological attributes that dictate microbial colonization and activity (Akther *et al.*, 2025; Compant *et al.*, 2010; Guo *et al.*, 2024).

The transition of free-living soil microbes into endophytes is mediated by complex selective filters, including root exudates, plant surface structures and immune responses (Yan *et al.*, 2025; Xu *et al.*, 2023). This continuum concept emphasizes that the endosphere and rhizosphere are dynamically linked habitats forming a critical ecological interface, rather than isolated entities (Akther *et al.*, 2025; Compant *et al.*, 2010) (Figure 10.1).

### 10.2.1 The Rhizosphere: Gateway to the Endosphere

The rhizosphere functions as the primary point of entry for microbes looking to invade plant tissues. Root exudates, including amino acids, sugars, secondary metabolites and phenolics, function as chemical signals, attracting beneficial microorganisms while deterring potential pathogens. This biochemical signalling helps the selective enrichment of microbial populations having traits essential for colonization, such as chemotaxis, adhesion and biofilm formation (Akther *et al.*, 2025; Compant *et al.*, 2010). Many endophytes originate from rhizospheric populations capable of successfully navigating this selection process and entering the root interior. Entry typically occurs through natural openings (e.g., root hairs), lateral root junctions or cracks formed during root growth (Yan *et al.*, 2025). Once entering the root, these microbes must adapt to the new microenvironments by regulating gene expression, producing cell wall–degrading enzymes under controlled conditions and evading host defence mechanisms (Guo *et al.*, 2024).

### 10.2.2 The Rhizoplane: A Transitional Microhabitat

The rhizoplane, the immediate surface of the root, acts as the transitional zone between the endosphere and the rhizosphere. It is characterized by intense competition for nutrients and high microbial density. Surface adhesion processes, often mediated by surface proteins and extracellular polymeric substances (EPS), are crucial for microbial colonization here. This biofilm-like structure offers protection from environmental stress and serves as the starting point for endophytic invasion (Akther *et al.*, 2025; Guo *et al.*, 2024). Within this zone, microbial communities use mechanisms such as quorum sensing and horizontal gene transfer to find which strains can efficiently enter the plant's internal tissues (Guo *et al.*, 2024; Yan *et al.*, 2025).

### 10.2.3 The Endosphere: A Selective and Protected Habitat

Inside the plant, microbes meet a uniquely nutrient-limited environment heavily regulated by host physiology (Akther *et al.*, 2025; Compant *et al.*, 2010). Unlike the carbon-rich rhizosphere, the endosphere offers restricted nutrient access, favouring specialized microorganisms that can set up an efficient mutualistic cohabitation. Endophytic bacteria (e.g., *Burkholderia*, *Bacillus*, *Pseudomonas*) and fungal genera (e.g., *Fusarium*, *Trichoderma*) have evolved various adaptations to survive here (Guo *et al.*, 2024). These adaptations include inhibiting plant immune recognition, synthesizing phytohormones (like indole-3-acetic acid (IAA)) and producing volatile chemical compounds that stimulate host growth (Compant

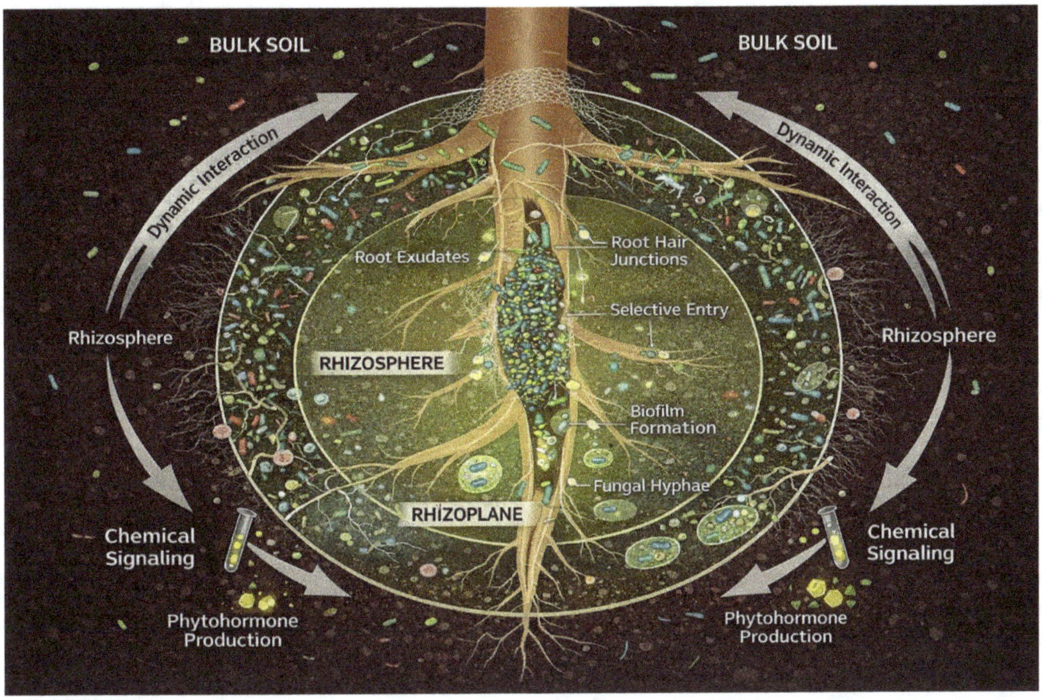


Figure 10.1 Rhizosphere–endosphere continuum.

et al., 2010; Xu et al., 2023). Consequently, the endosphere is a highly selective biological niche where microbial survival is intrinsically linked to environmental adaptability and host compatibility (Akther et al., 2025).

### 10.2.4 Ecological and Functional Integration

The rhizosphere–endosphere continuum functions as an interconnected ecological system where microbes continually traverse this gradient, forming a dynamic network that influences both soil processes and plant health (Akther et al., 2025; Yan et al., 2025). The equilibrium between endophytic and rhizospheric populations is subject to environmental factors such as nutrient availability, soil pH and moisture (Guo et al., 2024; Akther et al., 2025). Similarly, the developmental stage and genotype of the plant significantly affect the microbial communities along this continuum (Yan et al., 2025; Xu et al., 2023). Understanding these cross-border interactions is essential for designing effective strategies that use beneficial endophytes for improved environmental sustainability and enhanced crop performance (Compant et al., 2010; Akther et al., 2025).

## 10.3 DIVERSITY AND TAXONOMY OF ENDOPHYTIC MICROBIOMES

Endophytic microorganisms form a remarkably diverse assemblage, including bacteria, fungi, archaea, actinomycetes and even certain protists, all living asymptomatically within plant tissues. The composition of these internal communities varies significantly based on the plant species, the specific tissue colonized and the prevailing environmental conditions (Long et al., 2025; Sharma et al., 2023; Crucitti et al., 2025).

This diversity reflects host physiological status, genotype and ecological factors such as temperature, soil type and the availability of microbes in the surrounding environment (Lei et al., 2025). Many endophytes are facultative, shifting between endophytic and free-living lifestyles in response to environmental cues, while others are obligatory symbionts that have co-evolved closely with their hosts. This co-existence of various microbial forms highlights the ecological flexibility and capacity of endophytic communities to occupy specific niches within the plant (Long et al., 2025; Lei et al., 2025).

### 10.3.1 Bacterial Endophytes

Bacterial endophytes are among the most thoroughly studied groups within the plant microbiome. The dominant bacterial phyla found include Proteobacteria, Firmicutes, Actinobacteria and Bacteroidetes (Long *et al.*, 2025; Sharma *et al.*, 2023; Deepa *et al.*, 2024). Commonly described endophytic genera include *Bacillus*, *Pseudomonas*, *Burkholderia*, *Azospirillum* and *Enterobacter* (Long *et al.*, 2025; Sharma *et al.*, 2023).

Many of these bacteria are renowned for their ability to promote plant growth via mechanisms such as nitrogen fixation, phosphate solubilization, siderophore synthesis and the production of phytohormones. Certain species, like *Azoarcus* and *Herbaspirillum*, colonize intercellular spaces and vascular tissues, directly supporting nitrogen metabolism and stress tolerance (Sharma *et al.*, 2023). Due to their closer proximity to soil microbial sources, bacterial endophytes often show distinct spatial partitioning within the plant and a high degree of host specificity (Long *et al.*, 2025).

### 10.3.2 Fungal Endophytes

Fungal endophytes are seen in every plant species examined to date.They belong to various taxonomic groups, primarily Basidiomycota and Ascomycota, encompassing both non-clavicipitaceous and clavicipitaceous forms (Long *et al.*, 2025; Crucitti *et al.*, 2025; Lei *et al.*, 2025). Examples include *Neotyphodium* species (Clavicipitaceae) that induce systemic infections in grasses, producing alkaloids that protect the host from herbivores. In tropical species and woody plants, non-clavicipitaceous endophytes like *Fusarium*, *Colletotrichum*, *Trichoderma* and *Penicillium* are prevalent, contributing to environmental adaptation and host resistance (Long *et al.*, 2025; Deepa *et al.*, 2024; Lei *et al.*, 2025). These fungi show the context-dependent nature of endophytic interactions, alternating between latent pathogenic and mutualistic states based on environmental stress or host condition (Crucitti *et al.*, 2025).

### 10.3.3 Archaea and Other Rare Endophytes

Recent studies have found archaeal lineages, including Euryarchaeota, Thaumarchaeota and Crenarchaeota, in the leaves and roots of various wild and crop plants, a finding contrary to earlier beliefs that archaea were absent from plant tissues (Long *et al.*, 2025).Though their specific functions are still being found, there is evidence suggesting roles in ammonia oxidation, methane cycling and enhancing stress tolerance in challenging environments (Long *et al.*, 2025).

Additionally, actinomycetes, particularly *Streptomyces* species, are noteworthy for producing bioactive secondary metabolites that promote growth and protect plants from infections (Sharma *et al.*, 2023). These rare and often unculturable species highlight a profound depth of microbial diversity that cannot be captured by conventional culture-based methods (Long *et al.*, 2025).

### 10.3.4 Methods for Studying Endophytic Diversity

The combined use of culture-dependent and culture-independent methodologies has significantly advanced the understanding of endophytic microbiomes (Long *et al.*, 2025; Sharma *et al.*, 2023). While conventional culture methods are still valuable for isolating and functionally testing specific strains, they are limited. Molecular techniques, such as 16S ribosomal ribonucleic acid (rRNA) gene sequencing for bacteria and internal transcribed spacer (ITS) region sequencing for fungi, have revolutionized the classification and identification of unculturable organisms (Deepa *et al.*, 2024). Current metagenomic and amplicon-based studies provide comprehensive community profiles, while techniques like fluorescence in situ hybridization (FISH), stable isotope probing and confocal microscopy reveal metabolic activity and spatial organization within plant tissues (Long *et al.*, 2025). These technological advancements have catalysed a change in thinking from focusing on individual species to understanding the comprehensive functional networks and community dynamics within the plant microbiome (Lei *et al.*, 2025).

### 10.3.5 Factors Influencing Endophytic Composition

The structure of endophytic communities is influenced by a multitude of abiotic and biotic factors (Long *et al.*, 2025; Sharma *et al.*, 2023; Lei *et al.*, 2025). Host genotype, tissue type and plant developmental stage strongly figure out which microbes show persistent associations (Crucitti *et al.*, 2025; Sharma *et al.*, 2023). For instance, root endophytes often differ from leaf or stem populations due to the distinct metabolic gradients and confined microenvironments they represent (Long *et al.*, 2025).

Environmental variables such as soil pH, moisture, nutrient availability and human agricultural inputs further modulate microbial assemblages (Long *et al.*, 2025; Deepa *et al.*, 2024). Furthermore, geographical and climate factors can influence biogeographical patterns in endophyte diversity, suggesting that selective ecological forces drive the co-evolution of plant–microbe symbioses (Crucitti *et al.*, 2025). A thorough understanding of these factors is crucial for predicting microbial responses to environmental changes and for engineering endophytic communities for beneficial outcomes (Long *et al.*, 2025; Sharma *et al.*, 2023).

## 10.4 MECHANISMS OF ENDOPHYTE–HOST INTERACTIONS

Establishing an endophytic association is not a passive process but requires a finely tuned sequence of structural, biochemical and molecular interactions between the host and the microbe. Unlike harmful bacteria, endophytes must colonize internal tissues without provoking strong defence reactions (Mushtaq *et al.*, 2023; Rather & Masroor, 2025).

Successful colonization needs mutual recognition, the maintenance of physiological balance and the suppression or modification of the host immune system (Brader et al., 2014; Singh *et al.*, 2024).This careful coordination ensures the endophyte has a stable niche and access to nutrients, while the host gains ecological or physiological advantages, such as stress tolerance or growth promotion (Rather & Masroor, 2025; Kandel *et al.*, 2015) (Figure 10.2).

### 10.4.1 Entry and Colonization Pathways

Endophytic colonization begins with recognition and attachment at the root surface (Singh *et al.*, 2024; Mushtaq *et al.*, 2023). Plant roots release various exudates, including organic acids, flavonoids, phenolics, amino acids and sugars, that function as chemical attractants for compatible microorganisms (Brader *et al.*, 2014; Singh *et al.*, 2024). Microbes respond by using surface structures like flagellin, lipopolysaccharides and adhesins to adhere to the rhizoplane. Entry into the plant tissues is typically achieved through natural apertures like stomata, lateral root junctions and root hairs, or through wounds caused by environmental stress or growth (Singh *et al.*, 2024).

Some bacteria help entry by releasing plant cell wall–degrading enzymes (e.g., endoglucanases, pectinases, cellulases), which allow access without causing clear tissue damage (Singh *et al.*, 2024; Brader et al., 2014). Once inside, microorganisms form endophytic populations that can travel systemically throughout the plant, colonizing the apoplast (intercellular spaces) and in some cases, the xylem vessels. Fungal endophytes often use hyphal penetration, guided by mechanical pressure and enzymatic activity (Rather & Masroor, 2025). Specialized infection structures, such as appressoria, are sometimes used by fungi to gently penetrate the epidermal layer, while others coordinate entry with changes to the host cell wall via chemical signalling (Brader *et al.*, 2014). These nuanced invasion tactics highlight the development of mutual tolerance and compatibility between endophytes and hosts (Rather & Masroor, 2025).

### 10.4.2 Molecular Signalling and Recognition

Communication between endophytes and plants is mediated by complex molecular signalling pathways that ensure reciprocal recognition and prevent the activation of a strong immune response (Singh *et al.*, 2024; Brader *et al.*, 2014). Plants have pattern-recognition receptors (PRRs) that detect microbe-associated molecular patterns (MAMPs), such as chitin fragments or flagellin. Consequently, beneficial microorganisms must change or conceal these signals to prevent severe defence reactions (Singh *et al.*, 2024; Kandel *et al.*, 2015). Endophytes employ several strategies to achieve this:

- Production of effector proteins that inhibit the expression of host defence genes.
- Release of antioxidant enzymes (e.g., peroxidases and catalases) that detoxify reactive oxygen species (ROS) produced during plant immune activation.
- Modulation of phytohormones like jasmonic acid, salicylic acid and ethylene to adjust host immunity (Kandel *et al.*, 2015; Mushtaq *et al.*, 2023).

Concurrently, plants can detect beneficial microbial signals, including volatile chemical molecules, lipo-chito oligosaccharides and siderophores, which trigger mild, non-destructive immune responses known as induced systemic resistance (ISR) (Brader *et al.*, 2014). ISR effectively

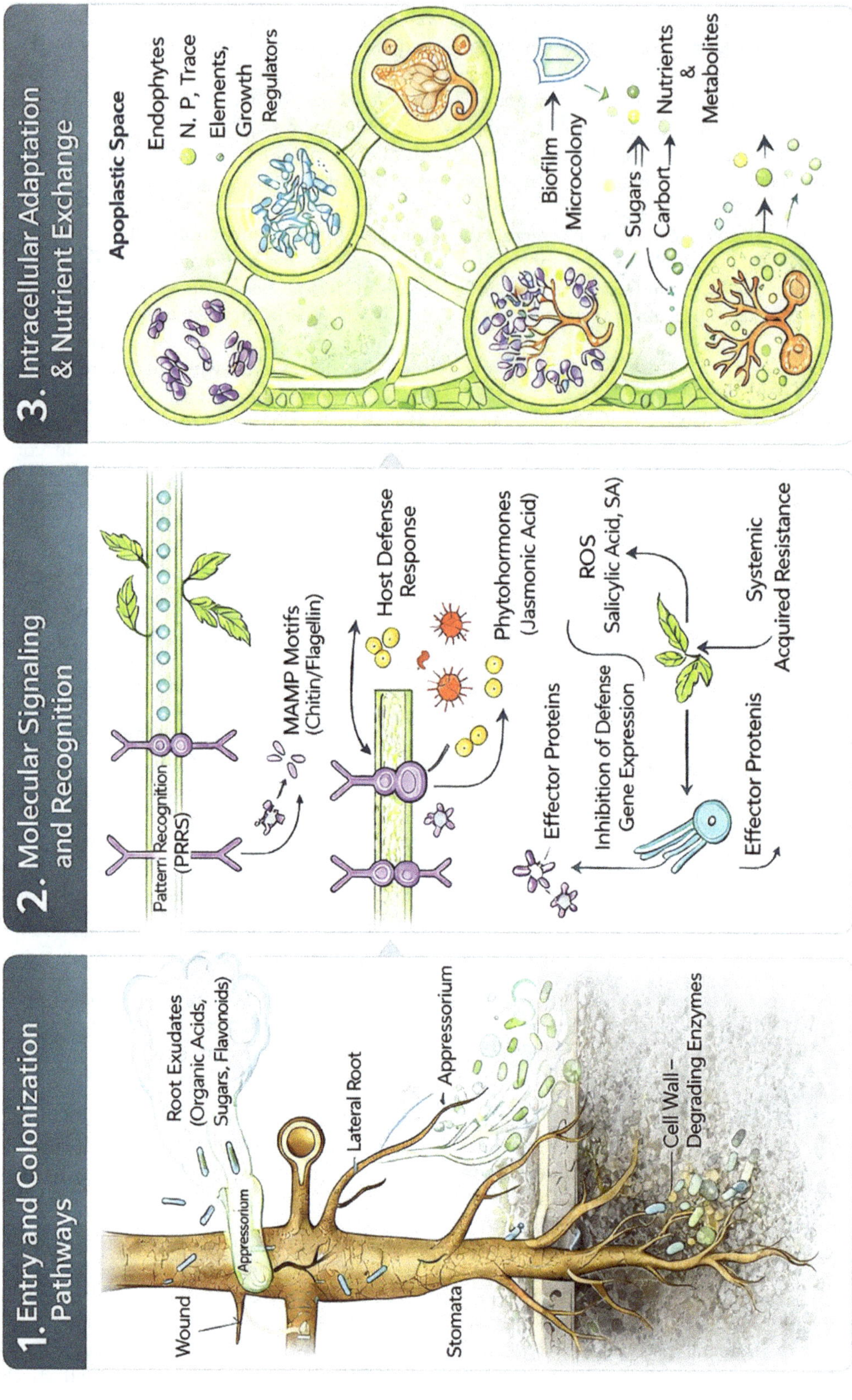


Figure 10.2 A mutualistic journey of endophytes–plant interaction.

conditions the plant defensively, preparing it to react more rapidly to later stress, and is a foundational aspect of mutualistic defensive conditioning (Kandel *et al.*, 2015; Singh *et al.*, 2024).

### 10.4.3 Intracellular Adaptation and Nutrient Exchange

Upon entry, endophytes must adapt to a highly regulated, nutrient-limited interior environment (Brader *et al.*, 2014; Rather & Masroor, 2025). Avoid triggering host defence, they support low population densities and change their metabolic pathways to use plant-derived carbon sources, such as amino acids, sugars and organic acids (Mushtaq *et al.*, 2023). Many endophytes form biofilm-like microcolonies in the intercellular gaps, providing stability and defence (Kandel *et al.*, 2015).

Symbiotic nutrient exchange is often bidirectional: the endophyte provides trace elements, phosphorus or nitrogen, along with secondary metabolites like growth regulators or vitamins, while the plant supplies carbon molecules (Rather & Masroor, 2025; Brader *et al.*, 2014). This metabolic complementarity enhances plant fitness and is often likened to a "microbial organ" used in concert with the plant tissues. In fungal endophytes, intracellular accommodation may involve forming specialized structures like haustoria or arbuscules that ease nutrient exchange without killing the host cells. These structures resemble mycorrhizal interactions, showing an evolutionary convergence in positive plant–microbe associations (Rather & Masroor, 2025).

### 10.4.4 Significance of Mechanistic Understanding

A thorough understanding of the mechanisms underlying plant–endophyte interactions is vital for managing agriculture and ecosystems using microbial inoculants (Singh *et al.*, 2024; Kandel *et al.*, 2015; Brader *et al.*, 2014). By deciphering immune regulation, molecular signalling and colonization pathways, researchers can find candidate strains with reliable field performance and design microbial consortia that function consistently across diverse environments (Rather & Masroor, 2025).

## 10.5 ECOLOGICAL AND FUNCTIONAL ROLES OF ENDOPHYTES

Endophytic microbes are often termed "hidden helpers" due to their many direct and indirect contributions to promoting plant growth. Their functional roles extend across plant growth promotion, stress tolerance, disease suppression and ecosystem-level nutrient cycling (Moturu *et al.*, 2025; Sowjanya *et al.*, 2025; Batista *et al.*, 2018).

### 10.5.1 Endophytes as Plant Growth Promoters

Endophytes enhance plant growth through multiple-nutrient and hormone-mediated mechanisms. Many bacterial endophytes fix atmospheric nitrogen, solubilize phosphate and produce siderophores, improving nutrient availability in nutrient-poor or degraded soils. They also synthesize key phytohormones such as IAA, cytokinins and gibberellins, which stimulate root and shoot development (Frontiers, 2021; Riddech *et al.*, 2022; Moturu *et al.*, 2025).

Additionally, several strains produce ACC deaminase, reducing stress-induced ethylene levels and helping plants keep growth under drought, salinity or flooding (Moturu *et al.*, 2025). Beyond direct physiological effects, endophytes influence rhizospheric microbial communities by releasing organic acids and volatile compounds that change root exudation and attract beneficial symbionts (Batista *et al.*, 2018). Through these coordinated nutrient, hormonal and microbial interactions, endophytes support balanced plant growth and contribute significantly to sustainable agriculture and ecological restoration (~~Frontiers~~, Salvi et al. 2022).

### 10.5.2 Stress Tolerance and Environmental Adaptation

Endophytes play a crucial role in enhancing plant tolerance to abiotic stresses such as drought, salinity, heavy metals, temperature extremes and UV radiation (Moturu *et al.*, 2025; Riddech *et al.*, 2022). Bacterial endophytes, including *Enterobacter cloacae*, *Pseudomonas fluorescens*, *Bacillus subtilis* and *Burkholderia* spp., improve osmotic balance, water-use efficiency and antioxidant activity by increasing osmoprotectants (e.g., proline, glycine betaine) and regulating stress-responsive genes, including aquaporins (Sowjanya *et al.*, 2025). Fungal endophytes like *Trichoderma spp.* and *Piriformospora indica* activate jasmonic acid and salicylic acid-signalling pathways, providing systemic stress resistance (Frontiers, 2021).

Endophytes also support phytoremediation by detoxifying or immobilizing contaminants.

Metal-tolerant strains associated with *Populus* and *Brassica* species help sequester or volatilize heavy metals such as cadmium, lead and arsenic, while other endophytes degrade organic pollutants (Frontiers, 2021). Additionally, endophytes from extreme environments enhance UV protection and thermotolerance through pigments (melanin, carotenoids) and heat-shock proteins (Moturu *et al.*, 2025). Collectively, these mechanisms enable plants to thrive in marginal or contaminated ecosystems, contributing significantly to environmental resilience and habitat restoration (Frontiers, 2022).

### 10.5.3 Disease Suppression and Immune Modulation

Endophytic microbes function as internal biocontrol agents, protecting plants from diseases through antibiosis, competition and immune modulation (Moturu *et al.*, 2025; Batista *et al.*, 2018). They produce antimicrobial compounds such as lipopeptides, alkaloids, phenolics and cell wall–degrading enzymes (e.g., chitinases, glucanases) that inhibit pathogens (Ridddach *et al.*, 2022). By competing for nutrients and ecological niches, they limit pathogen establishment (Batista *et al.*, 2018).

Many endophytes also trigger ISR via jasmonic acid and salicylic acid pathways, priming plants for rapid defence responses (Frontiers, 2022). Bacterial endophytes like *B. subtilis*, *Bacillus amyloliquefaciens*, *P. fluorescens* and *Pseudomonas putida*, as well as fungal endophytes such as *Trichoderma harzianum*, effectively suppress soil-borne pathogens (Frontiers, 2021; Moturu *et al.*, 2025). Collectively, endophytes provide sustainable, environmentally friendly disease suppression and reduce reliance on chemical pesticides (Frontiers, 2022).

### 10.5.4 Nutrient Cycling and Ecosystem Processes

Endophytes play a significant role in plant nutrient acquisition and in shaping broader ecosystem nutrient cycles (Batista *et al.*, 2018; Moturu *et al.*, 2025). Nitrogen-fixing bacteria such as *Herbaspirillum*, *Burkholderia*, *Azospirillum* and *Gluconacetobacter* convert atmospheric nitrogen into plant-usable forms, reducing dependence on synthetic fertilizers, especially in non-leguminous crops (Frontiers, 2022). Many endophytes also solubilize phosphate through organic acids and phosphatases and release siderophores that enhance iron availability, strengthening plant nutrition under limiting conditions (Moturu *et al.*, 2025).

Fungal endophytes ease the decomposition of lignin and cellulose, accelerating carbon turnover and influencing soil organic matter dynamics (Batista *et al.*, 2018). Their effects on plant productivity, litter quality and root-associated signalling help shape plant–plant interactions, microbial community structure, carbon sequestration and overall ecosystem stability. Through these integrated functions, endophytes function as key drivers of biogeochemical cycling and environmental resilience (Frontiers, 2022).

### 10.5.5 Contribution to Ecosystem Functioning

Endophytes connect plant health to environmental stability, affecting ecosystem-level processes. In forest ecosystems, endophytic communities contribute to carbon sequestration by increasing plant longevity and biomass. In agricultural systems, they promote sustainable production by fostering biodiversity, lowering chemical inputs and improving soil fertility. Endophyte-driven stress tolerance also strengthens ecosystem resilience to land degradation, climate variability and droughts. They further contribute to microbial inheritance through vertical transmission via seeds, ensuring the continuation of ecological functions across plant generations.

### 10.5.6 Endophytes in Environmental Biotechnology

The functional versatility of endophytes has led to their increasing use in biotechnological applications, including the following:

- Biofertilizers: Endophytic consortia are formulated to enhance nutrient intake and growth.
- Biopesticides: Selected or changed endophytic strains are used for pathogen reduction.
- Phytoremediation aids: Microbes that immobilize metals or degrade organic contaminants in plants.
- Industrial bioproducts: Secondary metabolites, enzymes and bioactive compounds with applications in agriculture and medicine.

Thus, endophyte-based biotechnology provides sustainable solutions for food security and environmental management, bridging the gap between industry and ecology.

## 10.6 APPLICATIONS AND FUTURE PROSPECTS

Endophytic microbes are a powerful, yet underutilized, biological resource for enhancing sustainable agriculture, bioremediation and climate resilience. Their unique ability to colonize internal plant tissues, form stable associations and modulate plant physiology from within positions them as next-generation bioinoculants that surpass conventional rhizospheric models (González-García *et al.*, 2024; Medari *et al.*, 2025; Pathak *et al.*, 2024). Integrating endophyte-based strategies into environmental management and industrial biotechnology holds the potential to increase productivity while reducing ecological footprint and chemical inputs (Das et al. 2025; Pathak *et al.*, 2024).

### 10.6.1 Agricultural Applications

Biofertilizers and Plant Growth Promotion: Endophytic bacteria, including *Bacillus*, *Enterobacter*, *Burkholderia* and *Azospirillum* species, are recognized for stimulating plant development through biological nitrogen fixation, phytohormone synthesis and phosphate solubilization (González-García *et al.*, 2024; Medari *et al.*, 2025). Unlike traditional rhizospheric inoculants, endophytic strains can colonize interior tissues, offering consistent and long-term benefits even under varying field conditions.

Endophyte-based biofertilizers improve nutrient-use efficiency, enhance root development and increase crop yields while minimizing dependence on synthetic fertilizers (González-García *et al.*, 2024; Medari *et al.*, 2025). Commercial formulations having endophytic bacteria (e.g., *P. fluorescens* and *B. subtilis*) are currently in use for cereals, legumes and horticultural crops (Pathak *et al.*, 2023).

Biocontrol and Plant Protection: Endophytes serve as internal biological barriers against pests and infections. They produce lytic enzymes, volatile chemical compounds and antibacterial substances that inhibit a broad spectrum of fungal, nematode and bacterial diseases (Leichtweis *et al.*, 2021; Pathak *et al.*, 2023). *Beauveria* and *Trichoderma* species have been commercialized as endophytic biocontrol agents due to their dual roles in disease suppression and plant growth promotion (González-García *et al.*, 2024). Internal colonization ensures persistence, providing reliable protection even in challenging climate or soil conditions, compared to rhizosphere-bound biocontrols (Medari *et al.*, 2025).

Abiotic Stress Management: As climatic stress intensifies, drought, salinity and temperature fluctuations pose significant challenges to agriculture. Endophytes mitigate these stresses by boosting osmotic balance, regulating phytohormonal pathways and strengthening antioxidant systems (González-García *et al.*, 2024; Pathak *et al.*, 2023). Salt-tolerant endophytes, such as *E. cloacae* and *Halomonas*, improve wheat and rice performance in saline soils; fungal endophytes from arid-zone plants confer drought tolerance through accumulation of compatible solutes and regulation of stress-related gene expression (Leichtweis *et al.*, 2021).

### 10.6.2 Phytoremediation and Environmental Restoration

Endophytes are essential collaborators in phytoremediation, partnering with plants to clean up contaminated environments. They metabolize or immobilize harmful chemicals, enabling plant survival in soils rich in organic contaminants, metal-pollutants or salinity (Leichtweis *et al.*, 2021).

Heavy Metal Detoxification: Endophytic fungi and bacteria utilize redox enzymes, efflux transporters and metallothioneins to convert heavy metals into less toxic forms or immobilize metals in roots, reducing transport to edible parts (Leichtweis *et al.*, 2021; González-García *et al.*, 2024).

Organic Pollutant Degradation: Certain endophytes carry catabolic genes that help plants degrade pollutants (chlorinated chemicals, PAHs, toluene); collaborative degradation by endophytes with plant metabolism accelerates remediation (Pathak *et al.*, 2023).

Ecosystem Rehabilitation: Endophyte-assisted plant establishment can improve soil structure and fertility in degraded environments, supporting revegetation in salt, desert or mining areas (González-García *et al.*, 2024).

### 10.6.3 Industrial and Pharmaceutical Applications

Endophytic microbes produce bioactive metabolites with medical, agricultural and industrial uses. Notably, *Taxomyces andreanae*, a fungal endophyte from *Taxus brevifolia*, produces taxol, a leading anticancer drug (Leichtweis *et al.*, 2021). Bacterial endophytes from medicinal plants also synthesize phenolics, terpenoids and alkaloids with pharmacological potential, while endophyte-derived enzymes (cellulases, proteases, lipases) are used in bioprocessing, biofuel production and food technology (Leichtweis *et al.*, 2021; Pathak *et al.*, 2023).

### 10.6.4 Climate Change Mitigation and Carbon Sequestration

Endophytes indirectly support climate resilience by promoting plant growth, reducing chemical input dependence and enhancing carbon sequestration via increased biomass (Leichtweis *et al.*, 2021). Their stress tolerance allows plants to flourish in marginal areas, promoting carbon uptake; some endophyte consortia regulate methane and nitrous oxide oxidation, connecting microbiomes to greenhouse gas cycling (Pathak *et al.*, 2023).

## 10.7 CHALLENGES AND FUTURE RESEARCH DIRECTIONS

Despite significant advancements in the understanding of endophytic microbiomes, many questions about their ecological significance, function and diversity remain. The complexity of plant–microbe–environment interactions, limitations in current methodologies and the absence of standardized frameworks continue to pose challenges for both applied and basic research (Anand *et al.*, 2023; Ramatsitsi and Manyevere, 2025; Sandrawati *et al.*, 2020). To fully unlock the potential of endophytes in environmental and agricultural systems, the field must transition from descriptive research to mechanistic, predictive and translational microbiome science (Rather & Masroor, 2025; Heinig *et al.*, 2013).

### 10.7.1 Methodological and Technical Limitations

Isolation and Cultivation Challenges: A major hurdle in endophyte research is the difficulty of cultivating the majority of microbial species from plant tissues. Traditional culture-based methods recover only a small fraction of the microbial diversity; many endophytes exist in slow-growing or dormant phases not supported by standard media (Elsarkasi, 2025; Sandrawati *et al.*, 2020; Anand *et al.*, 2023). Advanced methods like microfluidics, co-culturing and diffusion chambers are improving access to these uncultivable taxa (Elsarkasi, 2025).

Omics and Data Integration Issues: While high-throughput sequencing has revolutionized microbiome research, challenges in interpretation and data integration persist (Anand *et al.*, 2023). Unbiased extraction, amplification and sequencing protocols, coupled with robust reference databases, are needed for improved comparability and repeatability in plant-associated microbiome studies (Rather & Masroor, 2025).

### 10.7.2 Ecological Complexity and Context Dependence

Endophytic communities are highly dynamic and context-dependent, influenced by soil type, climate, host genotype and developmental stage (Anand *et al.*, 2023; Ramatsitsi *et al.*, 2025). The same microbial strain may have neutral, negative or positive effects based on the environment and host. Understanding factors driving microbiome stability and assembly requires systems-level modelling and field research (Sandrawati *et al.*, 2020; Ramatsitsi *et al.*, 2025).

### 10.7.3 Translational Challenges and Field Performance

Endophytes often show variable efficacy in field applications due to environmental variability, native microbial competition and formulation instability (Sandrawati *et al.*, 2020). Developing reliable inoculants requires refining delivery methods, carrier materials and ensuring compatibility with diverse field conditions (Heinig *et al.*, 2013; Anand *et al.*, 2023). Regulatory and biosafety frameworks must also be addressed for commercial use (Rather & Masroor, 2025).

### 10.7.4 Ethical, Regulatory and Societal Dimensions

The use of genetically engineered or synthetic microbial consortia raises ethical and biosafety concerns. Harmonized regulations, public engagement and transparent risk assessment are necessary for responsible innovation and wide acceptance (Rather & Masroor, 2025).

### 10.7.5 Future Research Directions

Future research should focus on the following:

- Microbiome assembly rules, host genetic traits and compatibility (Rather & Masroor, 2025; Anand *et al.*, 2023).

- Integration of multi-omics and experimental ecology using synthetic communities and gnotobiotic systems (Ramatsitsi *et al.*, 2025).
- Microbiome engineering and CRISPR-based boosting of beneficial traits (Elsarkasi, 2025).
- Seed microbiome engineering and strategies for climate-smart agriculture (Sandrawati *et al.*, 2020).
- Data standards, microbial collections and open genomic repositories for collaboration (Rather & Masroor, 2025).

## 10.8 CONCLUSION

As key players in plant growth, stress tolerance, disease suppression and ecosystem-level nutrient cycling, endophytic communities are a crucial but traditionally disregarded part of environmental microbiomes. Their diversity, complex host interaction mechanisms and functional adaptability go well beyond the rhizosphere, as this chapter proves, creating a dynamic continuum that significantly influences plant health and ecological resilience. The depth and complexity of these hidden microbial networks are being revealed by advances in multi-omics and imaging technologies, highlighting their promise as sustainable tools in environmental restoration, biotechnology and agriculture.

In order to use endophytes as dependable bioinoculants in the future and promote resilient ecosystems, climate-smart agriculture and a better knowledge of plant–microbe–environment interactions, it will be crucial to integrate mechanistic insight with applied innovation.

## REFERENCES


Akther, S.M., Hu, J., Miller, G., and Shi, W. (2025). Foot traffic on turf primaily shaped the endophytic bacteriome compared to the rhizosphere and bulk soil. *Frontiers in Microbiology*, 16:1488371. doi: 10.3389/fmicb.2025.1488371

Batista, B. D., et al. (2018). Beneficial relationships between endophytic bacteria and medicinal plants. *Frontiers in Microbiology*, 9, 282.

Berendsen, R. L., Pieterse, C. M., & Bakker, P. A. (2012). The rhizosphere microbiome and plant health. *Trends in Plant Science*, 17(8), 478–486.

Brader G, Compant S, Mitter B, Trognitz F, Sessitsch A. (2014). Metabolic potential of endophytic bacteria. *Curr Opin Biotechnol*. 2014 Jun;27(100):30–37. doi: 10.1016/j.copbio.2013.09.012

Bulgarelli, D., et al. (2013). Structure and function of the bacterial root microbiota in wild and domesticated barley. *Cell Host & Microbe*, 13(3), 260–271.

Compant, S., Clément, C., & Sessitsch, A. (2010). Plant growth-promoting bacteria in the rhizo- and endosphere of plants: Their role, colonization, mechanisms, and potential application. *Soil Biology and Biochemistry*, 42(5), 669–678.

Compant, S., Samad, A., Faist, H., & Sessitsch, A. (2019). A review on the plant microbiome: Ecology, functions, and emerging trends in microbial application. *Journal of Advanced Research*, 19, 29–37.

Crucitti, D., et al. (2025). Endophytic microbiota diversity in the phyllosphere of Sicilian olive trees. *Environmental Microbiology Reports*, 17(3), 455–465.

Das, D., Sharma, P.L., Paul, P. et al. (2025). Harnessing endophytes: innovative strategies for sustainable agricultural practices. *Discov Bact* 2, 1 (2025). https://doi.org/10.1007/s44351-025-00011-z

Deepa, N., et al. (2024). Unraveling the functional characteristics of endophytic bacteria in Pelargonium graveolens. *Microbial Ecology*, 87, 321–335.

Elsarkasi, R. F. (2025). Advances and challenges in the isolation and culture of endophytic fungi from therapeutic plants. *Bulletin of Faculty of Science, Zagazig University*, 47, 61–76.

Fatima, T., Hashmi, I., & Khan, A. L. (2022). Omics approaches in endophyte research: Expanding our understanding of plant–microbe interactions. *Frontiers in Microbiology*, 13, 861547.

Frontiers (2021). Beneficial relationships between endophytic bacteria and medicinal plants. *Frontiers in Microbiology*, 12, 620.

González-García, M., et al. (2024). Harnessing bacterial endophytes for environmental resilience and agricultural sustainability. *Science of the Total Environment*, 927, 171832.

Guo J, Ning H, Li Y, Xu Q, Shen Q, Ling N, Guo S (2024). Assemblages of rhizospheric and root endospheric mycobiota and their ecological associations with functional traits of rice. *mBio*. 2024 Mar 13;15(3):e0273323. doi: 10.1128/mbio.02733-23

Hardoim, P. R., et al. (2015). The hidden world within plants: Ecological and evolutionary considerations for defining functioning of microbial endophytes. *Microbiology and Molecular Biology Reviews*, 79(3), 293–320.

Hardoim, P. R., Van Overbeek, L. S., & Van Elsas, J. D. (2008). Properties of bacterial endophytes and their proposed role in plant growth. *Trends in Microbiology*, 16(10), 463–471.

Heinig, U., et al. (2013). Commercialization bottlenecks in industrial production of endophyte-derived pharmaceuticals. *Journal of Industrial Microbiology & Biotechnology*, 39, 173–180.

Kandel, S. L., Joubert, P. M., & Doty, S. L. (2015). Bacterial endophyte colonization and distribution within plants. *Microorganisms*, 3(4), 944–953.

Lei, L., et al. (2025). Distinctive structure of endophytic microbial communities in rice roots and leaves. *Applied and Environmental Microbiology*, 91(1), e01483–24.

Liu, H., Carvalhais, L. C., Crawford, M., Singh, E., Dennis, P. G., Pieterse, C. M., & Schenk, P. M. (2017). Inner plant values: Diversity, colonization, and benefits from endophytic bacteria. *Frontiers in Microbiology*, 8, 2552.

Long, X. N., et al. (2025). Diversity and functional roles of endophytic and rhizospheric microorganisms in O. vulgatum. *Frontiers in Microbiology*, 16, 887654.

Medari, S., et al. (2025). Green guardians: Bacterial endophytes in protecting vegetable crops against environmental stress and pathogens. *Plant Protection Science*, 61(1), 35–47.

Mendes, R., Garbeva, P., & Raaijmakers, J. M. (2013). The rhizosphere microbiome: Significance of plant beneficial, plant pathogenic, and human pathogenic microorganisms. *FEMS Microbiology Reviews*, 37(5), 634–663.

Moturu, U. S., Nunna, T., Avula, V. G., Jagarlamudi, V. R., Gutha, R. R., & Thamminana, S. (2025). Multi trait plant growth promoting endophytic bacterial. *Journal of Pure and Applied Microbiology*, 19(1), 287–299.

Mushtaq, S., et al. (2023). Interaction between bacterial endophytes and host plants: Mechanisms and ecological significance. *Frontiers in Plant Science*, 14, 117814.

Pathak H.K., Chauhan P.K., Seth C.S., Dubey G., Upadhyay S.K. (2024). Mechanistic and future prospects in rhizospheric engineering for agricultural contaminants removal, soil health restoration, and management of climate change stress. *Science of The Total Environment* 2024; 927: 172116

Philippot, L., Raaijmakers, J. M., Lemanceau, P., & Van der Putten, W. H. (2013). Going back to the roots: The microbial ecology of the rhizosphere. *Nature Reviews Microbiology*, 11(11), 789–799.

Rai, M., Agarkar, G., & Rathod, D. (2014). Plant–endophyte interactions: Potential for biotechnological applications. *Critical Reviews in Biotechnology*, 34(4), 326–337.

Ramatsitsi N., Manyevere A. (2025). Roles of endophytic fungi in plant resilience under abiotic stress: A mechanistic review with implications for climate-smart agriculture. *Plant Signal Behav*. Dec 31;20(1):2578712. doi: 10.1080/15592324.2025.2578712

Rather, R. A., & Masroor, M. (2025). A theoretical exploration of fungal endophyte's role as plant symbionts: Mechanisms, diversity, and ecology. *Journal of Fungal Biology*, 23(2), 102–119.

Riddech, N., Ma, N. Y., Ho, P. N., & Sarin, P. (2022). A plant growth promoting of rhizobacteria and endophytic bacteria. *Journal of Pure and Applied Microbiology*, 16(3), 1553–1564.

Rodriguez, R. J., White Jr., J. F., Arnold, A. E., & Redman, R. S. (2009). Fungal endophytes: Diversity and functional roles. *New Phytologist*, 182(2), 314–330.

Salvi P., Mahawar H., Agarrwal R., Kajal, Gautam V., and Deshmukh R. (2022). Advancement in the molecular perspective of plant-endophytic interaction to mitigate drought stress in plants. *Front. Microbiol.* 13:981355. doi: 10.3389/fmicb.2022.981355

Sandrawati, I., et al. (2020). Constraints and advances in commercialization of endophyte-derived products. *Frontiers in Plant Science*, 11, 1171.

Santoyo, G., Moreno-Hagelsieb, G., Orozco-Mosqueda, M. C., & Glick, B. R. (2016). Plant growth-promoting bacterial endophytes. *Microbiological Research*, 183, 92–99.

Sharma, P., et al. (2023). Endophytic bacterial taxonomic and functional diversity in wheat seeds. *Journal of Plant Interactions*, 18(1), 77–90.

Singh, B. K., Trivedi, P., Egidi, E., Macdonald, C. A., & Delgado-Baquerizo, M. (2020). Crop microbiome and sustainable agriculture. *Nature Reviews Microbiology*, 18, 601–602.

Singh, R., Kumar, P., & Srivastava, S. (2024). Current advances in the functional diversity and mechanisms of endophyte-host plant interactions. *Frontiers in Microbiology*, 15, 246793.

Sowjanya, U. S., et al. (2025). Multi-trait plant growth promoting endophytic bacterial. *Journal of Pure and Applied Microbiology*, 19(1), 287–299.

White, R. A., et al. (2019). The known and unknown microbial diversity in the plant microbiome. *Plos One*, 14(9), e0221962.

Xu D, Yu X, Chen J, Liu H, Zheng Y, Qu H, Bao Y. (2023). Microbial Assemblages Associated with the Soil-Root Continuum of an Endangered Plant, Helianthemum songaricum Schrenk. *Microbiol Spectr*. 2023 Jun 15;11(3):e0338922. doi: 10.1128/spectrum.03389-22.

Yan W, Yuan MM, Wang S, Sorensen PO, Wen T, Xu Y, Wang H, Jiao S, Chen J, Shangguan Z, Deng L, Li Z, Zhong Y. Microbial community dynamics in the soil-root continuum are linked with plant species turnover during secondary succession. *ISME Commun*. 2025 Jan 29;5(1):ycaf012. doi: 10.1093/ismeco/ycaf012.

# INDEX

CONTENTS

For Product Safety Concerns and Information please contact our EU
representative GPSR@taylorandfrancis.com
Taylor & Francis Verlag GmbH, Kaufingerstraße 24, 80331 München, Germany

www.ingramcontent.com/pod-product-compliance
Lightning Source LLC
LaVergne TN
LVHW062238200926
848484LV00006B/1568